T0192204

CAMBRIDGE LIBRARY COLLECTION

Books of enduring scholarly value

Life Sciences

Until the nineteenth century, the various subjects now known as the life sciences were regarded either as arcane studies which had little impact on ordinary daily life, or as a genteel hobby for the leisured classes. The increasing academic rigour and systematisation brought to the study of botany, zoology and other disciplines, and their adoption in university curricula, are reflected in the books reissued in this series.

Travels through Norway and Lapland

In *Travels Through Norway and Lapland*, Leopold von Buch (1774–1853), a German geologist and palaeontologist, recounts his expedition to Scandinavia in 1806–1808. This book, originally published in Berlin in 1810, and in this English translation in 1813, describes these large, sparsely populated regions at the turn of the nineteenth century. The translator's preface provides an important geo-political backdrop – the possibility of war in Norway and the machinations of Sweden, Russia and Great Britain over the future of this territory. Von Buch's observations, however, are firmly engaged with the scientific. He writes that his motivation for the expedition was to find out how the harsh climate influenced the land, and he records detailed information about the weather and the region's mineralogy and geological structure. He also describes the local population, providing a wide-ranging account of life in the remote reaches of Northern Europe.

Cambridge University Press has long been a pioneer in the reissuing of out-of-print titles from its own backlist, producing digital reprints of books that are still sought after by scholars and students but could not be reprinted economically using traditional technology. The Cambridge Library Collection extends this activity to a wider range of books which are still of importance to researchers and professionals, either for the source material they contain, or as landmarks in the history of their academic discipline.

Drawing from the world-renowned collections in the Cambridge University Library, and guided by the advice of experts in each subject area, Cambridge University Press is using state-of-the-art scanning machines in its own Printing House to capture the content of each book selected for inclusion. The files are processed to give a consistently clear, crisp image, and the books finished to the high quality standard for which the Press is recognised around the world. The latest print-on-demand technology ensures that the books will remain available indefinitely, and that orders for single or multiple copies can quickly be supplied.

The Cambridge Library Collection will bring back to life books of enduring scholarly value (including out-of-copyright works originally issued by other publishers) across a wide range of disciplines in the humanities and social sciences and in science and technology.

Travels through Norway and Lapland

During the years 1806,1807, and 1808

LEOPOLD VON BUCH
EDITED BY ROBERT JAMESON

CAMBRIDGE
UNIVERSITY PRESS

CAMBRIDGE UNIVERSITY PRESS

Cambridge, New York, Melbourne, Madrid, Cape Town,
Singapore, São Paolo, Delhi, Tokyo, Mexico City

Published in the United States of America by Cambridge University Press, New York

www.cambridge.org
Information on this title: www.cambridge.org/9781108028813

© in this compilation Cambridge University Press 2011

This edition first published 1813
This digitally printed version 2011

ISBN 978-1-108-02881-3 Paperback

This book reproduces the text of the original edition. The content and language reflect
the beliefs, practices and terminology of their time, and have not been updated.

Cambridge University Press wishes to make clear that the book, unless originally published
by Cambridge, is not being republished by, in association or collaboration with, or
with the endorsement or approval of, the original publisher or its successors in title.

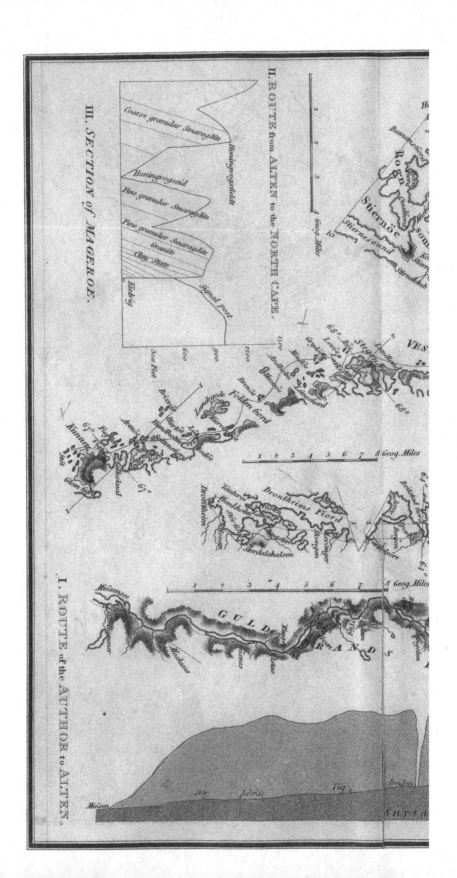

II. ROUTE from ALTEN to the NORTH CAPE.

Geog. Miles

Coarse granular Smaragdite

Honingvogseid

Fine granular Smaragdite

Fine granular Smaragdite
Granite
Clay Slate

Honingvogsfield

Signal post

Kiebig

III. SECTION of MAGEROE.

Feet

I. ROUTE of the AUTHOR to ALTEN.

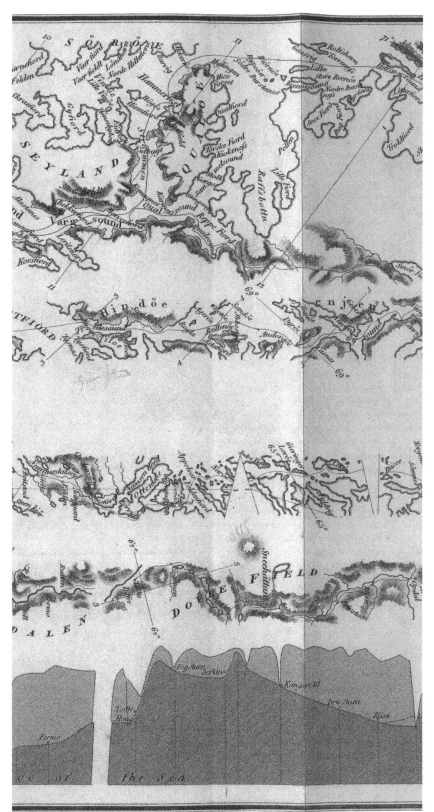

Published July 12th 1813 by Henry Colburn Conduit Street London.

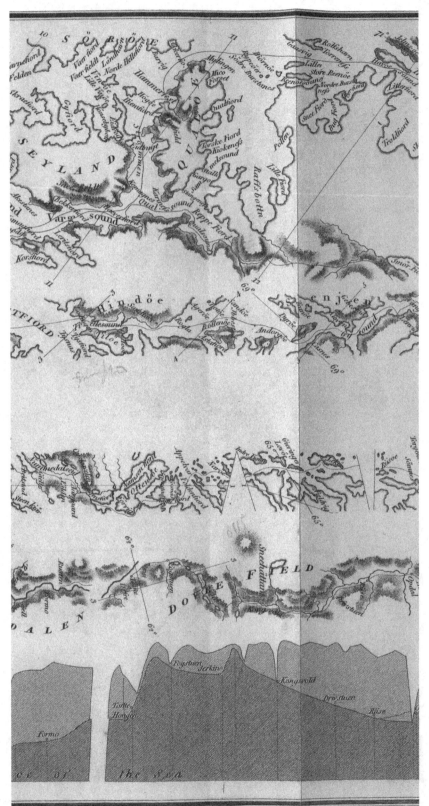

Published July 12th 1813 by Henry Colburn Conduit Street London.

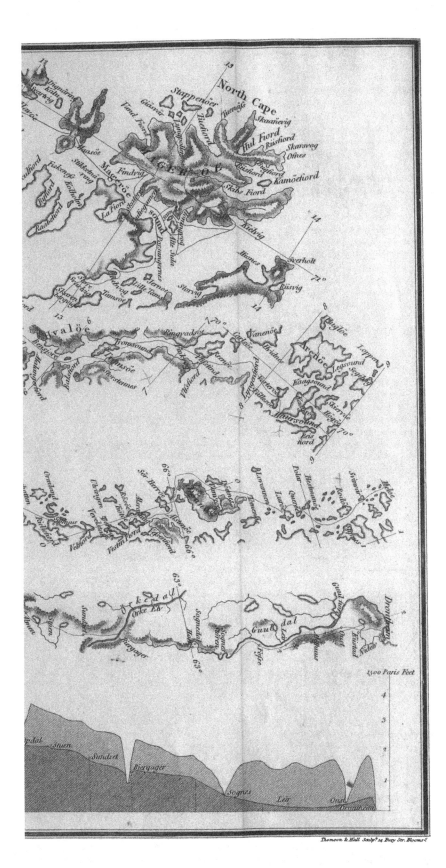

North Cape

G E R S Ö E

Maasöe

1500 Paris Feet

4

3

2

1

Thomson & Hall. Sculp.^r 14 Bury Str. Blooms.y

TRAVELS

THROUGH

NORWAY AND LAPLAND,

DURING THE YEARS 1806, 1807, AND 1808,

BY

LEOPOLD VON BUCH,

MEMBER OF THE ROYAL ACADEMY OF SCIENCES AT BERLIN.

TRANSLATED FROM THE ORIGINAL GERMAN

BY

JOHN BLACK.

WITH NOTES AND ILLUSTRATIONS, CHIEFLY MINERALOGICAL,

AND SOME ACCOUNT OF THE AUTHOR,

BY

ROBERT JAMESON, F.R.S.E.; F.L.S. &c.

PROFESSOR OF NATURAL HISTORY IN THE UNIVERSITY OF EDINBURGH.

Illustrated with Maps and Physical Sections.

LONDON:

PRINTED FOR HENRY COLBURN,

BRITISH AND FOREIGN PUBLIC LIBRARY, CONDUIT-STREET, HANOVER-SQUARE;

AND SOLD BY GEORGE GOLDIE, EDINBURGH; AND

JOHN CUMMING, DUBLIN.

1813.

TRANSLATOR'S PREFACE.

LITTLE apology can be necessary for laying before the public any additional information respecting Norway at the present moment.

That country will in all probability soon become the theatre of a bloody war, in which the British nation are pledged to co-operate. Our attention is therefore now very naturally directed to the nature of the country and the character of the people who inhabit it. But all the information we can obtain from the works which we already possess is extremely unsatisfactory. The romantic scenery of a part of the coast drew from the celebrated Mary Wollstonecroft, about twenty years ago, some of the most beautiful specimens of description in the English language; and the sweetness of the appearance of the Norwegian vallies is spoken of with more enthusiasm than we might expect by M. Malthus, in a note to the second edition of his well known work on Population, in which work we have a good deal of new information respecting the constitution of society in the north. All that we at present know of Norway, however, is in truth merely sufficient to whet our curiosity. From the intimate connexion which has long subsisted between that country and England, we might, on a first view, be induced to wonder at the scantiness of our knowledge; but the intercourse has been chiefly, if not altogether, confined to traders and seamen, a description of people from whom even if they possessed the requisite opportunities, and if their visits had not been confined to a few of the southern ports, it were vain to expect much satisfactory information.

But this deficiency is in a great measure supplied in the following pages. M. Von Buch travelled through Norway in the years 1806, 1807, and 1808 ; and he has given us the most minute and satisfactory descriptions of the appearance of the country, the nature of its productions, and the character and occupations of its inhabitants. In the course of his travels he was frequently exposed to great hardships and inconveniences, which he has related with a singular degree of modesty. He went across the country from Christiania to Drontheim in the spring of 1807, a season of the year when the struggle between summer and winter renders every mode of travelling equally disagreeable and dangerous, and when even the natives never think of venturing from home. In the course of the summer of that year he explored the country from Drontheim to the North Cape, and returned on the approach of winter through Lapland and Sweden to the south of Norway. The number of mountains he ascended, and of which he determined the heighths by the barometer, in the course of his journey, is almost inconceivable. Unfortunately the bulk of readers of travels are but little acquainted with the fatigues of ascending high northern mountains, and therefore peculiarly ill-qualified to appreciate the sufferings and the merits of a modest and enterprising traveller.

From the view of the country which M. Von Buch has given us, we may see the difficulty with which a conquest by Sweden, assisted even by the mighty power of Russia, and by Great Britain, will be attended. It is in fact hardly possible that such a country should ever be conquered. Russia has been for a long time back encroaching on Finmark, the most northern part of the Norwegian territory, and a war may place her in the possession of nearly the whole of that district. But it is extremely probable that the rest of the country, except with the consent of the natives, can never be separated from the crown of Denmark.

The country from which Norway has the most to dread is Great Britain. It is in our power to interrupt the extensive fisheries, to deprive it of some of its most essential supplies, and to drive many of its brave and hardy inhabitants to the use of bark-bread, and the other miserable substitutes for adequate food. We may add to their privations and sufferings, but we cannot conquer them. The prosperity and comfort of Norway are in a great measure dependent on England. In the harvest of 1806, M. Von Buch saw a number of ships laden with hay from England and Ireland in the Bay of Christiania; and he more than once informs us that on the intercourse with England the prosperity of whole provinces depends

It is not to be wondered at then that the inhabitants of Norway should in general feel a strong attachment for England, the country from whose friendship her chief advantages are derived, and from whose enmity she can only experience the most dreadful sufferings. M. Von Buch tells us that at Christiania every appearance which had the least tendency to justify the English was anxiously laid hold of. Every measure of a hostile or unjustifiable nature they imputed to the ministry, and every action of kindness they uniformly imputed to the nation at large. Are we then to reward this unoffending people, the only people in the world perhaps who are sincerely attached to us, by joining in a fruitless attempt to subject them to their hated neighbours?

For however much the Norwegians may dislike the Danes, their dislike of the Swedes is naturally much greater, and it is probable they will never consent to a subjection to Sweden. The two countries of Norway and Sweden have from the earliest times been engaged in almost perpetual warfare, and the Swedish attempts have uniformly failed. Such is the nature of the country of Norway, that by the pre-occupation of difficult passes the destruction of an invading army is frequently almost inevitable. Bodies of regular troops have been more than once destroyed in some of these passes by the peasantry. We have a curious account in this work of the destruction, by a few peasants in Guldbrandsdalen, of Colonel

Sinclair and nine hundred Scotch, who were marching through the country to join the army of Gustavus Adolphus. So late as 1788, the Swedes were overthrown by the Norwegians at the pass of Quistrum, who would have afterwards taken the rich town of Gottenberg but for the interference of the English ambassador.

M. Von Buch frequently speaks in high terms of the dignified character and independent spirit of the Norwegians, and the fondness with which even the peasantry dwell on the heroic actions of their ancestors. The celebrated Danish poet and historian, Holberg, who was a native of Norway, in his Civil and Ecclesiastical State of Denmark and Norway, observes: " With respect to the bravery of the Norwegians, I may boldly assert that no people in our times can equal them. I do not say this for the purpose of flattering my country-people, as authors are too apt to do. Whoever entertains such a suspicion, let him examine the annals which contain the actions of that people. Both Swedish and Danish historians agree in stating that no nation has ever yet obtained any advantage over the Norwegians. The Swedes, who boast much of their bravery, are forced to confess that they are equalled by the Norwegians. The latter, however, esteem themselves much braver than the Swedes; and they ground their opinion on the numerous victories which they have gained with a greatly inferior force over their brave Swedish neighbours." This is the fond account of a native; but, making every allowance for national partiality, we may safely pronounce the Norwegians one of the most brave and high-minded nations on the face of the earth. But on this subject the Translator has already perhaps enlarged too much.

M. Von Buch has related his travels in a lively and agreeable manner, but in rather a loose and negligent style. His meaning is, however, at all times, clear and obvious; and objections to his diction will rather be made by the critic than the person who reads merely for information and amusement, who is frequently more pleased with the negligent transcripts from a traveller's journal than with the more nicely constructed sentences which may be produced by the diligence of subsequent labour.

Of the manner in which the translation is executed, it is hardly fitting for the Translator to speak. He was anxious to understand his original, and to convey his meaning in English as accurately as he possibly could. He may be blamed by some, perhaps, for adhering too closely to the style of his original, and for not adapting the composition somewhat more to the English taste. With respect to the latitude allowable to a Translator, different opinions are entertained. Some recommend as close an adherence to the character and language of an original as possible; while others think, if all the information be conveyed, every deviation from the character or manner of an original by which it may be improved is not only justifiable but necessary. But a slight alteration of an expression frequently produces a most important change in the meaning; and attempts at improvement of an original it is believed are in general viewed with more suspicion than satisfaction.

To Professor Jameson, of Edinburgh, the Translator is under the greatest obligations. He has enriched the work with a number of notes and observations, which the learned will know how to appreciate. Without the assistance of that gentleman, the mineralogical portion of the work would have frequently been very inadequately executed. It must be highly gratifying to M. Von Buch to find that his book received this degree of attention from one whose name stands so deservedly high in the scientific world as Mr. Jameson.

From the same gentleman the Translator obtained the following account of M. Von Buch, with which he will close this Preface.

" M. Von Buch, the celebrated Author of these Travels, is a native of Prussia. He received his mineralogical education in the famous Mining Academy at Freyberg, in Saxony, under the illustrious Professor Werner. Very early he distinguished himself by indefatigable industry, great acuteness, and enthusiastic love of Natural History. During his residence in Saxony, he published several interesting papers in the Miner's Journal. His first separate publication was a mineralogical description of Landeck, in Silesia, printed in the year 1797. This little tract (for it did not exceed fifty pages quarto) was at the time of its publication the best mineralogical geography that had appeared in Germany. It was his first essay on quitting the school of Werner, and the work of his early youth. It has been translated into French by an eminent miner and mineralogist, M. Daubuisson, and we possess an excellent English version of it by Dr. Anderson, of Leith, printed in 1810.

" His next work, entitled 'Geognostical Observations made during Travels in Germany and Italy,' was published in the year 1802. This volume contains a geognostical description of Silesia. From the account given by Von Buch, it appears that the red sandstone of that country contains very important beds of coal. This fact is there well established, and should be known to the coal viewers of Britain, who to a man are of opinion that coal is never to be sought for in districts composed of red sandstone. In the same volume there are geognostical accounts of the salt countries belonging to Austria; of Berchtolsgaden and Salzburg; a comparison of the passages over Mount Cenis and the Brenner; and, lastly, observations on the remarkable district of Pergine.

" From this period, until the spring of 1806, when he left Germany for Norway, he was actively employed in examining many of the most curious and interesting countries in Italy, Switzerland, and Germany. During his absence in Norway, a second volume of his mineralogical travels was published, and which proved equally valuable with the first. It contains a geognostical description of the strata on which the city of Rome is built, from which it appears that they are entirely of aquatic origin, and that the craters described by Breislac are nothing more than old quarries. The second article is an account of Monte Albano near Rome; and the third contains an excellent description of Mount Vesuvius. But probably the most important part of this volume is the series of letters addressed to the late celebrated Professor Karsten, on the mineralogy of Auvergne. These letters contain the fullest and most accurate account hitherto published of that enigmatical country. He proves that it contains many undoubted extinct volcanoes, and although many of the lavas are very nearly allied to true basalt, he does not adopt the volcanic origin of that substance to its full extent. He is of opinion that basalt may be formed either in the humid or the dry way, and thus endeavours to reconcile the neptunian and volcanic theories.

" But of all his writings, the present work, his Travels in Norway and Lapland, is to be considered as the most generally interesting. It abounds in curious and important observations in regard to the climate of these remote regions, and he has shewn how the geographical and physical distributions of several of the most important vegetables that grow in the Scandinavian peninsula are connected with situation and climate. He has, in this department, added several facts to those already known by the admirable researches of the enterprising Wahlenberg. The mineralogical facts with which it abounds are most interesting, as will appear from the following enumeration of a few of them.

1. Norway and Lapland are principally composed of primitive and transition rocks: floetz rocks occur very rarely, and alluvial rocks are uncommon.

2. Granite, contrary to the general belief of mineralogists, is a rare rock in Norway and Lapland, it even occurs but seldom in Sweden, and it is to be considered as one of the least frequent of the primitive rocks in Scandinavia.

3. The granite frequently alternates with gneiss.

4. A newer granite sometimes occurs resting on mica-slate, as at

Forvig; or connected with clay-slate and diallage rock, as in the island of Mageroe.

5. Besides the gneiss, which is associated with the oldest granite, there is another of newer formation, which rests upon mica-slate.

6. Gneiss appears to be by far the most frequent and abundant rock in Scandinavia, all the other primitive rocks appearing in some degree subordinate to it.

7. In the island of Mageroe, and in other quarters of Norway, there appears a species of simple aggregated mountain rock, composed of compact felspar and diallage. This rock is the Gabbro of the Italians, and appears in Norway to be connected with clay-slate.

8. All the magnetic iron-stone of Scandinavia occurs in beds in gneiss, and not in veins, as has often been maintained by mineralogists.

9. The class of transition rocks in Norway contains besides grey-wacke, alum-slate, clay-slate, lime-stone, and other rocks well known to mineralogists as members of that class, the following rocks :

a. *Granite*, which sometimes contains hornblende.

b. *Syenite*, which contains labrador felspar, and numerous crystals of the gem named zircon.

c. *Porphyry.*

d. Amygdaloid.

e. Basalt.

f. Sand-stone

10. The transition lime-stone of Norway is sometimes granular foliated, like that which occurs in primitive country, and contains much tremolite.

His descriptions of the different tribes he associated with during his long, perilous, and dangerous journey, his delineation of manners, and pictures of society, are truly excellent. It was not to be expected that a man of Von Buch's feelings and taste could pass through Norway without being forcibly struck with the grandeur and rude sublimity of its mountains, and coasts, and rivers, and charmed with the softer and more beautiful scenery of a Christiania. His descriptions of scenery are therefore always animated and faithful. His personal fatigues, inconveniences, and dangers, are so slightly touched on, that it is only the experienced traveller who can discover, through the veil which he has cast over his relations, his frequent distressing anxiety of mind, the alarming dangers to which he was so often exposed, and the excessive and overpowering fatigue to which he was subjected. The various political relations of this country, now becoming so important, are clearly and forcibly stated. The hatred of the Swedes and dislike of the Danes, which prevails throughout Norway, are repeatedly mentioned. But since Von Buch left that country it is said that the dislike of the Danes has increased so much that now Swedes and Danes are viewed as equally obnoxious by the people of Norway. If they have any regard for a foreign nation, for all are by them considered as foreigners, it is for the British. The fishery at Loffoden, which occupies twenty thousand men, proves how important the acquisition of Norway must be to any power capable of calling forth its energies."

AUTHOR'S PREFACE.

———

THE principal motive which induced me to lay before the public the hasty observations made by me in the course of my travels in the North of Europe, was the consideration of the scanty materials possessed by us for obtaining a knowledge of these regions, so seldom visited by strangers. Mathias Sprengel, the Historian, was quite correct in affirming twenty years ago, that through Pontoppidan's Finmark Magazine, from which he was enabled to compile his description of Finmark, that country might be considered as a new discovery for us, as it was formerly in a great measure a *terra incognita*. Since the time of Pontoppidan, we have received very good information respecting Finmark from Thaarup*; and a very short, but excellent description of the same country from Sommerfeldt, in the Norwegian Topographical Journal. But many objects of utility and importance still required to be more accurately known; and with respect to the interior of the country, we have from all these descriptions but very little information. Before the appearance of Skioldebrand's and Acerbi's Travels to North Cape we had a very imperfect idea of the possibility of the union between Swedish and Norwegian Lapland to the North of Torneo; and Wahlenberg's beautiful Topography of Kemi-Lapmark, as well as his observations on the Sulitjelma mountain, in Salten, yet remain buried in the Swedish language. A journey along the northern coast of Norway, has, besides, so far as I know,

———

* Stalistik II. Theil II. Abtheil Kopenh. 1797.

b 2

never yet been published; for the travels in these regions promised by Father Hell do not appear to have been found among his papers. I hope the little which I have here published will not be deemed superfluous till we obtain better and more accurate information respecting the sea-coast of these Alpine countries. I am not afraid that I shall be too severely charged with negligence for the many errors which will probably be found in the following work; for it is the unavoidable fate of a traveller to be seldom able to convey accurate information. If those who are qualified to communicate more correct information should be stimulated through me to publish it, my errors will have evidently been of public utility.

While this work was in the press, a Treatise on the Segeberg, in Holstein, was published by M. Steffen in his work on Geognosy, in which he expresses a belief, by no means improbable, that the gyps formation of this rock is altogether peculiar, and newer than any of the secondary formations which have yet come to our knowledge. He has carefully compared it with the gyps of Montmartre at Paris, which has been proved in the most satisfactory manner by M. M. Cuvier and Brogniart to be new. If remains of land-animals should be found in the Segeberg or Boracite in the gyps of Montmartre, it would be impossible afterwards to dispute the identity of these two gyps formations.

CONTENTS.

CHAPTER I.
JOURNEY FROM BERLIN TO CHRISTIANIA.

Peat-moss at Linum.—Boulderstones in Priegnitz.—Hamburg.—Reimarus.—Roeding's Museum.—Segeberg.—Kielerfiord —University.—Packet-boat.—Voyage to Copenhagen.—Baltic.—Collection of Minerals of the University in Copenhagen.—Rosenburg.—Schumacher —Classen's Library.—Granite of the Quay.—The Sound —Journey through Holland.—Götha elf.—No Granite in the North.—Rock-channels (Scheeren-Natur).—Quistrum—Suinesund.—Fredericstadt.—Moss.—View from Christiania. - Page 1

CHAPTER II.
CHRISTIANIA.

Stone Houses.—Division of the Town into different Trads.—Meeting of the Country People at the Yearly Fair.—Development of Cultivation from the Town outwards.—Barter of various Articles.—Exports.—Timber Magazine.—Berndt-Ancker.—Division of Society in Christiania.—Theatre.—Falssen.—School Library.—Military Academy.—Country Houses.—Climate of the Town. - 27

CHAPTER III.
MINERALOGICAL EXCURSIONS IN THE COUNTRY ROUND CHRISTIANIA.

Transition formation at Christiania.—Orthoceratite in the Lime-stone.—Veins of Porphyry.—Sandstone.—Porphyry in Hills.—Kolaas, Krogsioven.—Sandstone at the Holsfiord.—Transition granite at Hörtekullen.—Granite at Strömsöe.—The Granite is confined by the Boundaries of the Transition formation.—Paradiesbacken.—Marble Quarries —Granite of After Vardekullen.—The Fiord of Christiania divides the Old and New Rocks.—Zircon-Syenite.—On the Porphyry and Clay-slate at Greffen.—Granite thereon.—Order of the Minerals at Christiania. - 45

CHAPTER IV.
JOURNEY FROM CHRISTIANIA TO DRONTHEIM.

Romerije.—Difficulty of the Journey in Spring.—Hedemarcken.—Extent of the Transition Mountains in Norway.—Gneiss and Clay-slate at Miösen.—Provost Pihl.—Fertility of the Province.—Hammer.—Greywacke at Fangsbierg.—King Oluf in

Ringsager.—View of Guldbrandsdalen.—Wolves on the Ice of the Lakes.—Clay-slate of Ringebo.—Bark Bread.—Colonel Sinclair in Kringelen.—Quartz at Viig.—Mica-slate in Kringelen.—Gneiss on the Heights.—Valley of Lessöe.—Gneiss of Rusten-berg.—Dovrefieldt Sneehätlan.—Fogstuen.—Snow-line.—Jerkin.—Straits of the Valley at Kongsvold.—Drivstuen.—Opdalen, End of Dovrefieldt.—Mica-slate on the Height.—Gneiss at the Foot.—First Firs.—Ocrkedal.—End of the Mountain Range at Stören.—Interesting Guldal.—Arrival at Drontheim. 69

CHAPTER V.

DRONTHEIM.

Reception in Drontheim.—Spirit of the Inhabitants.—Branches of Subsistence.— Röraas.—Manufactures.—Architecture and Appearance of the Town.—Fate of Wooden Towns.—Society of Sciences in Drontheim.—Provost Wille............ 108

CHAPTER VI.

JOURNEY FROM DRONTHEIM TO FINMARK.

Country-Houses at Drontheim.—Stördalen.—Cultivation at Levanger.—Värdalsöre, Gloves.—Tellegrod.—Clay-Slate in Stördal, Gneiss in Värdal.—Shell Clay at the Figa-Elv.—Steenkiär.—Beestadfiord.—Ship-Ellide.—Järn-Nätter at Eilden.— Embarkation at Lyngenfiord.—Appelvär —Näröens Market.—Riisoe.—Difficulty of living here.—Egg Islands. The Waves smoothed by Hail.—Lecko —Battle between Eagles and Oxen.—Marble in Gneiss.—Helgeland.—The Night diminishes the Strength of the Wind.—Climate of Helgeland.—Counsellor Brodtkorb.—Granite of Bevelstad.—Tiölöe.—Alstahoug, a Bishop's Seat.—Sörherröe.—Phœnicians in the North.—Stratification of the Islands.—Catching of Land Birds.—Luröe.—Kun-nen.—Glacier at Gaasvär.—Boundaries of the Spruce Fir Region.—Vigtil.— Hundholm.—Bergens Trade to Nordland.—Bodöe.—Lazaretto.—Expence of Wood in this country.—Grydöe.—Stegen.—Prästekonentind.—Mica-Slate —Laws of Stratification.—Simon Kildal, a Preacher.—Schools.—Westfiord.—Mälström.—Sal-tenström.—Lödingen.—Climate.—Gneiss.—Sun at Midnight in Mid-Summer's Day. —Northern Lights in Winter.—Does the Climate of the North grow worse ?—Fisheries at Vaage —Their importance.—Arrival of the Fish.—Method of catching them with Nets.—With Lines.—Angling.—Form of the Hooks.—Laplanders in Lödingen.— Intermixture with Finns.—Arrival in Finmark's Amt.—Cassness.—Faxefieldt.— Fishing and Agriculture cannot be united.—Klöwen at Senjen.—Strata of Tremolite.— Contempt of the Norwegians for the Laplanders.—Lenvig.—Tremolite —Colony of Bardon-jord.—Bensjord.—The whole of Tromsöe private property.—Saw-mills.— Arrival in Tromsöe.—Difficulty of the town's advancement.—Shell-beds.—Impression of perpetual Day.—Culture of Grain in Lyngenfiord.—The Laplanders are dan-gerous Neighbours to the Norwegians.—Lyngensklub, Lodde, Salmo arcticus.—

Lyngen's Glacier. — Maursund. — Alt-Eids Fiord. — Appearance and Climate.— Glacier of the Jöckulfieldts.—Smaragdite and Felspar at Alt-Eid.—Similar Rock in the Neighbourhood of Bergen. — Passage over Alt-Eid. — Langfiord. — Arrival in Alten. 116

CHAPTER VII.

FINMARK.

View of Alten.—Climate.—Quäns in Alten.—Origin of the Quäns.—Finns in Ag-gershuusstift.—The Norwegians inhabit the Islands, the Quäns and Laplanders the Interior.—Clay-Slate and Quartz at Alten.—Voyage to Hammerfest.—Object of the Institution.—Climate.—Tyvefieldt, Gneiss.—Trade and Fisheries of the Russians in the North.—Influence and Consequences.—Voyage to Mansöe.—Vivacity of the Lap-landers.—Danger of a Residence in these Islands.—Norwegian Dwellings on Ma-geröe.—Rein-Deer on Mageröe.—Winds.—Passage to Repvog.—Hooks of the Lap-landers. — Olderfiord. — Sea-Laplanders. — Passage to Reppefiord. — Merchants in Finmark.— Their Appearance and Influence. — Rage for Brandy among the Lap-landers.—Causes of it.—Return to Alten.—Thunder Storm.—Talvig.—Mountains.— Boundaries of Vegetation.—Clay-Slate.—Mica-Slate.—Akka Solki. 240

CHAPTER VIII.

JOURNEY FROM ALTEN TO TORNEO.

Departure from Alten.—Valley of Alten.—Dwellings of the Laplanders on the Moun-tains.—Rein-deer Flocks, and Huts of the Laplanders.—Zjolmijaure.—Names of the Laplanders.—Breaking up of a Lapland Family.—Diet of the Laplanders.—Rein-deer Milk.—Siaberdasjock.—Kautokejno.—Spreading of the Finlanders in Lapland. —Salmon Fishery in the Tana.—Course of the Kiolen Mountains.—Mica-slate of Nuppi Vara.—Granite of Kautokejno.—No Mountains between the White Sea and the Bothnian Gulph.—Departure from Kautokejno —The Kloker, the Interpreter of the Clergyman.—Fisheries in the Inland Lakes.—Boundaries between the Kingdom. —Disputes and Wars before they were settled.— Entry into Swed n.—Re-appearance of the Scotch fir.—Difference of the manner in which the Rein-deer Moss spreads in Sweden and Norway.—Lippajärfwi.—Palajoen Suu.—Striking of Salmon in the River Muonio.—Boundaries of the Norway Spruce.—Muonioniska.—The Lap-landers and Finlanders are different People.—Granite on the Boundaries between the Kingdoms.—Gneiss at Palajoen Suu and Muonioniska.—Water-fall of Gianpaika. —View of the River Muonio.—Colare Kengis.—Rappa River, Red Granite at Kengis.—Iron-work of Kengis.—Lapland Iron-stone Mountains.—Tärändo-Elv.— Departure from Kengis.—Pello.—Pullingi at Svanstein.—Makarenji.—Excellent Road to Torneo.—Cultivation of the Land.—Gneiss at Korpikylä.—Clay-slate at

Wojakkala.—Transition Formation at Torneo.—Spreading of the Finlanders in recent Times.—A Country is not depopulated by emigration.—Lists of Exports from Wester-Bothnia and Lapland.—View of Torneo.—Architecture of the Town.—Manner of Living 311

CHAPTER IX.

JOURNEY FROM TORNEO TO CHRISTIANIA:

Woods on leaving Torneo.—End of the Finn Population.—M. Grape, a Clergyman.—Calix-Elf.—Baron Hermelin's Undertakings in Luleo Lapmark.—Raneo —Luleo.—Piteo.—Splendid Church of Skelefteo.—Decrease of the Surface of the Sea in the Bothnian Gulph.—Wahlenberg.—Umeo, its Church.—Gneiss in Westerbothnia.—Angermann-land.—Linen.—Skulaberg.—Sundsvall.—Changes of the Gneiss.—Helsingeland.—Gestrikeland.—Gefle.—Course of the Limit of the Scotch Firs over the Surface of Europe.—Dal-Elf.—Upsala.—Observations of the Temperature.—Stockholm.—View of the Town.—Collection of Minerals belonging to the College of Miners.—Departure from Stockholm.—Söder.—Telje Canal.—Orebro.—West Gothland.—Ruins of Udewalla.—Storm at Swinesund.—Frederickshall.—Return to Christiania. 380

CHAPTER X.

RETURN FROM CHRISTIANIA TO BERLIN.

Drammen.—Holmestrandt.—Remarkable Range of Rocks at Holmestrandt.—Basalt, Porphyry in Sandstone.—They belong to the Transition Formation.—Jarlsberg.—Laurvig.—Beeches and Brambles.—Bridge over Louwen-Elf.—Porsgrund.—Cloister of Giemsiö at Skeen.—Boundaries of the Transition Formation at Skeen.—Beauty of the Zirconsyenite.—Limestone with Petrifactions.—Quartz.—Amygdaloid.—Zirconsyenite on Veedlösekullen.—Porphyry beneath.—Porphyry Veins in the Limestone.—Road from Konsberg to Skeen.—Zirconsyenite at the Skrimsfieldt—At the Luxefieldt —Beautiful Situation of Skeen.—Departure.—Söndel-elv.—Iron-Works of Näss.—Arendal.—Christiansand.—Sources of Inaustry.—View of the Town.—Windmills —Passage to Nye.—Helliesund.—Storm.—Situation of the Island.—Signa's.—Sea Crab Fishery at Helliesund and Farsund.—Gun-boats.—Unfortunate Endeavour to reach Jutland.—The Lugger Privateer Virksomhed.—Fresh unsuccessful Attempt.—Kumlefiord.—Danger of Grain Ships.—New Attempt, Passage to Brekkestöe.—Pilots.—New Attempt —Dreary Prospect of the Coast of Jutland.—Arrival in Lycken.—Difficulty of landing on the Northern Coast of Jutland.—Vendsyssel.—Aalborg.—Randers.—Aarhuus.—Desert Heath.—Flensburg.—Schleswig.—Kiel.—Berlin. 409

TRAVELS

THROUGH

NORWAY AND LAPLAND.

CHAPTER I.

JOURNEY FROM BERLIN TO CHRISTIANIA.

Peat-moss at Linum.—Boulderstones in Priegnitz.—Hamburg.—Reimarus.—Roeding's Museum.—Segeberg.—Kielerfiord.—University.—Packet-boat —Voyage to Copenhagen.—Baltic.—Collection of Minerals of the University in Copenhagen.—Rosenburg.—Schumacher.—Classen's Library.—Granite of the Quay.—The Sound.—Journey through Holland.—Götha elf.—No Granite in the North.—Rock-channels (Scheeren-Natur).—Quistrum-Suinesund.—Fredericstadt.—Moss.—View from Christiania.

THE principal motive which induced me to visit the polar regions was a desire to investigate, as far as a hasty journey would permit, the manner in which, in high northern latitudes, the nature of the land becomes gradually changed by the climate, till at last, the noxious influence of snow and ice is destructive to every thing which has life, and the mode in which the more solid part of the earth's body is composed, whether agreeably to the constitution of more southern lands, or in conformity to other laws.

With respect to the dismal and inanimate country between Berlin and Hamburg, the observations which can be made are few and insignificant;

B

yet few as they are, they require a person to be well acquainted with the topography of the country, which a traveller can hardly be supposed to be. The peat-mosses at Linum, twenty-three English miles from Berlin, and in the immediate neighbourhood of Fehrbellin, are, however, well deserving of notice. By the naturalist, in particular, they will be viewed with peculiar interest, as marine plants* are not unfrequently found in the mosses ; and this great bed of peat is one of the objects from which we are enabled to obtain a knowledge of the history and progressive formation of the earth.   Whether these plants are to be found throughout the whole depth of the bed, or only in the lower parts of it, is yet unknown to us.   In the latter case, we should be led to ascribe a uniformity of origin to peat-mosses, at a great distance from one another ; an important circumstance in geognosy.   In the vicinity of. Drontheim, the flat peninsula of Oereland consists also of a great peat-bed, of which the undermost strata are composed almost entirely of half decayed marine plants (the long leaves of the *Zostera* and others).   The uppermost strata on the other hand contains only bog plants *(Sphagnum palustre)*, and appear therefore removed from the influence of the sea.† Is the origin of the northern German peat-mosses in any wise connected with the numberless downs, wholly similar to those on the Dutch coasts, which run through the sandy wastes of the north of Germany ?

Pregnitz is well known to waggoners and coachmen, on account of the multitude of boulder, or rolled stones.   They assert that the Hamburg road in Pregnitz is covered with stones, gathered from the fields, and that as they disappear it is again covered by others ; and the uncomfortable jolting of the carriage between Kyritz, Perleberg, and Lenzen, seemed to confirm their account.   It is possible that stones may be more heaped together in this country than in other provinces, and on that account the fact is deserving of attention.   They are all, however, only very small pieces of red and grey granite, and I very seldom saw any of those large blocks so frequent in Mecklenburg, Pomerania, and the northern parts of the Mark of Brandenburg.

---

* *Fucus Sacharinus.*   M. de Humboldt observed leaves of this marine plant in Linmer peat-moss, from 8 to 10 inches in length, and from 1 to 1½ inches in breadth, as fresh and uncorrupted as they are found in the sea at Heligoland.   *Bergmännisch Journal*, 1792. I. 551.

† *Fabricius Reise nach Norwegen*, 264.

These are still more seldom to be seen in the sandy wastes near Luebthene in Mecklenburg, where all nature appears to consist only of sand and a few dwarfish firs.

The traveller welcomes the appearance of Boitzenburg on the Elbe. The large looking-glasses in many of the houses, and the general cleanliness every where visible, remind us of the industry of Holland; and the market-place, surrounded with trees, produces an effect very different from that of the ruins and wastes of Perleberg. Though this transitory impression be soon again effaced by the sandy heaths of Lauenburg, yet on leaving Escheburg, we easily perceive that we are approaching a large and thriving town, which extends the sphere of its activity to a great distance around. How beautiful and rich is the view from the height down the vale of the Elbe, and its islands! And what surprising animation in Bergedorf! It is almost inconceivable for such a small town. The numerous country-houses by the road side shew that the land here supports not only productive labourers, but rich consumers; and amidst such objects of contemplation, we advance with eagerness and joy of heart to a great town. The size of the windows, and the number of them in the exterior of the houses, although not always a model of the beauty of a façade, give rise from their invariable neatness to a pleasing idea of comfort and cleanliness; an impression admirably calculated to urge us on through the tumult of the streets of Hamburg to the inn, where we can again enjoy refreshment and repose.

To visit Reimarus, the philosopher in Hamburg, is a pleasing duty, which even in a hasty passage one is unwilling to forego. He is one of the most distinguished labourers in the field of geognosy. His small work on De Luc's Theory of the Earth is a remarkable example of the possibility of seeing with the eyes of others, and yet by the application of a strong and well-regulated mind, to exhibit new and peculiar prospects; but this however, so far as I know, has succeeded only with a Reimarus. A talent for careful observation and judicious reflection is rarely to be met with, even among those geognosts, to whom we owe a multitude of insulated but important observations. M. Reimarus possesses also several remarkable articles in his cabinet of minerals.

The very curious collection of Roeding ought to be visited by every person, however short his stay in Hamburg. The proprietor has made it a sort of public collection, for it may at all times be seen by any person for the sum of three marcks; and the town has appropriated a building for its reception in the neighbourhood of the wall. Quadrupeds are seldom prepared with such an appearance of life as they here are. Many animals of the continent are seen for the first time here, as also several from New Holland: the flying opossum *(fliegende beutelthiere)*, and a great variety of birds. The collection of shells possesses also great beauty and perfection; but the minerals are quite insignificant, and in bad order. The collector never had it at heart to increase his collection of them.

I left Hamburg on the evening of the 4th of July, 1806, and travelled by night through Wandsbeck and Arnsburg to Kiel. The environs of this place are flat and uninteresting; and it is only after passing Oldeslohe, that hills begin to rise, and thickets to diversify the uniformity of the fields. In the bottom of a long green, and well-wooded valley, the Castle of Travendahl makes its appearance, and immediately gives rise to recollections which may have surprised many a one here, quite unprepared for them. But we willingly associate ideas of peace with so friendly and pleasing a country. The way to Kiel runs along the heights, and after a few miles we ascend the Segeberg*, the figure of which resembles nothing in the whole North of Germany. The rock stands on the height like Hohen-Twiel; it is more than two hundred feet high; and the little town environs the hill, as Stolpen, in Saxony, does the Basalt mountain. That a gypsum-rock should rise in this insulated manner above the surface is certainly a remarkable phenomenon; and in this respect it well deserves to be more accurately examined. This rock has become lately more important in a mineralogical point of view, as *boracites* have been found in the gypsum here, which before were only discovered at Luneburg †.

---

* *Segeberg*, i. e. the Sege mountain, *berg* signifying mountain, in German. Such an explanatory list of the names of the most common natural objects occurring in the composition of words, as is given by Dr. Beaufort in his ingenious Memoir of a Map of Ireland, might be advantageously appended to every work of this kind.—*T*.

† Of this remarkable hill a very interesting and accurate description has been published by Dr. Steffens, in his Geognostical Essays. According to him, the Sege-berg is principally composed of gypsum, having the following characters: The principal colour is white, sometimes yellowish, reddish, more frequently greyish white and blackish grey. Internally it is dull, or weakly glimmering; the fracture seldom even, often splintery, and frequently small granular foliated; is translucent on the edges, soft, sectile, easily frangible, feels meagre, and is rather

The environs of Ploen have been oft and most deservedly celebrated. Would it be supposed that the lakes, the majestic beeches, the hills, the thickets, and high cherry-trees, recall to us the vales of Switzerland? And yet where are oaks and beeches to be found in Switzerland of the beauty and heighth of those on the road from Ploen to Pretz?

Long before I reached Kiel, the silvery edge of the bay lighted up the distant horizon, and vessels in full sail were seen entering from the north. Such a sight can never be viewed without filling the mind with elevation and hope. By ships, the most distant parts of the earth are connected together; and wherever men are connected together, there may we expect to find a more extensive circle of ideas, and a quicker progress in improvement. This was well known to the ancient Icelanders, who strongly recommended to their youth to travel into other countries, and in whose language *stupid* and *untravelled* are synonimous *; and therefore Svipdag, an Icelandic bard, lamented the wretchedness of his life, compelled to dwell on the mountains, in wastes and wildernesses, without ever going near other men, or others coming to him †. Men rise in character only among men, and never in solitude.

But in truth, we by no means find that activity in Kiel which the entrance of these ships seems to promise; and we are still less disposed to be struck with it, on quitting the bustle of Hamburg. But I received a peculiar and by no means unpleasant impression from hearing at the inn, Copenhagen, Cronberg, Elsineur, or Landscron, spoken of as if they were places in the neighbourhood, and but a few miles distant. For I knew very well from the map that Copenhagen was farther than Berlin from Kiel. But such is

---

heavy. In the hill it occurs in large globular distant concretions, and between these is interposed a greyish black, striped, foliated gypsum. The gypsum also contains layers of a kind of stinkstone, and crystals of boracite. It appears to belong to the same formation as that of Lüneburg, in Hanover, and which rests upon chalk. It is therefore newer than the fibrous or second floetz gypsum of Werner; and in position, &c. nearly agrees with the gypsum of Montmartre, near Paris.—J.

* *Heimskt er heimalit barn,* is an old Icelandic proverb that may lead us to an unexpected derivation of *Haemisch.*

† *Dáuflig er äsi vor at vära her uppi i fiöllom, i afdölom, oc ubygdum, oc komma aldrig, til annata manna, ne adrir til vor.* Ihre von Reisen der Isländer nach Constantinopel. Schlözer nord. Gesch. 557.

the effect of navigation. A packet-boat goes regularly every week between the two towns, by which means the distance in both places seems to disappear. Curiosity induced me to wait for the departure of one of these packet-boats; but the days which this delay occasioned me to pass in Kiel were spent with delight and instruction in the pleasant society of Professor Pfaff.

The University, notwithstanding the smallness of the town, is far from attracting much notice. The number of students seldom exceeds one hundred and forty; partly natives of Holstein, and partly natural-born Danes, who form a strong contrast to one another. All whom I conversed with agreed that the Danes have by far the quickest capacity; that the natives of Holstein obtain their knowledge with much more labour and difficulty; and that therefore many of them are averse from the toil of learning. They go mechanically through the round of studies; and when every thing has been repeated to them, they believe they have then finished their course. On the other hand, the Danes often mistake levity for genius; soon believe themselves at the very pinnacle of knowledge; stop short in the career of instruction, and not unfrequently attempt to cover over their ignorance with arrogance. Such errors are every where prevalent; but it would be highly advantageous for both nations, if, by their meeting together in this manner at Kiel, an amalgamation of the two characters could be produced.

A residence of several years in this place would not be unpleasant. The country is agreeable, the bay abounds with numerous and varied landscapes, and the beautiful environs of Ploen and Pretz are at a short distance. The long residence of the Court gave rise to a number of little improvements between the castle-garden and the sea, which furnish at present such walks to the inhabitants of this place as few towns possess. The town of Kiel a short time ago allowed the Queen of Denmark to lay out a delightful summer-house in this country, on an elevation in Duesternbro, where there is a most delightful view, partly over the bay, and partly over the town and its neighbourhood. The Queen went almost daily there in summer, followed by the King, to enjoy from the height the prospect over the country. How delightful is it to see such places chosen as their favourite residences by the rulers of a country!

On Thursday, the 10th of July, we were all assembled in the packet-boat, and yielded ourselves to the waves. All was life and motion, singing and rejoicing. A fresh and favourable breeze swelled the sails, and Kiel, with its castle, promenade, and contiguous villages, rapidly withdrew from our sight. In less than an hour after our departure, we got out of the Kielerfiord, and were in the open sea. The increasing motion, and the rocking of the ship, gradually deprived our singers of their voices: every one sought for a place of repose to enable him to withstand what for most was an unusual sensation; and I could now examine the general composition of this little colony. In the cabin, which was tolerably roomy, the society was very good, and mostly belonging to Copenhagen—people of polished manners, and cultivated minds. Among them we had the wife of a Counsellor of State, with her attendant. Most had little cabinets around the cabin, with what are called koejen; i. e. places for sleeping in the sides of the ship, two generally above one another. In the entrance of the cabin were several of these koejen without cabinets, and others were also in the inside of the cabin itself, so that we had from twelve to fourteen sleeping places hid in the ship's sides. Towards evening, and in the night, the company of the cabin had disappeared, which but a short while before was in motion in almost every part of the ship. Near the mast and the fore part of the vessel about fifty recruits were tumbling about, all of them young and well formed men, in bright red Danish uniforms. Most of them had deserted from the Prussians, and were to be delivered over to the regiment of Marines at Copenhagen, where they are disposed of in different ships, or sent off to the garrisons of St. Croix, or Tranquebar. The hold of the ship was assigned to them, and they durst not approach the cabin. The part of the deck surrounding the helm was reserved for the cabin-passengers exclusively; so that we had it completely in our power to select our company. The ship itself was a considerable transport vessel, called a brig, which, besides a large mainmast, was also provided with a small foremast. It was called *Den Nye Pröve* (The New Proof), was almost new, a fast sailer, and commanded by Captain Smith, a young, artless, and pleasant man. About mid-day we had already lost sight of the Holstein coast to the north; but towards the south it stretched far out like a faint blue stripe, on which no objects could be discerned. Langeland next appeared in the distance, like four or five islands together; for the bendings of the sea concealed from us the connection between the hills, which are about two hundred feet high. In the afternoon we saw Laaland and Femern over against us, with the Church of

Petersdorff, which shines at an uncommon distance ; both of them very flat islands, on which not a hill is observable.   The wind now was laid, the sea became quite calm, the sick gradually appeared again on the deck, and the ship scarcely seemed to move over the deep.

The night was beautiful, serene, and clear : we might have almost supposed ourselves on shore.   By break of day we saw the wood on the coast of Femern ; but the progress of the ship was almost imperceptible, although we had a light wind in our favour.   The captain thought he ought to lie more towards the land of Femern, as the current to the west, under Laaland, was too strong, and retarded the progress of the ship.   I should not have dreamt of the existence of currents in this part of the Baltic ; but the captain was quite right.   Our progress to be sure was not very quick, but it was still something.   I did not learn whether this western current under Laaland was accidental or constant.   One would think that it must have a strong dependance on the current in Oeresund, which in north-west winds flows into the sea, and in west winds flows out.

The island of Moen appeared to approach nearer and nearer to us, and we turned our looks with delight to the land ; for the island has a most inviting aspect on this side.   Green meadows spread themselves softly down towards the coast, crowned above with well-built country-houses, and cornfields, and here and there a thicket or a grove relieved the uniformity of the landscape.   We saw also herds by the water, and villages towards the shore. About four or five o'clock in the afternoon, we had proceeded round the extreme southernmost point of the island.   The row of steep chalk-rocks, of which we could not before see the smallest trace, then all at once made their appearance.   As far as we could now see round the island, these rocks, which are of a dazzling white colour, were in view, and rose to more than two hundred feet of perpendicular height.   The sea now also suddenly appeared as if alive.   Far and near, the vessels of the Sound floated around us.   We saw at least a hundred at once ; and from Kiel to this place we had scarcely seen one.   The advancing night deprived us of this animated spectacle. Moens coast still shone at a distance, and before us appeared the white rocks of Stevens-Klint, on the coast of Zealand.

On Saturday, the 12th of July, at break of day, Stevens-Klint lay far behind us ; and all that we saw of it was a light and almost imperceptible stripe.   The wind was very faint ; but the current in the Sound drove us quickly along.   Again a current, and in opposition to the one of yesterday! The towers of Copenhagen appeared now like masts in the distance, rising out of the water.   On the right, the blue stripe of the Swedish coast stretched out as far as the view could reach, and experienced eyes could distinguish Falsterbo and Malmoe.   We went close by Saltholm, but we did not see the island : the cattle and the mills, at a distance of only a quarter of a German mile from us, seemed as if not on the land, but in the water; such is the flatness of the island : and it appeared quite incredible to me, though I afterwards heard it confirmed, that there were considerable lime-stone quarries in this island, from which the whole of the lime-stone of the capital was derived.   But the mouths of these quarries lie below the level of the sea. We passed quickly by the animated Dragoe towards Amak, whose inhabitants fit out ships for the East and West Indies; and we immediately came alongside of the black hulks which serve as a battery to defend the entrance of Copenhagen.   Every Danish tongue was now let loose at this sight, and Provesten, the name of the battery, and the 2nd of April, 1801, rendered us almost heedless of Copenhagen, which lay stretched out before us.   Who would not have been warmed, and felt his heart swell within him, in the place where a few moored vessels opposed such a gallant resistance to the whole might of the English fleet, and Nelson, their victorious leader! If they had been supported with more firmness on land, they would have compelled him to retreat.

We sailed round Amak ; and there we saw the excellent stone battery of *the Three Crowns* in the water, which were determined to do against Nelson what the *Provesten* so many hours with heroic courage actually effected.   All was now peace here; country-houses among green shrubberies and gardens, stretched along the shore ; the town with its high steeples lay in all its extent before us, and in the distance the castle of Friedrichsberg.   At eight in the morning we were lying at anchor before the Custom-house.

What a fortunate passage !   In forty-nine hours from Kiel, with so smooth and unruffled a sea.   But the passage is seldom so favourable.   The packet, which was expected in Kiel from Copenhagen, was twelve days at sea, coasted Pomerania and Rugen, was driven on Fühnen, and instead of reaching the entrance of the Kielerfiord, was at length compelled to land at Heiligenhaven, on the western coast of Holstein, where the impatient passengers were left.

Yet the Baltic, which we crossed, is little more than a lake. For it has not here the depth which we suppose in a great sea. Our captain the whole night through kept sounding between Laaland and Femern, a great way out at sea, lest the vessel should ground. The bottom is never more than sixteen fathoms deep, and commonly not above eight, nine, or ten, which is certainly a very shallow sea. On the Genoa coast, and at a very short distance from the land, Saussure found a bottom first at Portofino, at eight hundred and eighty-six feet, and then at Nizza at a depth of one thousand eight hundred feet. The salt contained in the water of the Baltic is also very trifling. I drew up sea water when we were exactly in the middle, between Laaland, Femern and Moen; and I found in Copenhagen the specific gravity of this water, by means of very accurate scales, according to professor Tralle's problem, to be 1,00937 at sixty-six degrees of Fahrenheit. This was even a high result, for according to the experiments of Wilke at Landscron, the gravity of the Sound water during east winds was only from 1,0047 to 1,0060, but on the other hand it rose during north-west winds to 1,0189. In the open sea at Heligoland, the sea water weighs 1,0321, and even at Ritzebüttel 1,0216.* Probably the saltness of the whole of the Baltic may never be constant, but variable, as in the Sound, according to the winds and currents. How could salt be contained in this shallow sea, into which so many considerable rivers flow, which would expel the sea water, if the winds did not incessantly introduce new salt water from the Northern Ocean?

Few towns in Europe have experienced in modern times such a hard fate as Copenhagen. Traces of the great fire, which in 1794 consumed a great portion of the town, were yet discernible, and the castle of Christiansburg was still in ruins. But since that period, the yet more dreadful bombardment of the English in 1807 inflicted a wound on the town, and the whole country, which ages will not be able to heal. Copenhagen stands no longer now on the same proud eminence on which she stood previous to the breaking out of the English war. It is no longer, as it used to be, the capital of the only peaceful state in Europe, the staple of the whole commerce of the North.

---

* M. de Humboldt *Bergm. Journal*, 1792. 138.

During all these misfortunes, a protecting genius seems to have watched over the principal scientific collections in the town.   However near to the danger, they were always saved from the fire, as if they had been unconsumable.   The Great Royal Library remains yet insulated, beside the ruined walls of *Christiansburg*; and the excellent collection of the University exists in the only building, or rather the only rooms, in the midst of interminable ruins.   The flames also did not spread towards Rosenburg or Amalienburg, where their ravages might have been so fatal.

The collection of minerals belonging to the University is in fact very considerable, and, as might be expected, every thing belonging to the north is found here in extraordinary beauty.   Arendal's Epidote, of an extraordinary size; scapolite, crystals of yellow titanium.   These fossils are first seen here in perfection.   I never saw such beautiful and large zircon crystals from the Syenite of Friedrichsvaern as in this collection.   All the pieces are excellently kept, which is seldom the case in such large collections; and it may not be amiss to inform those who before were ignorant of that circumstance, that professor Wad, whose merit is so great, belongs to the Wernerian school.

The royal collection in Rosenburg is also one of the most remarkable and distinguished; not from the plan of the institution, for it is within these few years only, or more properly speaking, since the direction of professor Wad, that the enlargement and arrangement of the collection have been conducted on any thing like a scientific principle; but on account of the monstrous and collossal size of the specimens.   There are pieces of Kongsberger native silver, a foot in length, and from six to eight pounds in weight.   A mass of silver in its natural rock, said to be worth ten thousand dollars, is in truth by no means remarkable.   But the calcedony of Iceland is of a most extraordinary magnificence.   The drops of calcedony hang from the top to the bottom of the piece, like inch thick pillars behind one another. Many remain in the middle, and do not reach the bottom.*   The quantity of zeolite is immense.   A piece of amber from Jutland, placed on a velvet

---

* The Calcedonies of Iceland and Feroe are remarkable, not only on account of their magnificence and extraordinary beauty, but also for the various curious and interesting forms they exhibit, all of which, even the stalactitic, I consider to be crystalline shoots.—J.

cushion, is little inferior in size to that in the Berlin collection. Large pieces of geyser sinter, which the collection lately received from Iceland, were almost in the state of conchoidal opal. Besides these magnificent specimens, the collection possesses as great a treasure of choice and well-preserved northern fossils from Arendal as the cabinet of the University, since the beautiful collection of Manthey, the counsellor of state, was purchased and brought to Rosenburg. This department is so complete, that it is almost unequalled. It were to be wished that the same provision was made for those who wish to obtain by means of ocular inspection a knowledge of the composition of the mountains of the Danish dominions, which are better known. But we look in vain for specimens of rocks in any collection of Copenhagen. The appearance of the lime-stone used in Jutland or in Faxo; the stone in which the Kongsberg mine was formerly wrought; or the figure of the large rocks on the west coast of Norway at Bergen or in Nordland, we seek in vain to discover either from a series or from individual specimens; and yet it would be so easy, and at the same time so princely, to form in a royal collection something like a picture of the interior of the whole dominions.

Professor Schumacher has also made a very beautiful and complete collection of Norwegian fossils; among them he has many things which are not to be found in other cabinets. I should doubt the possibility of shewing more beautiful and distinct specimens of the *Leucite* of Friedrichsvarn. The crystals are as large as the leucite of Albano; we can easily recognize the double octahedral pyramid with four terminal planes; and the white colour gives it a still stronger resemblance to the Roman leucite. They rest insulated on Hornblende in the Syenite, which contains zircon in such abundance. Yet in France there is a conviction that these crystals are not leucite, but analcim (Werner's cubicite). The difference between these two fossils externally is indeed very trifling, and consists principally in the greater hardness of the analcim, and its less frequent tendency to foliated fracture. But in the chemical analysis they differ more. Leucite contains twenty-four pro cent kali, and analcim, on the other hand, ten pro cent of soda. I cannot omit noticing that the appearance of this ample collection of Arendal put me always in mind of the fossils of Vesuvius. The analogy between them is great. Here and there new and wholly unknown fossils were contained in primitive stones; and those which were known appeared in forms seldom hitherto observed. But in both places they are numerous, and heaped together to a degree which we seldom find in an existing bed; and were we to find all that the country of Arendal produces in such uncommon perfection at a distance from their first beds, and heaped together on the declivity of a volcano like Vesuvius, we should be as embarrassed as we now are at the appearance of so many druses of Nephelin, Meionite, Vesuvian, Hornblende, and Felspar, in the granular limestone on the sides of Vesuvius. The first beds of these masses may therefore have been the same as a bed in micaceous slate, or gneiss like that of Arendal; and in this case, it must be sought westwards in the sea, or in Sardinia and Corsica; for towards the west, the primitive rocks are to be found on the Italian coasts.

The exotic articles possessed by M. Schumacher in his collection are numerous, but of no great importance.

The treasures of the Great Royal Library are well known. But the excellent collection of books of Classen, which no stranger in Copenhagen can examine without envying, is much less known than it ought to be. General Classen bequeathed not only his books to the public, but also a sufficient fund along with them, part of which was dedicated to the erection of a suitable building for the reception of the books, and the remainder to provide a revenue for the increase of the collection. He chiefly possessed historical books. But the directors of the new institution had the good sense to give up this department entirely to the Great Library, and to confine themselves solely to natural history, the arts, and travels. They wisely judged that in this way alone it was possible for them to attain any thing like perfection; and every one who wishes for other books, may find them with more certainty in the Great Library. The consequence is, that in Classen's library we not only find the most expensive botanical works and original travels, but also a more complete collection of even the most fugitive German and foreign publications connected with these departments, than is any where else to be met with. This library possesses a yearly revenue of four thousand rix dollars; which is more than that of the Great Library, and indeed than that of most of the public libraries of Europe.

In Copenhagen therefore there is no want of assistance from books; and in this point of view it is worthy of the capital of an extensive state.

That the town is in general beautiful and well built, all travellers are unanimous in stating; and of this we become soon convinced. After every fire, which has consumed whole streets, they have been always rebuilt more beautiful and wide, and altogether on a more convenient plan, so that the town bears in many places no resemblance to what it was before 1728, 1794, and 1807. Another species of magnificence cannot fail to strike the inhabitant of a flat country, which is little noticed in the descriptions of Copenhagen. The streets are almost every where at the sides paved with large oblong granite flags, and many of the canals are wholly lined with flags of a monstrous size. I conjectured at first, that they had been brought from some quarry in Norway; but I was assured by M. Wad that this immense quantity was all derived from large masses on the coast of Zealand. This is a remarkable circumstance, and deserving of attention. If so many large blocks are found in Zealand, they must necessarily have made their way over the sea; for there are no granite mountains in Zealand. In whatever manner these blocks may have been driven over to Zealand, we may easily conceive that they have been brought in the same manner to Mecklenburg, Pomerania, and Brandenburg. Large granite and gneiss blocks are even seen on the smaller islands; for instance, there are many on Femoe at Laaland. These are still further proofs that all the granite of the plains in the north of Germany, however great the distance, has been torn from the northern mountains, and not from the hills of Saxony or Silesia. We are not in possession of facts sufficient to enable us to develope the wonderful revolutions of nature, by which this may have been effected; but every observation brings us nearer to the causes, and in a few years perhaps they may be discovered.

In the great works begun shortly before the breaking out of the English war, for the purpose of defending the fortress of Cronenburg from the violence of the waves, large blocks were conveyed from Norway; they were brought from thence for the first time; and from the account of Lieutenant-Colonel Hammer in Bergen, they were not divided by nature into regular parallelopipeds, but into magnitudes and forms, which however required little labour at Nattling in Quindherred Prastegield, at the great Hardangerfiord, and about ten or twelve miles south from Bergen. It is a pity that this source of gain for Norway should have been put a stop to by the war; for the probability is, that large blocks of gneiss and flags would have been at last brought to Copenhagen from Hardangerfiord;

that Copenhagen would have thus been improved in beauty; and that the industry of Norway would have been somewhat quickened.

The season of the year was now far advanced, and to any one who intended to ascend the high mountains of Norway, there was no time to be lost. I had therefore but a few days to remain at Copenhagen, and was under the necessity of setting out for Christiania as quickly as possible. The Kongsberg silver waggon was then setting out, and I availed myself of such an excellent opportunity of travelling quickly and expeditiously through a country the language of which was not understood by me. According to its first destination, this was not a waggon for passengers, but was dispatched through Sweden to Kongsberg to convey the produce of the silver mines to the Mint at Copenhagen. It was necessary, however, that an overseer should accompany the silver to Copenhagen, and natural that he should allow another traveller to occupy the empty place in his carriage. From this a sort of regular diligence has at length arisen, which has contributed not a little to increase the intercourse between Copenhagen and Christiania. It has given rise in both places to mutual wants and inquiries, of which formerly even an idea was hardly entertained. In winter, the silver waggon made its appearance in Copenhagen, not so much loaded with silver as with delicious birds of all sorts for the kitchens of the capital: wood and black grouse, ptarmigans, and hazel-grouse *(Tetrao)* were sent in incredible quantities, to the great advantage of the mountaineers of Norway. On the other hand, Christiania was supplied with many little wants from the southern countries, as well as with journals and books, which were always uncertain and late of arrival by sea-carriage. The war, which put an end to this intercourse, first demonstrated the extent of its utility to a country whose advancement is so much retarded by the want of internal and external intercourse.

This diligence left Copenhagen every three weeks; and its departure was always two or three days beforehand announced in the newspapers. In fourteen days the journey is generally finished.

I set out from Copenhagen about five o'clock in the morning of the twenty-fourth of July, and I found the silver-waggon filled with good and agreeable society, so that for the first miles towards Elsineur I repented very little the choice which I had made, particularly as the person in charge of the waggon (generally an uncultivated man) not only takes care of its progress, but also provides food and beds for his passengers, and settles with them on arriving at Christiania. We proceeded quickly, and arrived by ten o'clock at Elsineur.

I had often heard of the grandeur of the view of the vessels sailing in the Sound, when descending from the heights above Elsineur: but my expectations were exceeded: nothing could equal it. The Sound is covered with sails, and the vessels ascending and descending seem to pass with the swiftness of an arrow. Cronenburg, with its steeples and walls, appears in the most picturesque manner to oppose their progress; but they clear it and disappear in a moment, as if swallowed up by the earth. There is something enchanting and over-powering in the perpetual repetition of the light and waving motion of such large and proudly extended masses. We soon got nearer to them, for the preparations for embarkation retarded us only half an hour. We tacked across the Sound, during which time we passed the half of all the European flags, amongst which, however, the American,* a skyblue studded with golden stars, was by far the most frequent.

In an hour's time we reached the shore of Helsingborg, and trod the Swedish shore.

At my very first step out of the boat, I saw, with no small astonishment, a work which might have done honour to the greatest state. They had begun the building of a mole of hewn granite, twenty feet at least in breadth, which already advanced far into the sea, and was continued with great diligence. A large, secure, and complete harbour was projected at this place, which the Sound on both sides stands very much in need of; for even in Elsineur, where so many vessels put in, the only security against the wind and waves is a row of stakes covered with wooden scaffoldings. What would Sweden not be, if the spirit which conceived such a plan had never forsaken its monarchs?

We divided ourselves in Helsingborg among several light Swedish carriages, such as are generally used for travelling in that country; and we proceeded the same afternoon to Engelholm. What was there in Helsingborg to detain us! The town is dull, and so small, that we look for it sometime after we leave it. We travelled over a flat country, through

---

* A nice critic might here object to the classing the American among the European flags.—T.

several considerable villages, and reached Engelholm at nightfall. Close to the town we were not a little surprised by the appearance of one of the most daring wooden bridges that ever was constructed. It is a hanging bridge, and is, at least, sixty feet in width, and fifty feet above the surface of the river. This elevated bridge is surrounded with beeches, and has a very picturesque effect. It is entirely new, and was just then completed. But how little does this impression correspond with that we received from the town ! It is only one street; the houses are only one story high; and the market-place is a green field. The inhabitants do not exceed a few hundreds.

On the 21st of July we proceeded, for more than four English miles, through a flat country, with a majestic prospect of the range of rocks of Kullen, which advance far into the sea, till we came to Margaretetorp: there we ascended the range of small hills of two or three hundred feet in height, which sepa- rate Schonen from Halland, and which are called Halland's Os : a thick beech wood ascended to the top, through which we had now and then an extensive view of the sea. The road here is highly romantic : the range is, however, by no means broad, and before we had completed four miles more to Karup, we were again on the plain.

Here, for the first time for twelve months, did I again see continued rock. With the exception of Möens chalk-rocks and the insulated Segeberg, from Berlin to this place such a sight was impossible. They were small and insig- nificant rocks of granite, or rather gneiss, with mica sparingly intermingled, and not lying in scales one piece above another, which now and then pro- jected on the declivity towards Karup. From these, however, the immense quantities of blocks and boulder stones cannot be derived with which the whole descent from Margaretetorp is covered, for in that case they would probably be equally divided on both sides of the mountain, as well more large and angular. All the way from Helsingborg to Fleningen great blocks lay scattered on the surface, and they are there at a considerable distance from these low hills. They probably belong to the blocks with which Zealand, Mecklenburg, and Pomerania are covered. This was also the opinion of Professor Wad, who told me that we ought not to look to Norway for the origin of these masses, but rather suppose them to come from Schonen through Smoland to Sweden ; and this is confirmed by M. Haussman, who travelled so attentively through Sweden. In Smoland, he says, all is desolation. The blocks lie above one another like rocks, and the whole province is covered by them. Though the masses of the north of Germany should not have been derived directly from Halland and Smoland, yet these are always additional helps to guide us to the beginning of this great revolution of nature.

It is true that it is only local, and confined to the north; but as is the case with all observations in geognosy—if we succeed in connecting all the facts relating to the spreading of boulder stones in the north in the order of cause and effect, light will be thrown on a number of similar phenomena in other parts of the earth, and we shall then be enabled to ascend to the general causes which concurred in producing all these different phenomena perhaps at the very same period.

In Laholm, I saw the water of the river precipitated over the rocks in a small noisy cascade. The novelty of this view renders it always uncommonly agreeable. What Swede would have ever remarked this little waterfall? But for four hundred and fifty English miles on the road from Denmark, we see here, for the first time, water precipitated over a rock.

The country all the way to Halmstadt is singular, and has a melancholic and dismal appearance. Small rocks of gneiss rise on every side of the road, not more than twenty feet in height, and surrounded with large blocks like fallen towers. Black heath covers the level ground between the rocks, or pure sand makes its appearance through the dark covering; and a wretched cottage is now and then seen at great intervals. Though Halland may probably never have been more populous, yet it had not always this dismal appearance. For the *Knyttlinga Saga* informs us that the whole country was formerly a continued wood of beech and oak, which was applied to the feeding of swine; and this was found so advantageous, that *Knud the Great, Suend Ulfsön's son*, demanded the possession of the whole wood as a royal domain, for the purpose of feeding the royal swineherds.* Although Halmstadt is the capital of the whole province, it does not much diminish this dismal impression; and the country does not change till nine miles further on between Quiville and Sloinge. A small range of hills then commences similar to Hallandsos, and every thing immediately assumes an animated and interesting appearance. The trees have all a fresh and green aspect, and they grow vigorously all the way up the hills. The

---

* *Suhm. Kiöbenh. Gelesk Skrif. XI. 110.*

cottages are also numerous along the declivities, and among the thickets of ash, beech, oak, and sycamore. Gittinge, about the middle of the heights, and completely buried among the trees, is highly in the Idyl style, and brings us in mind of Switzerland; but all this disappears with the hills. At Mostrup and Sloinge in the plain, and all the way to Falckenberg, nothing is to be seen but blocks of gneiss, rocks and heath: and as to Falckenberg, what a town! only one street, and hardly a hundred houses in it. Four of the houses are covered with tiles, and these are public houses; the rest are wretched huts covered with straw. I went along the street to the harbour, and there I saw a single vessel, a galley, and the sea in the distance, where many thousand ships of the Sound pass near this place. Falckenberg's advantageous natural situation for the herring and other fisheries has never been profited by. The harbour seems good, and is at the mouth of a considerable river. We were very tolerably accommodated here, however, at the public houses.

The houses of Falckenberg are covered with straw: this is undoubtedly through poverty, as tiles or boards would be dearer: but what is here the effect of want, would farther north be deemed more than superfluous; for all the peasants of these countries, who live beyond the sixty-first degree of latitude, consider a straw roof as a most unwarrantable piece of prodigality, sufficient to draw down the vengeance of Heaven. Straw is there like corn, a noble gift of God for the maintenance of man and beast; and straw on the roof, is to the inhabitant of Norway, or Westbothnia, or Jämteland, such a sight as a roof covered with bread would be to a German boor. In Falckenberg we see, for the last time, houses covered with straw, on our way to Norway; and it is singular enough that the country about Halland possesses a sufficiency of straw for the purpose of thatching. Yet there is no stir in the harbour!

Warberg is a better, larger, and more beautiful place, and several ships at anchor were lying here. It is a sort of fortification, with a castle beside it, towards the sea-shore, in which state prisoners are generally kept. General Pechlin, well known in the recent Swedish history, died a few years ago in confinement here.

I know not whether it was in consequence of the incessant rain, which drenched us the whole day through, or whether it was the natural impression of the country, but from Warberg onwards, I frequently believed myself on the Height of St. Gotthardt, every thing on the way appeared so bare, solitary, and rocky: no trees, no habitations, and nothing but heath was to be seen. But *Ljunghedar* (Heath), the Swedes themselves acknowledge as the character of North Halland*; and therefore the appearance of the country may not be much better in more favourable weather. Even Gottenburg, the second town of the Swedish Kingdom, does not, till we are quite close to it, give any sort of animation to the country, as great towns in general do. We did not enter it, but passed the night at Reberg, an English mile from the town, on the common Norwegian road.

Our way next morning lay through the valley along the large stream of Gotha Elv for three miles, sometimes under rocks, and sometimes over hills, with the view of the broad stream, on which ships frequently made their appearance. The cottages are now every where insulated, and never after this place collected in villages. Altogether, the prospects are highly diversified, agreeable, and perpetually changing. At the old fort of Kongelf we crossed both arms of the river at a ferry. The arm, which farther down runs through Gottenburg, was neither rapid nor broad; but the arm which runs under Kongelf is very rapid, and by far the largest. In breadth it does not yield to the Mayn below Frankfort. Its rapidity, however, renders it a dead stream: it is dreaded by vessels who proceed down the more placid arm to Gottenburg. We left this really delightful valley for the first time at Holmen, in view of the great and noisy sluice of Trollhättan, and advanced over the bleak hills again towards the sea. Beside the advantage of the beautiful prospects, this valley completely freed me from a prejudice, which, misled by descriptions, I had too long entertained. How often had I heard and read that granite almost every where makes its appearance in Sweden; that the rocks of Gottenburg consist entirely of granite; and that the sluices of Trollhättan are of granite. I looked on many rocks in Halland with astonishment: they appeared to me not only striped, but even slatey; not granite but gneiss; and between Quiville and Sloinge, not unfrequently beds of hornblende and felspar, mixed in stripes, beds such as can scarcely be said to occur in granite. I saw also beds dipping to the south; but in granite the stratification is not distinct. Nearer to Falckenberg all the blocks appeared always striped gneiss, although their number had very

---

considerably increased. For from the rocks of the little hills, about a quarter of a mile from the road, descended an uninterrupted waste of blocks, as rough and wild as streams of lava generally appear. The rocks of Kongsbacka and Warberg suggested more and more the idea of gneiss. Yet I believed that all these might be only accidental changes of the granite, and that on a closer examination they would perhaps appear less remarkable; but from Gotha Elv, onwards, there can remain no longer any doubt as to the nature of the stone. It is throughout, and in the whole extent, no where granite, but clear and decided gneiss. The whole is distinctly slatey; the mica lies in scales one above another, and never in single folia as in granite. Hence the beds are easily followed: on both sides of the valley they regularly dip towards the west and north-west, and preserve this direction from Kongelf to Holmen. There I was at last convinced that granite is probably in Sweden and the whole of the north a great rarity, and that it has never properly been distinguished from gneiss by Swedish and foreign mineralogists. M. Haussmann travelled through all Sweden, and at the end of his journey he affirmed that granite of the older formation, such as we see in Saxony, Silesia, at the Hartz, in Lower Austria, and Lower Dauphiny, never was seen by him in Sweden*. Many perhaps will be inclined to think that this distinction between gneiss and granite does not deserve so serious a consideration, and is not of sufficient importance to justify an opposition to so many deserving men, who have always continued to mention granite as existing in Sweden. But if we do not endeavour, along with the increase of our geological experience, to regulate it with an equally increasing accuracy, we shall long be strangers to the real knowledge of the structure of the earth. The difference between gneiss and granite is very considerable every where, and of course also in Sweden. Such a stratification, such a slatey structure, and the beds of hornblende, lime-stone, and ores, which appear the same in Sweden as in other countries, are not to be found in granite. On the other hand, we may always expect fewer foreign beds in real granite, where quartz, felspar, and mica are granularly aggregated, than in gneiss, where the aggregation is slatey †.

---

* *Moll. N. Jahrb. der Berg. und Hüttenkunde I.* 18.

† In different parts of Scotland, as in the Island of Coll, Long Island, Rona, &c. occur the same kinds of gneiss as those which have been in Sweden confounded with granite.—*J.*

As we descended from Grohede towards Uddewalla, new scenes and most wonderful prospects were opened to us. The country appeared more and more cut into small rocks; and dales, like canals, run between them. Farther on, small sea-bays forced their way through these narrow and steep fissures, and winded in a singular manner round the rocks. From the heights the country looks as if it were split into innumerable pieces. And as we descended the narrow openings, and looked through the small rock-islands, the views were dark and gloomy, as in many parts of the Lake of Lucerne, or Upper Austria, but in miniature, for the rocks are but a few hundred feet high. This is the true *Skiär*, almost exclusively peculiar to the north. The vallies or fissures between the rocks become deeper and deeper; the sea covers also more and more the islands, till at last they appear only like shoals, occasionally above the surface; and at a few miles in the sea they can only be discerned during the furious motion of the waves among these deep rocks in a storm. The Norwegians and Swedes have very appropriately named these rock-islands *Skiär*. They have borrowed it from *Skiäre*—to cut through; and, in fact, the tops and ridges of rocks not only cut the surface of the sea, but the whole coast for miles through; and though the surface of the sea were raised a few hundred feet higher, the land would still be divided by the water into fissures and islands, similar to those which at present run along the coast. A person descending from an eminence to one of these shores would with difficulty believe himself close to the sea. The deeply indented bays *(fiorde)* resemble large rivers, or small mountain-lakes, surrounded by steep rocks.

Our way down the narrow valley to Uddewalla lay through majestic trees, which gave a promise of a better climate below among the cliffs than between the dry rocks of the heights. Shortly after, we saw the town stretched out before us on both sides of the river. It was the largest and most thriving town that we had hitherto seen in Sweden. All the houses were of two stories, and had a new appearance. They were all new in reality, for the town was wholly burnt to the ground a few years before. There was life and agitation in the streets, and it was evident that the arm of the sea had not here penetrated so far between the rocks to no purpose. We had scarcely left the town, when we saw from a small eminence vessels lying in a little bay, concealed among the rocks. In this bay they find deeper water than at Uddewalla, and lie fast among the cliffs as in a dock. We arrived at Quistrum late in the evening, which consists of three or four well-built public-houses at the edge of a highly advancing sea-bay, and is almost wholly concealed under a thick row of ash trees.

The Danes are fond of speaking of Quistrum; for here the Norwegian army under Prince Charles of Hesse gained the only engagement of any consequence in the short war with Sweden in 1788. Quistrum is a narrow pass. The road is sunk between both sides of the height, and runs for some distance between steep rocks towards the sea. The Swedes endeavoured to prevent the advance of the Norwegians to Gottenburg, by securing the pass; but the attack was so quick and unexpected, that they were compelled to retreat. The way was now open to the Norwegians, and the rich town of Gottenburg would probably have fallen into their hands, if the English Ambassador had not arrived at the camp, and by his serious admonitions prevented the advance of the Danes.

On the 28th of July, we soon reached the heights, and lost all traces of the romantic views of the valley of Quistrum. It seemed to me frequently as if we had ascended a considerable height, when we ascertained by the barometer in the Desert of Svarteborg that we were not more than three hundred and ninety three English feet above the level of the sea. The rarity of trees here is quite astonishing; and the cottages have a naked and by no means agreeable appearance. Bahuslehn lies too near to the sea, and is too much exposed to the west winds from the Northern Ocean; and sea winds never allow trees to grow to any considerable height. In the neighbourhood of Hogdal we saw for the first time in the evening the romantic *Skiär* country, the sea-bays, the deeply-indented rocks, which now retreating and now advancing, displayed the most singular diversities of light and shade. We were close upon the confines of the kingdom. We had only one height to ascend, from whence we saw the Swinesund in the narrow valley deep under our feet.

Its appearance brought keenly to my recollection the view of the deep valley through which the Doux descends, about a mile from *la chaux de Fond*, in Switzerland, a situation which I have always accounted one of the most remarkable. But we cannot see at the Doux what appeared before our eyes in the descent into this valley: a large three-masted ship in full sail, in a narrow and deep mountain vale, in which the dark water appeared only a large stream—a most surprising sight indeed! The ship came from

Frederickshall, and went out into the sea. We could not possibly have conceived, from the view before us, where we should either have looked for the sea, or the town, which could send out such large vessels. It appears as if both kingdoms were here separated from one another by a great rent.

As we advanced up the steep ascent on the Norwegian side, we now and then had a view of the batteries and walls of the Fortress of Frederickshall over the dark pines on the Heights of Swinesund. The hills seemed to unite, and the fortress to rise out of a dark wilderness; but after attaining the top of the height, an open plain sprinkled with hills was spread out before us. We advanced speedily onwards, and found at Westgaard, about two English miles from Swinesund, a good, agreeable, and clean inn, provided with several of the comforts which are often found wanting in many a large town.

The beautiful country of Swinesund is not also without its importance to the geologist. I was almost inclined to believe in Hogdal that I had been too hasty, when I thought at the Gotha Elv that the existence of granite in these countries was unfounded; for Hogdal's stone was to appearance distinctly granite. Felspar occurred in pretty large crystals amongst the mica and quartz. This continued all along the Swedish descent to Swinesund, till the water-edge. But the very first stones on the Norwegian side appeared in the most distinct manner to be gneiss. They were throughout fine slatey, with quartz beds, and frequently beds of small granular hornblende. At Westgaard this gneiss appeared rich in mica, and the mica was scaly foliated. The granite of Hogdal is also wholly enveloped in gneiss; and we can hardly concede to it a proper and individual existence. It is probably only a variety of the gneiss, and wholly subordinate to it.

On the following day our road was admirably calculated to enable us to survey the country over which we had travelled. There were indeed no villages here, but farm-houses *(gaarde)* every where; and almost all of them were large and well built, and presented along the declivity an agreeable object to the eye. From the road we have a perpetual and enlivening view of distant bays and islands in the sea. About mid-day we passed through Frederickstadt, one of the principal fortifications of the country, and a very small, though not altogether an unthriving town. In time of peace, a whole regiment lies in the garrison here, and vessels come close to the walls. Although the number of inhabitants it was found in 1801 did not exceed one thousand eight hundred and thirty-seven persons : this is by no means a small number for a town in the north ; and the half of all the Swedish towns are not more populous. We made no long stay, but proceeded to descend the Glommen, the principal stream of the north, over which we passed at a ferry. We found the stream every way deserving of its fame. It runs with great rapidity, and below the town is at least as broad as the Rhine at Coblentz, notwithstanding the separation of a considerable arm from the main stream, about a mile farther up at Thunöe. The banks of the vale of the stream are insignificant, as well as all the heights which we passed. We soon arrived at Moss, where a considerable river runs foaming over rocks in the middle of the town, dashing down from one wheel to another, till at last, at the edge of the Bay of Christiania, it drives the bellows of a great iron-work. An immense quantity of deals lie in heaps beside the twenty saw-mills, which are set in motion by these wheels at the water-falls ; and stunned with the noise of the saws, the falling water, and the large iron hammers, we proceeded through the works to a high fir-wood on the other side of the town. Deals and iron ! We had seen collected in one place what maintains and enriches the whole of the south of Norway. How large and majestic is the appearance of the firs here ! Even the woods alone through which the road passes, and the timber of which is not cut into deals, give us a clear idea that these trees never reach their greatest growth, except in high latitudes. The trees are indeed of a singular beauty.

We were drawing near to the sixtieth degree of latitude. We were now so far north, that I expected the country to present altogether a new appearance, particularly as all the way from Moss we had been ascending considerably, and the country seemed a broad and extensive sweep of hills.

We had no doubt left behind us many a large tree in Schonen and Halland, which could no longer shoot up here. The horse-beech tree *(carpinus betulus.* Afven Bok) is not once to be seen beyond the limits of Schonen ; and white and black poplars penetrate but a small way into Halland. We lost the majestic beech *(fagus silvestris)* at the Gotha Elv, beyond which it is not to be seen. I was therefore the more surprised to see the most beau-tiful oaks at Aas, and in the country round Corsegard. Who would have expected to see them so high beyond the beeches ? And in the evening, at Skytsjord, where we stopt, I saw a high cherry-tree, with excellent ripe

E

cherries. Roses were also beginning to bloom, a month, it is true, later than in the North of Germany, but still they were in flower.

Skytsjord lies five hundred and fifty-three English feet above the sea. The distance to Christiania was fourteen English miles, and the way lay through deep vallies among the hills. These miles we passed in a most delightful summer morning, a favour which seemed to be conferred on us by Heaven that the view of the wonderful country round Christiania might be enjoyed by us in all its glory. What variety! What astonishing forms of objects, looking down from the height of Egeberg! The large town at the end of the bay, in the midst of the country, spreads out in small divergent masses in every direction, till it is at last lost in the distance among villages, farm-houses, and well-built country-houses. There are ships in the harbour, ships behind the fascinating little islands before the bay, and other sails still appear in the distance. The majestic forms in the horizon of the steep hills rising over other hills, which bound the country to the westwards, are worthy of *Claude Lorrain.* I have long been seeking for a resemblance to this country, and to this landscape. It is only to be found at Geneva, on the Savoy side, towards the mountains of Jura; but the Lake of Geneva does not possess the islands of the Fiord, the numerous masts, and the ships and boats in sail. Here we have the impression of an extraordinary and beautiful country, united in a wonderfully diversified manner with the pleasure derived from the contemplation of human industry and activity.

We descended by numerous serpentine windings the steep height of the Egeberg, through the remains of the old Town of Opslo, and through a continued row of houses along the Bay to Christiania, which we reached about mid-day of the 30th of July.

CHAPTER II.

CHRISTIANIA.

Stone Houses.—Division of the Town into different Trades.—Meeting of the Country People at the Yearly Fair.—Development of Cultivation from the Town outwards.— Barter of various Articles.—Exports.—Timber Magazine.—Berndt Ancker.—Division of Society in Christiania.— Theatre.— Falssen.—School Library.—Military Academy.—Country Houses.—Climate of the Town.

WHAT makes Christiania the capital of Norway, is not merely the presence of the principal constituted authorities and public bodies of the country, nor is it the superiority of its population, for Bergen contains double the number of inhabitants; but it is rather the extensive influence of this town over the greatest part of the country, the various connections of the inhabitants partly with the capital of the kingdom, and partly with foreign countries, and the social mode of life and cultivation of these inhabitants.   Whatever change takes place in any part of Europe, is in the same manner as in Germany keenly felt and eagerly followed: but this is not the case in Bergen.   Many means of assistance, which are generally looked for in a capital, and where men meet actively together in great bodies, are to be found united in Christiania much more than in Drontheim, and still more than in the narrow-minded Bergen: as for Christiansand it is too small.

Whoever is acquainted with northern towns, will discover, from the exterior of Christiania, that it is a distinguished, a thriving, and even a beautiful town; for the streets are not only broad and straight, and nearly all intersect one another at right angles, which gives a gay and animated appearance to the whole: but almost all the houses are built of stone ; and wooden log houses are, for the most part, banished to the remotest streets of the suburbs.   When a Norwegian descends from his hills to the town, he stares at these stone houses as an unparalleled piece of magnificence; for perhaps he never saw before, in the interior of the country, a single house of stone: and those who have lived sometime in Drontheim or Bergen, where stone houses are rarities, and wholly concealed among the wooden houses, are willingly disposed to consider the houses in Christiania a very great luxury: they attribute to them a beauty which they do not in themselves possess, and they involuntarily connect with it the idea of a general prosperity, of a brisk trade, and of the superiority of this town over every other.

In this case, however, they would not judge altogether correctly, for it is not optional with the inhabitants to build as they do, as log houses have been long prohibited by the government in the circumference of the town; and the wisdom of the prohibition has been confirmed by experience. There is not a town in Norway which has not been once, at least, burnt to the ground. The fire rages terribly among the dry boards. Whole streets burst into flames at once, and it is in vain to think of either extinguishing the fire or saving the property. How much has Bergen suffered from fire, where the houses are closely crowded together among the rocks! How much Drontheim and Skeen! Moss was twice, in the course of the year 1807, devastated by fire; and in Sweden, Gottenburg, Uddewalla, Norkiöping, Gefle, a slight inattention lays the whole town in ashes; and what cost centuries to build is annihilated in a few moments. Christiania hears also the alarm drum as often as other Norwegian and Swedish towns; but since its origin, during nearly two whole centuries, it has never lost entire streets, and seldom more than ten houses at once.

If it were not for the prohibition, the inhabitants would, in general, soon return to wooden houses; and the greater cheapness as yet, and greater quickness of erection, would overbalance in their minds the idea of safety of life and property. The government itself, with no great consistency, thought proper, in 1806, to erect a large, beautiful, and excellent military hospital of logs, on an eminence at one of the ends of the towns: a considerable fabric, which appears full in view all the way from Egeberg. With this royal building in sight at every corner of the town, we are less disposed to suspect that the building with stone was not perfectly free on the part of the inhabitants. It is a pity that so few of the houses will bear a narrow inspection: some of them are neatly built; but these are rare. Even the rich Chamberlain, Berndt Ancker, who was surrounded with such extravagant luxury, left behind him no buildings to do honour either to his native town or himself.

Formerly the proprietors of houses seem to have deemed it a very great ornament to mark the initials of their name, and the year of erection, with great iron hooks, on the outside of the houses. It deforms the houses very much.

The town is by no means uniform, but is divided into several small towns, the boundaries of which may almost be laid down with certainty; and in these the exterior, the houses, trades, and manner of living, are very different from one another. In great towns we are accustomed to see this; but in a town like Christiania we are hardly prepared to expect it. There is an exact boundary between the part of the town occupied with the inland trade and that where the foreign trade is carried on.

The straight streets, which cross at right angles, run up from the harbour, but do not extend all the way to the country. The capitalists, the wholesale dealers, the ship owners, those who hold government offices, find more room here than elsewhere for their large houses; and the consequence is, a greater stillness, and almost a dead silence in these streets. They are called the quartale, and every person in the quartale, according to the way of thinking here, is considered richer, finer, and more polished than the inhabitants of the other streets.

On the other hand, there is more stir in that part of the town which runs out into the country. The houses are more closely crowded together, and every bit of ground is carefully occupied. Whatever comes from the country must pass through these streets. All the artizans, shopkeepers, and retailers, who wish to dispose of their commodities to the country people, draw near to them; and signs and posts without number invite the entrance of purchasers. I have often considered, with astonishment, the multitude of small shops and booths. How is it possible, said I to myself, that so many people can derive a living in so small a town from the same trade? I looked over the lists, and found, that of nine thousand and five inhabitants, which Christiania contained in 1801, including the garrison, one hundred and ten were shopkeepers, two hundred and twenty retail dealers, and two hundred and forty-two master artizans. In what other town, with the same population, shall we find even the quarter of this number?— But let a person wait for the weekly market, and still more for the annual fair, or winter, which connects every place together, and he would then be almost tempted to believe that different nations were collected together in this place; for the Swedes, Danes, and

Norwegians, assuredly do not differ more from one another than the inhabitants of various vallies, who assemble from all parts to the annual fair. This is one of the most interesting spectacles for every stranger who visits Norway, and for every person who wishes to examine human nature, and to trace by what routes and associations man gradually advances in the progress of cultivation towards his destination.

For several days before the annual fair, which is held on the thirteenth of January, the town is filled with country people from all quarters; and figures make their appearance, such as before were not seen in the streets. The strong and robust inhabitant of *Guldbrandsdalen*, in his long coat of the seventeenth century, and with his little red cap on his head, walks by the side of the comparatively elegant boor of *Walders*, who, in features and dress, is as unlike him as if he came from beyond the sea. The rich proprietors from *Hedemarken* pass along as if they were of the inferior order of townspeople; and their coats of home-made cloth are cut in an antiquated fashion, as is usual in country places. From *Oesterdalen*, on the Swedish boundaries, appears a higher class of men; but we may easily see, from their carriage, that it is borrowed from their neighbours. On the other hand, we see the rough and almost stupid native of *Hallingdalen*, in a true national uniform, and the sturdy men of *Oevre Tellemarken*, still more rough and stupid. They alone yet continue to wear the broad northern girdle round the waist; which the native of Tellemarken embroiders and ornaments in quite a different manner from the other; and in this girdle they fix a large knife like the Italians, which was formerly as often used by them for attack and warfare as for conveniency. They wear a short jacket, with a sort of epaulette on it, and a small cap on the head: thin short leathern breeches contain in the side pockets all the wants of the moment, and almost always the important small iron tobacco-pipe. Every step and movement of these men is characteristic and definite. They have only one object in view, and nothing which surrounds them can deaden the eagerness with which they pursue that object. The boor of *Foulloug* and *Moss* is far from having this distinct character. Nearer to the town, his business is also more various, and he looks around him with attention and caution to discover any little advantages which may bring him easier and more securely to his end: he no longer lives insulated in his valley, relying on his own individual physical strength, but has become, through common interest and connections, a part of a nation,

This has been effected by the capital : it, and it alone, has effected this diversity among the country people, and it proves itself to be a capital in gradually burying, and even altogether changing, and extirpating, all nationality through so great an extent. Who would believe that in the times of Harald Haarfager, or Saint Oluf, the people in Guldbrandsdalen lived and dressed as at present? Who would suppose that the people of Oesterdalen, and the people of Hedemarken, possessed many remains of those times? But to be convinced that all these changes have proceeded from the town, we need travel but a very short way. An inhabitant of Guldbrandsdalen, in his long-bottomed coat, and monstrous stiff and indented flaps over his coat pockets, has quite a strange appearance when he appears in the streets of Christiania ; but the form of the dress and the men change upon us imperceptibly when we travel through their vallies.

In the suburbs of the town we find the same fashions that were prevalent in the quartale three or four years before; and there they again follow the fashion shortly before set by Paris and London. The peasant nearest the town, particularly in the neighbourhood of the streets leading to the country, takes a pattern from the coat he sees worn in the suburbs. He seldom penetrates farther into the town, and to the quartale he is altogether a stranger. It appears as if he changed his nature and habits with his dress; and this is natural enough; for it is only through more important connections he acquires the knowledge of this new fashion. In the clothes of the boors of Hedemarken and Foulloug, there is not the smallest trace of the national dress. The same fashion prevailed twenty-five years ago in Germany, and probably also in Christiania. As we ascend the country, the cut becomes older and older, but the dress of their ancestors is always perceptible ; and when we come to the strange dress in Guldbrandsdalen, what else is it but the regimental uniform of the times of Eugene and Marlborough? It is the same with the women ; they change perhaps slower and later ; but they must also at last yield to the influence of the town. " When we see a woman from Guldbrandsdalen in her full dress," said the noble and intelligent Chamberlain Rosenkrantz in Christiania to me one day, " we imagine ourselves standing before our old northern grandmothers, as they are occasionally to be seen in our antiquated family portraits."

If Hallingdalen, Walders, and especially Oevre Tellemarken, have yet retained in their exterior something exclusively peculiar to the country, they owe it to the remoteness of their vallies, and the difficulty of communication with the town. They are consequently seldom to be seen in the towns on the coast.

That the national character is in this manner limited to a few remote districts ; and that the towns have so powerful and extensive an influence on the surrounding country, and render the Norwegian a quite different being from what he was in the time of *Snorro Sturleson*, is lamented by many, and those among the most exalted characters, as a national calamity ; and they earnestly wish that it were possible to arrest the further progress. But why ? Are men to remain for ever stationary like insects ? Do they imagine that they have gained the golden fleece with that degree of virtue which can be practised in remote vallies ? And though this virtue may have somewhat of a national physiognomy, shall we concede to it any thing more than a relative excellence? And can, or should this excellence endure through the length of time ? It is certainly great and becoming to assert ones freedom boldly and vigorously in remote vales : but what if this freedom is never endangered ? Through social institutions, a still higher freedom may be acquired. Virtue has no national physiognomy, but belongs to all men, and to all ages. If it is ever produced by a particular national character, if the Norwegians, the Germans, the French, and the English, have each their particular virtue, however respectable this virtue may be, it is not pure ; it is not like the medicinal spring which restores health to the infirm, though superfluous to the strong.

We may therefore congratulate ourselves, and consider it as a fortunate circumstance, that we thus see a gradual change spreading from Christiania to the remotest provinces. Though evils formerly unknown may follow in the train, let them be weighed against the mass of newly developed good, and let us never forget that a free and happy man is a much more respectable and distinguished being than a free and happy *Samoiede*.

How different is the appearance of the more upland vales, from what it was before the town secured to the inhabitants a constant sale for their commodities ! How many conveniences, nay, almost necessaries of life, they can now have in exchange for their produce, to which formerly they were strangers ! And how many places of the country may now be turned to account, which were formerly doomed to remain uninhabited and waste ! It is certainly a great pleasure to meet on the days of the annual fair whole caravans of country people with their full-loaded sledges on all the roads leading to the town. They bring such a quantity of tallow, cheese, butter, and hides, with them, that we can hardly conceive how they can find a market for them in the town. But every landlord and householder waits for the time of the sledges : the boors are seldom embarrassed in the disposing of their tallow ; and they have it generally in their power to set their own price on their commodities. Yet in October, shortly before the commencement of the snow, thousands of oxen are driven to Christiania to supply the inhabitants with the necessary winter provisions. They take in return, corn, malt for beer at weddings and holidays, iron and ironmongery, and also, perhaps, fish, and some other small articles, which serve more for comfort than necessary support. This is the true division which nature and climate have made of the land :—grazing in the highest perfection among the hills, and grain from the town. Men are collected together in societies, that every situation may be applied to what is most suitable to it, and that the surplus may be exchanged for wants which other places can more easily supply.

The corn is mostly brought from Jutland, Fladstrand, Aaalborg, or Flensburg, partly in large ships by great capitalists, and partly in such small vessels, yachts, and even large boats, that we cannot help wondering how they durst expose themselves to the hazard of so boisterous a sea. But the passage is made in a single night, and the sale and profit are certain. That in time of peace the best and heaviest corn from the Baltic was always to be found in the harbour of Christiania is alone a sufficient proof that the town and country possessed means which enabled them to procure more than the necessaries of life. Those means were deals and iron, in return for which English gold flowed into Norway, and perhaps into no place more than Christiania; for the deals exported from Christiania have always been in high repute. It appears an easy matter to divide a tree at a saw-mill into deals and planks ; and the saw-mills themselves look exactly like those in other places; yet the greater prosperity of Christiania is entirely owing to the circumstance, that the deals exported from it are more skilfully sawed than elsewhere. The scrupulous and precise Englishman rejects the deals of Drontheim, and sends them to his less fastidious neighbours in Ireland, though the price of those of Christiania and Frederickstadt is much higher. This is not so much owing to the superior quality of the tree, as to the uniform thickness of the plank, and the accurate parallel of its two planes, and several other minute circumstances, that are only known to the sawer and to the delicate English merchant, but which nevertheless decide the happiness and misery, the wealth and poverty, of whole districts.

The activity and stir is great and striking in winter, when numberless sledges descend from the mountains with planks, and proceed with them to the great Timber Magazine. They are all accumulated in this place, which includes the whole space between the town and the suburb of Waterland, and stretches so far towards the bay, that the vessels seem to touch the piles of planks. Notwithstanding the extent, this magazine at the end of winter has the appearance of a great town of boards; and we lose ourselves in the multitude of avenues and streets leading through them. The noise of the country people bringing the timber continues without interruption so long as the snow lasts. They deliver over their boards to the overseers, who mark on their backs with chalk in letters and figures the place to which the boards were brought, and the number of them. It is a singular enough sight to see these boors hurrying away with all possible expedition to the counting-houses of the merchants in the Quartale, with this original species of obligation on their shoulders. By stopping on their way, or engaging in any other business, they might rub out the marks on their coats, and thus extinguish for ever all evidence of the debt. When they appear before the treasurer at the counting-house, they have no occasion to say a single word. They present their shoulders, and are immediately paid. The brush which the treasurer applies to his shoulders is the boor's acquittance.

There may be perhaps some twenty houses which have thriven by the timber trade: some of them are even rich. The great fortune which the ingenious Chamberlain Berndt Ancker acquired in a short time, principally by this trade, notwithstanding his expensive mode of living, and the immense sum of more than a million and a half of Danish dollars which he left behind him at his death, are certainly remarkable circumstances. His house is still standing; for he left his property to trustees, and destined the revenue to charitable purposes. It appears as if he could not bear the idea of breaking up this large sum, and that he wished it to remain a perpetual monument of what his talents enabled him to acquire. As the revenues must be expended in general charity, it is a pity that he did not take a pattern from the institutions of the worthy Pury, at Neufchatel, which still continue to have. such a beneficial effect on that place. The objects of Ancker's charity are widows and orphans, the poor and needy, and the fitting out of travellers to foreign countries; but all this is very indefinite, and instead of producing good, must waste and dry up the stream, by turning it into numberless channels.

If the power of controling the expenditure of these revenues were vested in the town, as is the case at Neufchatel, a regular stone harbour would probably have been gradually erected, in the room of the present tottering and filthy wooden quay *(Bryggen);* the town would have perhaps received a decent town-house, which it at present wants; and the pavements and streets would have been, in point of cleanliness and ornament, suitable to a great town. The fine supplies of water would not have remained at the crossings of streets, as at present, collected in wooden, but in stone reservoirs; and many other excellent improvements would have been adopted for the good of the town. What assists the town is returned over the whole country; and widows and orphans, the poor and needy, would have been easier provided for out of the great superfluity arising from it. Increased activity every where diminishes their numbers, which immediate pensions have a uniform tendency to increase. The memory of Ancker will always live in Christiania, from so many benevolent institutions; but in this way his honour would have extended throughout Europe, and the eternal gratitude of all Norway would have been secured to him.

The wealthy inhabitants of the town are engaged, from their extensive connections in trade, in numberless and difficult pursuits; but they contrive admirably to lighten the burdens of life by the pleasures of society. The prevailing tone of conversation here is what one would by no means have expected; for we frequently meet with the delicacy and polish of a capital with the high pride and independence so eminently peculiar to the Norwegians. We are more agreeably surprised still to find that this cultivation is no foreign and short-lived plant. Many of the most polished among the inhabitants, whose society would be an honour to any person, have seldom, perhaps, extended their travels beyond the country around Christiania; and the visits of others have been so short in foreign countries, that they would never have been what they are, if their manners had not been formed before leaving home.

Hence we observe in this what takes place in all capitals, where the art of social intercourse has made any considerable progress—the division of society into several classes, tolerably distinct from one another. That these divisions were effected, or in any considerable degree influenced, by riches, titles, influence, or personal connections with the state, I could never discover: they arose rather from a diversity of tone. Hence the boundaries of these divisions flow almost imperceptibly into one another, whatever may be the difference between the extremes. It is a proof of the refinement of manners in a town, when all are not united in one mass. The mind possessed of refinement ascends naturally to the top, and every thing like a common union in society is artificial, constrained, and cannot be permanent, because the parts which compose the union are heterogeneous. These divisions, marked out by nature, are no ways inimical to public spirit, or patriotism, as has been often proved by the example of England, and in miniature by the Canton of Schweitz, where shepherds and lords co-operate together in so singular, yet harmonious a manner.

I have often thought that the decided predilection of the Norwegians for the theatre may have had no small influence on their character. It is certainly surprising to find no town in Norway without a theatre. The most polished of the inhabitants play in a manner in public before the better sort of people, frequently tolerably, and often excellently. I saw several well-known persons in Bergen perform their different parts with the fervency and truth which belong only to the most skilful professional actors. Drontheim, Christiansand, and Frederickshall, have all of them their theatres; and when I was at the small town of Moss, I heard a very earnest deliberation respecting the means of constructing a theatre there also. Christiania has no less than two, and the whole winter through two different societies of Dilettanti tread the boards for the amusement of themselves and their fellow-citizens. The most beautiful and delightful music is spread and generally diffused, not merely by little occasional pieces, but by the representation of operas. Though the expression of the higher passions in tragedy requires a continued practice and study which the acting of Dilettante will not admit of, yet I shall always remember with lively pleasure the splendid representation of the national tragedy of *Dyvecke*, a piece certainly however praised beyond its deserts, in which the first families of the town distinguished themselves equally by their magnificence and their skill. They had an excellent poet among them, who seems to have given a good direction to their taste, and who entered with great enthusiasm into the management of their theatre. This was M. Falssen, Counsellor of State, President of the highest Tribunal in Christiania, and one of the three Members of the Government Commission for Norway, during the Anglo-Swedish war. The town was deprived of him by a sad accident in the winter of 1808; but his influence will long continue in circles dedicated to joy and festivity, through his sweet poetry, his translations of so many excellent French pieces for the theatre in Christiania, and still more through his original and affecting comic-opera of *Dragedukken*, with the lively music of Kuntze, in Copenhagen; and the Norwegians ought long to remember, that to the passionate but energetic official paper *Budstikken*, edited by him, they owed their courage and their confidence in the beginning of the Swedish war, when their own strength was unknown to them. His mind appears to have been of too vehement a cast for the cold blood of his fellow-citizens: it consumed itself.

The Gymnasium in Christiania, which bears the modest appellation of school, may be mentioned with distinction as a public establishment for education. Its merits are proved by the abilities of the teachers, and the progress made by the scholars. It supplies to a certain extent the want of a university in Norway, which has been so often warmly, but however reasonably, always fruitlessly demanded by the Norwegians, as a literary centre in the interior of a remote kingdom, which constitutes more than a third part of the whole monarchy. The school, which is situated in the best part of the town, is a large building, and has a serious and dignified external appearance. It contains, besides the rooms adapted for tuition, several collections, which are not very distinguished, and the library, which is not more ornamental than useful and profitable to the town. This library is open to the citizens, and contains perhaps not many rare, but a number of useful works. It owed its origin chiefly to the collection of Chancellor Deichmann, who died about twenty years ago, and who distinguished himself by his works on the modern history of Norway. This patriotic individual bequeathed his library to the town of Christiania, well judging that it would there be productive of the greatest benefit. In the same spirit several other more recent libraries have been incorporated with it, for which they are partly indebted to an Ancker; and they now continue unremittingly to procure the most important productions of the German and Danish press, so far as the school-funds, which are by no means scanty, will allow them. How few towns of the same extent, or in the same situation, can congratulate themselves on such a library! And as it is not suffered to remain idle, we can hardly doubt that it will greatly contribute to the diffusion of knowledge.

The excellent Military Academy, which directly fronts the school-house, is an object no less remarkable. It is certainly one of the best institutions in the Danish state, and has been the means of supplying the Danish army with a great number of useful and accomplished officers. It is a pleasant sight to see the hundred cadets, who generally receive an education here, either assembled together, or in the streets. Their vivacity, their blooming complexions, and their dignified behaviour, dispel at once every idea of constraint; and we soon see when we enter the building that it is a much nobler institution than similar schools for cadets generally are; yet the institution is almost wholly supported by the contributions of wealthy individuals. The academy is indebted for the house (an elegant little palace), and perhaps the most beautiful in the town, to the liberality of the Ancker family, by whom it was formerly inhabited; their instruments and books are legacies; and only two years ago it received from the chamberlain, Peder Ancker, the rich library and instruments which devolved to him on the death of his brother, Berndt Ancker. By these means they have been enabled from a mathematical school, which was the origin of the institution, to convert it into an academy, in which the young officers, besides the mathematical sciences and drawing, are diligently instructed in history, natural philosophy, natural history, and foreign languages. During several days of the week they practise leaping, climbing, rope-dancing, swimming, and other exercises, which professor Treschow in Copenhagen very appropriately calls the luxury of education; but a good officer will perhaps not regret the time he spent in such exercises. It is an excellent regulation, that the cadets neither lodge nor eat in the house; they are boarded with respectable people of the town, for the purpose of avoiding the monkishness of a secluded education. They wish to bring the young people as much as possible into contact with the world, and to break them at an early period of the narrow-mindedness which so circumscribed an occupation as that of a soldier has a necessary tendency to produce. The correctness of these principles has been confirmed by experience, even in the short space of a few years. So long as, the state of Denmark deems it necessary to keep up a great army, and to dedicate so much of its attention to that object, it were heartily to be wished that all the Danish officers found such a school for their formation as the Military Academy in Christiania.

We may easily conceive that such a beautiful country as the environs of Christiania does not in vain display its charms to the wealthy inhabitants, and that they will be disposed to quit the town in summer for the health and pleasure of a country life. The multitudes of country-houses round the town is in reality so great, that their appearance puts us in mind of Marseilles. A country-house is an essential piece of luxury in Christiania; and as a merchant in Hamburg does not suppose he can appear without his coach and his horses, so the country-house is the first expence of a rising citizen here. These small places are called *Lükken* in Christiania. Why they are so called I could never learn; and what is singular, this appellation is exclusively peculiar to this town. Many of their places are indeed very diminutive— a little house with a small meadow; but they have all an enchanting situation; and there is a perpetual variety of prospect from the height of the amphitheatre, of the Fiord, the town, and the hills. Whatever may be the number of these *Lükken*, we may boldly assert that there is not one of them without a prospect peculiar to itself. Most of them have not much to recommend them except this prospect, as little has been done for the surrounding grounds. But this they cannot be blamed for. The great desire to possess a small piece of ground in the neighbourhood of the town has raised the price of them so immoderately high, that it is seldom in the power of the possessor of one to ornament any part of it. A Lükke worth eight or twelve thousand dollars seldom exceeds the size of many a garden in Berlin; and a meadow worth a thousand rix-dollars may be overlooked at a glance. The occupation as a meadow is essentially necessary to the support of the town; for the country is not sufficiently inhabited to allow the market to be constantly supplied with every thing that house-keeping requires. Every family must keep their own cow; and the long winter requires great stores; hence a dry year, unusual warmth and drought in June and July, not unfrequently occasion great want and embarassment; and although the upland vales of Ringerige or Walders send some hay to the capital, it is by no means equal to the consumption. Assistance is then looked to from abroad, and hay is commissioned from England and Ireland. I could hardly believe my eyes, when I saw in the harvest of 1806 a number of ships loaded with hay in the mouth of the bay of Christiania. Is this hay exported to the Baltic or Jutland, to countries fertile in corn? No, I was answered, it is hay from England, commissioned to supply the wants of the householders in Christiania and Drammen. It is well with the country that possesses means and opportunities to supply its wants in such a manner; but it is still better with the country which by its own industry can produce what nature in the ordinary course of things refuses. And why should it not? When we see *the Aggers Elv*, a considerable stream close to the town, falling in noisy cascades from wheel to wheel, from saw-mill to paper-mill, and again to saw-mill; when we see numerous little streams descending from the wood-covered hills; and when we view at Frogner a considerable rivulet running through the midst of these possessions, before it falls into the Fiord at the west end of the town, a stream which in the greatest heat of summer is never dry, it is surprising that all these supplies of water have not been long ago made to fall from Lükke to Lükke, and to spread in a thousand various channels over the parched hills, as has been so beautifully done in the Emmenthal and Valais in Switzerland, and with so much art even in Norway itself, in the valley of Lessöe, and in Leerdalen below Fillefieldt. For this an agreement of all the proprietors among themselves is no doubt necessary, and it may be attended with some difficulty; but are we not to consider it as a want of public spirit that such an agreement has never taken place? And are we not entitled to suspect some error in the government, which, with such an excellent opportunity, prevents the inhabitants from finding their individual interest in the general good.

The possessors, in truth, show no want of individual industry. Bare rocks are yearly thrown down and converted to meadows, and many a place is now attractive which was formerly repulsive from its sterility. The small possession of Frydenlund, about an English mile from the town, formerly nothing but dry slates hardly covered with moss, has become, through the incessant labours of the indefatigable lady of General Wackenitz, one of the sweetest and loveliest places imaginable. And what has been effected by the noble and active Collet on his possession of Ulevold, will, in point of agriculture, long serve as a model for Norway.

Whoever takes a delight during his stay in Christiania in exploring the beauties of the surrounding country, must not neglect to visit the charming Skoyen, the country residence of Ploen the merchant, in point of situation, the crown of all the rural places in the neighbourhood of the town. The whole magnificence of nature is here unfolded to us: the Fiord, the town, and the hills, appear all entirely new, as if we had never before seen them. We never weary in looking down upon them, to follow the beautiful light spread over them, and to rivet our eyes on the picturesque forms of the hills of Bogstadt and Bärum. And again, what rural beauty, what charming solitary prospects, when we lose ourselves among the woods and dales that border on Skoyen! Here alone we live with nature! In Bogstadt, the magnificent seat of the Chamberlain Peder Ancker, we may please ourselves with viewing the way in which a rich individual may create and beautify a residence to give delight to a cultivated mind; and in Ulevold we may gratefully recognize the endeavours of the noble possessor to diffuse joy and benevolence around him.

This high cultivation and the beauty of the country around the town deceive us into a belief of a better climate than the place actually possesses. The appearance of the objects down the bay puts us so often in mind of Italy, that we would willingly associate the idea of Italian heat with them. It is confidently, however, believed by many, that the climate of Christiania is at all events better than might be expected from its high latitude. But that is not actually the case. By much too unfavourable an idea is entertained in other countries of nature under the sixtieth degree of latitude. Where oaks thrive, fruit-gardens may be cultivated with advantage and pleasure; and accordingly in Christiania not only apples and cherries, but even pears and apricots, grow in the open air: plums, however, do not succeed; and peaches and vines, as well as several sorts of pears, must be dispensed with. As to trees, the high ash thrives admirably, and it is a peculiar ornament to the country. Limes grew vigorously and beautifully; and sycamores and elms are among the most common trees of the woods. The aspen tree, *(Populus tremula)*, the alder, and the birch, grow always larger and finer; they are the true trees of the north; and the warmth of Christiania is even in some measure too great for their highest perfection: at least, the asper and birch seem here to love the shade very much.

Neither does the winter appear here much earlier than in the North of Germany: the snow is hardly expected to lie before the beginning of December; and continued frost is very rare in November. It is, however, sufficient to cover the harbour of Christiania with ice in the end of November, and the shipping is then for some months altogether at a stand. The inmost part of the bay, between the numerous islands and points, resembles a lake, and is therefore soon frozen. The *Bonnefiord*, an arm of more than fourteen English miles in length, is fully frozen, and in the main arm the ice extends frequently for nine English miles down the bay. The vessels are then frozen in, and lie in the harbour the whole winter through as if on land. People pass and repass between the yachts, galleys, and brigs, as through streets, and the land and water appear no longer separated. This continues for a long time. The fine season gradually makes its appearance. The snow has been long all melted on the hills of Christiania by the sun and the warm rains, and every thing has assumed a green and animated appearance, before the ships are disentangled from the thick ice. About the twenty-fourth of April the waves begin, at last, to beat against the moles of the harbour. The ship-owners then frequently lose all patience: for a few miles farther out in the Fiord, the ships of Dröback, Laurvig, and even Frederickstadt, have been long out at sea before the vessels at Christiania exhibit the smallest motion. They at last remove the obstacles by force, and break the ice. This is a most interesting moment. I heard once in February, that several ships wished to break through the ice, and I knew that they had at least a German mile to proceed through the hard ice to the nearest open water: I immediately ran to witness the Herculian undertaking; but I was not a little astonished to see the ships advanced a great way through the ice, and still continuing in motion, though slowly, as if they were in open water. The whole work is, in fact, much easier than one would be led to imagine. About fifty men stand opposite one another like an alley; and the space they allow between them corresponds to the breadth of the ship which is to be moved through. They cut along the solid mass of ice as far as their line extends, and then they separate, by cuts across from the one line to the other, immense rectangles of ice, perhaps more than twenty feet in length. A wooden plank is next placed in the cut so opened: the men then all proceed over to the opposite side; some of them press the rectangle of ice with all their might below the water: in the same moment, all the others lay hold of a number of ropes fastened to the board in the opposite cut, and shove the immense loosened mass of ice, with one effort, below the ice which is firm. They then begin to loosen another rectangle. The work proceeds so quickly, that the ship which follows hardly ever stops, and in the space of a few hours makes its way through a covering of two feet of ice for almost five English miles from Christiania to the open water. In this way several English ships of the line wrought their way in the winter of 1808 from Gottenburg through the ice into the open sea. Hence we may easily see that where the art of working through ice is properly understood, ships which are frozen in do not always necessarily fall into the hands of an advancing land army.

When the ice has left the vicinity of Christiania, the warmth increases with indescribable rapidity, and May, instead of being a spring month, is completely summer. On the third, fourth, and fifth of May, 1808, I observed that the thermometer at its highest rose to 70° Fahr. In the middle of the month all the trees were in leaf, except the ash (ask, fraxinus excelsior); and towards the end of the month the thermometer was daily at noon 19 or 20. In the beginning of July garden stuffs were every where to be had: the mean warmth of the month rose to upwards of 65°, and at noon it was generally 81, nay, even sometimes 86 degrees. They commenced their harvest before August, but September was not fully over before they began to think of stoves in the town.

We have as yet no series of observations of the temperature of Christiania continued for several years: it is therefore impossible to ascertain this temperature with sufficient accuracy to compare it with the warmth of other countries. We have, however, accounts of the greatest heights to which the thermometer has risen in summer, and the lowest to which it has fallen in winter. But we ought to be convinced that these examples are by no means sufficiently decisive to enable us to come to any sure result. We owe a series of observations, beginning with May, 1807, to the Lady of General Wackenitz: she made use of a very good Reaumur's mercury thermometer, which was placed towards the north on her estate at Frydenlund But the summers of 1807 and 1808 were uncommonly warm, and the winters extremely mild; so that the mean of her observations was somewhat higher than we dare lay down as an average for this country. She found the medium of both years.

In January	⊹ 0,43		October	3,244
February	— 1,358		November	1,874
March	— 1,375		December	1,619
April	— 4,805		Mean	5,292 R.
May	8,98			
June	13,155		In January 1809 it was	
July	15,243		only	— 7,2
August	15,897		Febr. 1809	— 3,33
September	9,224			

But if we take the lower temperatures of January and February, 1809, into account, the mean falls to 4,96 Reaumur. We shall not probably be far from the truth if we take 4,8 Reaumur for the mean annual temperature of Christiania. This is by no means too high for the latitude of the town, which besides is not far distant from the Western Ocean. These results correspond also tolerably well with the temperature of Copenhagen, as well as of Stockholm and Petersburg, both of which last places lie almost in the same latitude with Christiania, but much farther eastwards; which makes the Winter much more long and severe. The following short table will give us an idea of this correspondence.

	Copenhagen, according to Bugge*.	Stockholm, according to Wargentin.	Petersburg, according to Euler.
January	— 0,7	— 4,14	— 10,4
February	— 2,2	— 3,05	— 5,8
March	— 0,2	— 1,8	— 1,9
April	4,0	2,93	2,3
May	8,5	7,51	8,1
June	12,6	12,6	12,2
July	15,	14,3	15,
August	13,6	13,14	13,
September	11,7	9,3	8,5
October	7,5	4,74	3,2
November	2,9	1,26	2,5
December	0,7	1,59	4,1
Average	6,15	4,63	3,1

* According to an average of thirty years observations which M. Bugge communicated to me.

CHAPTER III.

MINERALOGICAL EXCURSIONS IN THE COUNTRY ROUND CHRISTIANIA.

Transition formation at Christiania.—Orthoceratite in the Lime-stone.—Veins of Por- phyry.—Sandstone.—Porphyry in Hills.—Kolaas, Krogskoven.—Sandstone at the Holsfiord.—Transition granite at Hörtekullen.—Granite at Strömsöe.—The Granite is confined by the Boundaries of the Transition formation.—Paradiesbacken.—Marble Quarries.—Granite of After Vardekullen.—The Fiord of Christiania divides the Old and New Rocks.—Zircon-Syenite.—On the Porphyry and Clay-slate at Greffen.— Granite thereon.—Order of the Minerals at Christiania.

AMONG the few specimens in the Copenhagen collections which illustrate with any thing like precision the mineralogical geography of Norway, I had remarked several black and thick Lime-stones from Eger; Clay-slate from the neighbourhood of Christiania; and the produce used by the alum-works at Opslo. I was therefore tolerably well prepared to find in this part of Norway the Transition formation, with probably all the minerals which are peculiar to this formation. But I found still more—stones which were never supposed to be in the Transition mountains, but which were here seen with such a distinctness of stratification, that not a doubt could remain as to their relations in this respect: if their true nature had been properly known, Chris- tiania would have been left with a conviction, that from this country geology may derive the most important acquisitions, and that in a mineralogical point of view, it is the most important of the whole of the north.

Porphyry in immense mountains, reposing on lime-stone full of petrifications *(auf Versteinerungsvollem Kalkstein);* a syenite over this porphyry consisting almost entirely of coarse-granular felspar, and in the same manner a granite not different throughout in its composition from the granite of the oldest mountains. Granite *above* transition lime-stone! Granite as a member of the transition formation!

Perhaps I should have long hesitated to acknowledge these very unusual and almost wholly new relations, if M. Hanssman had not, with his usual accuracy and penetration, examined the greater part of these countries before me, and confirmed and rectified the opinion which I entertained. His excellent treatise, in which he developes the whole of these relations, is known to mineralogists.*

In coming from Sweden, we must be very close to Christiania before we can have a suspicion of the change in the internal composition of the country. The *Egeberg*, which so beautifully commands the Plain of Christiania, consists wholly of a fine slatey gneiss, and when we get almost to the bottom of the hill, the black slate folia appear, and immediately afterwards the deep pits of the alum-work. These are the beds, which on account of their blackness, have been frequently considered as indicative of coal, and which have occasioned so many thousands throughout every part of Europe to be expended in fruitless labours. Such also has been the case here, and always with the same unfortunate result.

These slates admit of being more successfully used for alum, as they contain iron pyrites in small beds of an inch and more in porportion to their blackness. These iron pyrites are discomposed by smelting; the sulphur is turned into acid, and unites with the alumina of the slates. The black beds are separated by beds of clay-slate, and in these a number of strong flat masses, oval at the ends, at least several feet in diameter, lie beside one another, which are easily separated. They are thrown out of the pit as useless, and laid in heaps on the road side, where their appearance as flattened balls is singularly striking. These masses are more solid than the surrounding bed, and their fracture is small grained, uneven, or earthy. They are not unfrequently traversed by small black veins of calcareous spar; and perhaps themselves contain a good deal of calcareous earth besides the mass of clay-slate.†

---

* Bar Moll. Neue Zahrbuecher der Bergund Hüttenkunde I. Bd. 1 Lief. 34.

† The globular masses contained in clay-slate are to be viewed as distinct concretions. In the transition slate of the Pentland hills, in the vicinity of Edinburgh, I have observed similar appearances, but more on the great scale than in the slate of Christiania: the primitive clay-slate between Dresden and Freyberg I also observed in some places disposed in globular distinct concretions. Even the sand-stone of the coal formation, particularly when mixed with slate-clay, exhibits a similar appearance, of which I have given an account in my mineralogical description of Dumfriesshire.—J.

From these pits onwards the gneiss no longer makes its appearance in the neighbourhood of Christiania. The clay-slate becomes more distinct, and alternates frequently and in thin strata with thick blackish and dark smoke grey lime-stone. Wherever a stream or water-channel descends from the hills, these strata appear behind one another, and put us strongly in mind of the neighbourhood of Hoff, in Bareuth, or the Hartz, at Rübeland, and many other places where clay-slate and lime-stone appear exactly in the same manner. Nature resembles herself; she is the same in Norway as in southern countries; and her laws are general throughout the earth's surface.

These, it may be said, are trifles; but they acquire importance from this very generality; and may serve perhaps one day as a key to great and elevated views. How great was my joy, when, at the steep falls of the Aggers Elv, above the lower saw-mills, I discovered the orthoceratites, which so particularly distinguish throughout all Europe this formation, and this formation alone. They are many feet in length, divided into compartments, and for the most part at the edge and the walls of the compartment changed into calcareous spar. They are by no means unfrequent; several of them generally lie in various directions through one another. Pectinites, and several other not very distinguishable petrifications, appear frequently between them. That these orthoceratites do not appear accidentally for once in the lime-stone at the Aggers Elv becomes soon evident, when we examine with any degree of accuracy the lime-stone in its farther course. These wonderful productions never fail to make their appearance every where. At Raae, Soulhoug, and Saasen, on the west side of the lake of Fiskum, in the district of Eger, says the learned Provost Ström*, many thousands of orthoceratites lie above one another. He describes them with accuracy as in the section, on the one side plane, on the other curved, and generally with a pipe through the whole of their length; and he also confesses, that, though during his stay at the sea coast in Söndmör he acquired a knowledge of many marine animals, yet he never saw any thing in life similar to them. It is the same with almost all the petrifications in the more ancient mountains. The resemblance to present existing forms becomes gradually lost, in proportion to the age of the stones which unfold these organic remains†.

---

* Egers Bescrivelse.

† This interesting observation in the text, in regard to petrifications, was first made by Werner, to whom we owe nearly all the most important facts in regard to the distribution of organic remains in the crust of the earth.—J.

Not far from these, at the same lake of Fiskum, are to be found in considerable number the singular creatures which were formerly, and even by Ström, believed to be impressions of an unknown fish. M. Brunnich, however, proved in an excellent treatise, which, like that of Ström, has valuable drawings subjoined to it, that these remains must also be shell-fish, and probably oniscus*.

The lime-stone is never very thick in the hills of Christiania; and I know not, in fact, if many beds are to be found of more than one foot in thickness. The clay-slate is always the thickest; and it also penetrates in some degree into the lime-stone beds, and divides the lime-stone into balls and knots. Hence there is generally very little solidity in this stone, to the great annoyance of the diligent cultivators of the hills; for the stones of the dry masses of rocks which they throw down and level cannot be used for building, and can only be employed in walls where the balls and nuts of lime-stone give somewhat greater solidity to the clay-slate. They call this sort of rock here skiallebjerg (lamellar rock), because it falls asunder in single lamellac.

What the skiallebjerg cannot supply the people with, nature compensates them for in another manner, hitherto peculiar to the country round Christiania. This is by the numerous and great veins of *porphyry* which every where traverse the clay-slate here. If a small rock is seen isolated on a height, on examining it we infallibly find the remains of a porphyry vein, which rises out of the clay-slate: similar rocks in the distance mark the progress of the vein, and at the foot of the hill we see it traced in deep excavations in the clay-slate. The porphyry is alone used as a building-stone, and the soft slate at the side of it is allowed to remain. We enter conveniently into the excavated space, and can with ease follow the nature, thickness, direction, and position of the porphyry vein. The multitude of them is almost innumerable: on every hill new ones break out, and create confusion, when we wish to follow the same vein throughout its course to its termination. They frequently traverse the clay-slate at right angles, and dip under very great angles, often almost perpendicularly; and in the direction and inclination there is also an infinite diversity, and many of these veins must necessarily traverse one another. Their thickness is from ten to

---

* *Kiobenhavns Selskabs Skrifter.* Other well-known petrifications of this country, which Brunnich calls trilobiten, were described by him several years before.

fifteen fathoms and upwards; and veins of less than a fathom in thickness I never remember to have seen. All these spaces are however filled with a sort of porphyry, which is completely similar to that which, as a widely extended formation, and in high mountains, we find at only a mile's distance: a remarkable example of the filling up of the veins, with the formation which covers these repositories, and an important fact with respect to the theory of veins in general!

But, in fact, however various the composition of the porphyry in the veins may be, there is a similar bed to each in the porphyry mountain lying around it. We might, however, often be induced to believe that the porphyry in some veins is more highly crystallized than in others. It then bears a striking resemblance to the *colmünz stone (paterlestein)* in the Fichtelgebirge, in Bayreuth. It has the same solidity, the same contents, and almost the same size of the grain. Felspar runs through it in small but very long prisms, and alternates in an almost fine granular mixture with a black fossil, which we have some difficulty in recognising for hornblende. Epidote (pistacite) appears also in small green parts, an almost essential ingredient of the mixture; and the felspar is frequently coloured green with it. Cubes and points of iron pyrites are generally found interposed in great abundance, as also octahedral magnetic iron-stone pretty frequently. The whole peninsula of Tyveholm, which penetrates into the Fiord, in sight of the town, consists almost entirely of this mass: it is the remains of the vein which enters the bay at this place, and perhaps it is the same with that which afterwards again makes its appearance in the small island of Hovdöe. Agger's Church, on the north side of the town, owes also its high situation to a similar porphyritical greenstone vein. This even contains ore in small veins—blende and lead glance. There is still seen below the church the pits of the former trial works, which were sunk with the view of following these ores.

When the mass of the veins resembles the porphyry of the hills, the base is dark smoke grey, fine, but thick, splintery, and semi-hard. The imbedded felspar crystals are also extremely long, perhaps ten or fifteen times the extent of the breadth, and they follow one another in curved rows, like water in a whirlpool. Small indistinct crystals of epidote, often perceptible, merely from their grass-green colour, surround almost every felspar crystal. They also lie in it. Such is the appearance of the veins, which are numerous in the hills of Enehoug, between Christiania and Opslo. On the other hand, the felspar crystals in other porphyries are every where large and rhomboidal. They shine at a distance, and on that account are wonderfully striking. When we leave the suburb, Pebervigen, and take our course along the water-edge, rhombs of felspar glimmer upon us from each block. This is what we are not accustomed to in felspar, and we imagine at first that we have lighted on new and entirely unknown fossils. But this very porphyry, with the singular rhombs of felspar, is by no means rare in the porphyry mountains, and still less rare as veins. How often are these rhombs to be seen through the clay-slate at Haagenstadt, and Granevold, in Hadeland, and also in the woods below Hakkedalen! This porphyry may therefore be called rhomb-porphyry, and the other may be called needle-porphyry, in which the felspar appears in extremely long but very thin prisms, resembling needles, which traverses the base in so many directions.

Masses of such a nature, and in such abundance, passing through limestone, containing petrifactions (versteinerungskalk), have never yet been discovered in other places. These porphyries must renounce all pretensions to a place among the primitive rocks; and as organic remains are contained in the stone which surrounds the porphyry, how much more of similar remains might not be contained in that which is of later formation? These may have been preserved by the granular and crystalized portions of its mixture*.

I imagined a short excursion over the porphyry-mountain of Krogskoven to the west of Christiania would most effectually solve the nature of the connection between the porphyry of this mountain and that in the veins; I left Christiania therefore on the 15th of September, 1808. The road lies at first towards the Fiord, and follows for above four English miles the course of its windings till beyond Lysager, when it begins to ascend the heights by

* Several years ago I observed not only veins but also beds of porphyry in the great tract of transition rocks that traverse the south of Scotland. In my mineralogical travels through Scotland, published in 1801, and in my account of the island of Arran, published as early as 1798, I described particularly different kinds of porphyry occurring in veins, and large superimposed mountain masses over red sand-stone, containing coal and lime-stone full of petrifactions. These facts appear not to have been known to Von Buch; for I find, on examining the German translation of my travels, that the facts are mis-stated. It is there said the sand-stone is newer than the porphyry.—J.

which Christiania is on all sides enclosed. I found also here not unfrequently porphyry veins traversing clay-slate and lime-stone; and where the two highways to Ringerige and Drammen separate, I remarked one in particular of great thickness. The felspar crystals were white, and did not appear rhomboidal, but rectangular. Similar porphyry is not unfrequent also in the mountains.

The lime-stone and the clay-slate between the Fiord and the hills stretch N. E. and S. W. (h. 3. 4), and dip under an angle of 60° towards the northwest. I believe that is the most determinate direction of the strata in this neighbourhood. It continues always the same up the hill, so long as we see lime-stone and clay-slate; whereas nearer to Christiania the direction of the strata is so extremely mutable, that we are at last compelled to give up every hope of finding a determinate rule for them.

After ascending several hundred feet up the hill, I reached a valley that opened between perpendicular porphyry cliffs through the mountain. The two roads from Christiania and from Bogstadt to Bärum meet at this place. The lime-stone also terminates here; and before the porphyry could be laid on, a fine granular and fine miraceous grey sand-stone, a species of greywacke, made its appearance. Then came the porphyry, almost immediately in large and almost perpendicularly ascending rows of rocks, from which the blocks that had fallen off were scattered wildly about.

I turned off from the road towards the nearest and highest of those hills, the Kolaas, which in a great circuit commands the whole basin of Christiania. The same greywacke sand-stone made its appearance at the steep declivity towards Haslum Church, in grey and red alternating beds, and several hundred feet in thickness all the way to the bottom. When the porphyry begins to cover the sand-stone, it immediately forms such frightfully steep and perpendicular rocks, that we with difficulty discover small ravines and underwood to assist us in ascending. The first beds were rectangular porphyry, as at Lysager below; but rhomb porphyry shortly after made its appearance, of which by far the greatest mass of the hill consists. Here it was clearly demonstrated that the porphyry of the mountain could not be primitive any more than the porphyry of the veins. It lay without the smallest doubt upon sand-stone, and this again rested on transition lime-stone.

Kolaas, between Haslum Church and Bärum's Work, would be the highest porphyry hill of this country, if Bogstadaas did not somewhat surpass it in height. Asker Varde Kullen is also still higher, but it consists of granite, and not of porphyry. The barometer gave Kolaas a height of 1157 Paris feet above the Fiord*.

The great road to Ringerige runs from the height down a pleasant and extensive mountain-vale, surrounded by high hills; and an enlivening variety of farm-houses, copses, corn-fields, and woods, appears along the declivities. This is the Lommendal. In the bottom, where the water makes its way through a narrow fissure between the porphyry rocks, lies the celebrated iron-work of Bärum, the excellent iron and tasteful productions of which are exported to a great distance. It derives only charcoal and water from this mountain; the iron ore is brought from the island of Langöe, at Krageröe, and from the mines at Arendal. Several strata were discoverable on the declivity opposite the work, exactly like the green-stone of Tyveholm, and below Agger's Church. They were, however, wholly surrounded by rhomb porphyry, which gave the distinguishing character to all the rocks in the neighbourhood of the work.

The felspar shines forth along the whole declivity of the Lommendal. Where the mountain, however, rises somewhat more abruptly towards Viig, some of the strata are vesicular and porous: the vesicles run upwards from the lower side of the stratum at right angles, are long and small, and many of them filled with white calcareous spar. There is then very little felspar in it; and the base assumes a pure tile red colour. I never saw hornblende as an ingredient in the composition, nor even quartz. But at the top of the mountain, not far from Midtskoug, a small vein of quartz of two inches in thickness, with druses in the middle of it, runs through the porphyry. On

---

* Barom. Christiania, h. 7. 28 I. 1. Lin. Therm. 11.
    Kol-aas   h. 11. 26—10. 5   8.

The Norwegian language is rich in names for the different forms of mountains. *Aaas* (Ohs) is a very long extended row of small hills; Kullen is an insulated prominent head; Nuden a round and less prominent hill; Egg, a sharp ridge, an edge; Hammer, a rocky cape, which juts out either into the sea, or the plain; Bakke is a little hill; Fieldt, on the other hand, is the highest mountain rising beyond every ordinary human habitation; Tind, a point or peak on the mountain, the *horn* of the Swiss, and the *aiguilles* of Savoy; Fond, an ice-hill; Brä or Yökul, among the Laplanders geikna (jäkna), a glacier. In Christiansandstift, a distinguished height visible at a great distance is called hcien or hei—eidsheien.

the height, the porphyry in general undergoes very little change. Its basis is for the most part of a reddish brown colour, and the fracture is very compact, and generally splintery, opaque, and semihard. The felspars are smoke grey, and glistening. The former becomes grey from the weather, and the felspars white, by which both attract our attention the more. Here and there beside the felspar round small white calcareous spar nuts like trochites make their appearance. There is, however, no hornblende bed in the whole mountain. The height is a sort of table land, intersected with flat vallies; and where it is crossed by the roads, it may be at least nine English miles in breadth; and it is covered with a thick and continued wood of spruce firs, interspersed with Scotch firs, elms, and alders, in which bears and wolves roam about in great numbers. In the middle, on a situation a little elevated, is the only house to be met with on this road, called Midtskoug, the middle of the wood; and at a few steps distance from it there is a signal-post on a small hill, which appeared to me to designate the most elevated point of the whole country. From this place we in fact command a most extensive prospect over the dark wood. I found its elevation 1255 English feet above the sea *   But this small range of mountains rose still higher northwards, till Gyrihongen, which commands a view of both the districts of Hadeland and Ringerige, and which may be considered as the highest eminence of this country; yet it can hardly amount to 2130 English feet.

The mountains begin to fall four English miles and a half beyond Midtskoug; and the road rapidly descends to the plain of Ringerige. I had heard a great deal of this road; but 1 was greatly surprised when I saw it. It is a fissure between perpendicular rocks of an enormous height *(himmelhohen)*, and it falls with great suddenness: all is black and dark. The plain of Ringerige, strongly lighted up, is spread out in the depth below with its farm-houses, churches, lakes, and the majestic distant shapes of the hills of Walders. It appears as if we were looking down through an immense black tube; or as if the cleft of Aldersbach were placed on the height of Heuscheune, and we were to see through the cleft the Glatzer-plain spread out before us. The road is

---

* 15 Sept. h. 7. Christiania Bar. ....28. I. 4.   8 Lin.
      h. 4. Midtskoug Sign................27. 0. 6.—14
      h. 5. Commencement of the Sand-Stone 27. 6. 2.
      h. 6. Sundsvold, at the Steensfiord.....28. 2. 4.

called Krogsklewen, and is so steep, that large stones must be laid down to prevent carriages and horses from rolling down to the bottom.

In the middle of the declivity the sand-stone again makes its appearance, considerably below the porphyry: the line of separation is long and beautiful, and easily followed. We can cover the place with two fingers where the two formations separate. The first strata of the sand-stone are conglomerations of pieces of the size of a pigeon's egg : all is quartz, and throughout neither granite nor pieces of gneiss : strata of finer sand-stone then follow the whole way down the descent to Sundsvold, at the great lake of Ringerige, which is here called Steensfiord, deeper down Holsfiord, and Tyrifiord towards Drammen. Below Kolaas, there was a stratum of coarse pieces of quartz, every way similar to that just described, above the finer sand-stone. The change from the one formation to the other is very striking, even externally : the sand-stone rises upwards with a gentler slope ; but so soon as the porphyry makes its appearance, it rises like a crown of completely perpendicular rocks on the height, parallel with the Steensfiord towards Hadeland, as far as the eye can reach. The separation of two formations is seldom so beautiful and so accurately to be followed.

From Sundsvold upwards, the sand-stone here attains the thickness of 740 English feet. It was not so high below Kolaas ; but what the sand-stone gains in height is lost by the porphyry, for from Bärumsverck to Midtskougen, the porphyry height is fully 1124 English feet ; but it does not attain more than 532 feet from Midtskougen to the beginning of the sand-stone below Krogsklewen. The whole porphyry mass declines also from Ringerige and Hadeland, towards Christiania-fiord.

It was my intention to follow the porphyry at the lake down to the Holsfiord, and to go round the whole porphyry mountain of Krogskoven as far as the valley of Lier; but these rocks sink so precipitately into the lake, that in many places there is no possible way between them and the water. I was compelled to go across the lake in a boat to a peninsula, that for a length of nine English miles separates the two arms of the lake of Tyri and Holsfiord from one another. I reached land again at the place called Horn, where there are beautiful views of the majestic rocks which terminate in such a variety of forms the opposite part of the lake. The sand-stone now and then breaks through, and forms small promontories in the water ; and the stones are used in erecting a house and offices, and forming a meadow around.   The views are truly Swiss.   Between the farm-houses of Näss and Vevsrud, the sand-stone is at last wholly banished by the porphyry.   The rocks sink perpendicularly, and hence the eastern shore is wholly unin-habitable and desert.

The first stone that I met with at Horn, on the western shore, was black compact lime-stone and clay-slate, the same strata of which the hills of Christiania are composed.   The lake consequently lies on the division between the lime-stone and sand-stone; that the latter, however, reposes on the former is sufficiently proved by facts at Kolaas and Bärumsverck.

At Sör-Drag, half a mile from the end of the lake, the whole declivity is covered with large pieces of gneiss abounding in mica.   The gneiss was no doubt here quite in the neighbourhood, but I did not see it.   It is of importance to remark this, for it determines the western boundaries of the newer transition formations in Norway.   Modums Prästegieldt, which bounds these districts, is wholly formed of gneiss and mica slate, and lime-stone and clay-slate do not again make their appearance till beyond the range of great mountains on the shores of the Western Ocean.

Near the end of the Holsfiord, and not far from Hörtegaard, the lime-stone alternates with black flinty slate, with black hornstone, with single beds of white conchoidal hornstone, and with thick slatey clay-slate, re-posing in thin strata over one another.   They lie all probably under the lime-stone, and stretch nearly E. N. E. and W. N. W. h. 5, and dip 40° towards the north; but their extent is not great.   Even before coming to Gaard, red granite makes its appearance from under them: the separation is clearly seen: I did not follow it, but ascended the great Hörtekullen, a mountain which forms such a frightful precipice over the valley of Lier, and immediately over Hörtegaard, that I hardly ever saw any thing to equal it. It was excessively difficult to ascend it, even obliquely.   We see from its round summit to a great distance ; and even at Christiania it appears to rise above the hills that are nearer : it is seen in a most striking manner from the beautiful country-house of Skoyen.   Its height is considerable for so steep and abrupt a hill.   I found it to be 1257 feet above the sea.*

---

* 16 Sept. h 8. Horn, at the Holsfiord barometer.... 28. I. 5. 1 Lin. therm. 10.
    h. 10. Hörtekullen.....................27. I. 3. 1..........12.
    h. 4. at Lier-Elv, where Glitter-Elv falls in 28. I. 6. 1.   For

The black compact clay-slate continued all the way up the heights; for as I ascended the hill from the north side, and the strata incline also toward that side, I remained always on the same bed till the top; but in my descent on the south side I saw every moment a new bed, for the outgoings of the strata break forth towards the south. I had hardly descended 200 feet when the red granite made its appearance, and the clay-slate disappeared. The line of separation was also here so determinate, and seen for such a distance, that we might point out the junction an inch in breadth. And what was wonderful, the separation runs exactly in the same direction with the beds of clay-slate, nearly E. and W. (h. 5—6) 50° towards the north, as if the granite were only a bed in the clay-slate; and this in the midst of an entirely insulated summit, rising high above the neighbouring hills. If this granite differs much from clay-slate in the period of its formation, the same causes however of the descent of the stratification must have necessarily operated on both formations; and then it is singular enough how the upper plane of the granite was so very smooth, as if at this arrival of the clay-slate it had been spread out like a table.

At first I thought I discovered in the granite a separation into strata, which had the same direction and dip with the clay-slate; but farther down, with all my efforts, this was no longer discernible. The granite appeared a connected mass, full of fissures, but not stratified.

This granite was small granular, and composed of much beautiful flesh red felspar of somewhat less, though on the whole pretty frequent quartz, which was conchoidal, grey coloured, transparent, and granular, of different magnitudes, even small granular; and lastly of a small quantity of minute, insulated, and rarely grouped scales of black mica. No hornblende; nothing syenitic; likewise no beds of hornblende, and rarely a bed of fine granite, which has a grey colour, owing to the abundance of small scales of mica. It is a true granite, and has no resemblance to gneiss; but is it granite of the oldest formation? Is it the fundamental rock on which the gneiss almost universally spread over the north rests? Probably not. But at least we

---

For the topography of the mountains of this country, it is not unimportant to remark, that Hörtekullen exceeds in height Askfield in the valley of Lier, towards Bragernäss: but Ornefieldt, above the church of Liers, and Kroftekullen, above the Paradiesbakken at Giellebeck, are somewhat higher.

have here clay-slate and black lime-stone; and consequently the sand-stone, and entire porphyry formation of Krogskoven, reposing on it.

The beautiful vale of Lier, which we overlook in its whole extent from Hörtekullen, extends for nine English miles from the shores of the Holsfiord to the sea bay of Drammen. It is one of the sweetest valleys of the country, highly populous and thriving, from the neighbourhood of the three great towns of Bragernäss, Strömsöe, and Tangen, which bear the common name of Drammen, and it abounds in rich and varied prospects. It ends nearly at the place where the long town of Bragernäss begins; and from thence is separated from the vale of Drammen by a pretty high and steep chain of hills. The highest hill of this chain is called Soeldergaas: it lies above Krogstad near Eger, and is highly celebrated throughout the whole country for its extensive view. M. Esmarck determined its height to be 1819 English feet. M. Esmarck found this hill also to be composed of granite, but covered with porphyry at the top. Such is the wide extent of this granite.

But it goes much farther. Granite again makes its appearance behind the town of Strömsöe, on the other side of Drammens Elv, the largest stream in Norway, from its breadth and depth of water, and not only forms considerable hills there, but stretches itself over a great space. I did not lose sight of it till the foot of the small chain of hills about two English miles from Oestre, where it is concealed below lime-stone and slate. It continues to run with the range of hills for some miles in a parallel with the Drammensfiord, till that bay unites with the greater bay of Christianiafiord. In the steep and deeply cut hills, which to the south-west of Strömsöe separate the vale of Drammen from the lake of Eger, the granite breaks again frequently forth, and the newer formations appear to lie on it on beds of but small extent. Considerable hopes were entertained about the middle of the last century of working to advantage a silver mine in the mountain of Skouge, not far from Strömsöe. The ores occurred in veins in flinty slate, which belongs to the formation of transition clay-slate; but the granite was not far distant. As it was always looked upon as the fundamental formation, they calculated the depth to which they might sink the principal pit (Wedels Göpel, *Wedelseje gräbel*), before coming to the granite. The mouth of the pit lay 1234 Paris feet above the fiord; it was carried downwards for 93 fathoms, whereas, according to their reckoning, the granite was to make its appearance at 15 or 16 fathoms, supposing the surface

I

of the granite to run horizontally under the clay-slate. These relations were still more clearly unfolded in a canal for ventilation opened in the granite, and which was carried on to the flinty slate reposing on it, in which ore was contained.* The same granite makes its appearance frequently between the Werck and Eger. We can no more therefore doubt here than at Hörtekullen of its lying under the clay-slate, and consequently under the transition formation.

But it must be owned, however, that it has something peculiar here in its composition. It is almost all small granular flesh red felspar; the quartz pyramids which lie in it appear quite trifling compared with the abundance of the felspar; and the scales of mica are at most but seldom to be found. On the other hand at Strömsöe and Tangen, many long and black hornblende crystals lie among the felspar. It resembles very much zircon syenite. But this whole formation is *toto cœlo* different from gneiss. We never think here of gneiss. Whenever any trace of a slatey composition appears in this stone, it is characterised by flesh-red granular felspar, and the want of slatey mica becomes striking.

There is no hope of finding any place in the country about Drammen where the granite and gneiss come in contact; for this granite, and it is a very remarkable circumstance, never stretches beyond the limits which have in Norway been assigned to the extent of transition mountains. But of all the members of this formation, clay-slate and lime-stone are usually those which stretch themselves out to the greatest distance around; and hence the gneiss of the bases of the mountain is usually bordered by them, and not by granite. This singular limitation of the granite might give rise to a very well-founded suspicion, that it may also itself be a part of the transition formation, if we did not know how much farther in reality granite and zirconsyenite, which resembles it so much, spread themselves over subjacent porphyry and transition lime-stone.

On my return from Bragernäss (or Drammen) to Christiania, I went at first along the usual high road—a most interesting tract! So long as the way runs across the Vale of Lier, we are pleasingly taken up with the contempla-

---

* Bescribelse af Jarlsberga Sölvhaltig Blye-og Kaaberverck af Faje i norsk Topographisk Journal Heft **XXV.**

tion of the country-houses along the side of the hills, the prospects, and the farm-houses; and when it rises steeply, all at once, from the vale to the highest eminence on the road between Drammen and Christiania, the prospects are so numerous, they succeed one another with such rapidity, and there is such a variety in the objects which solicit our notice, that we willingly stop in our progress for the sake of more fully enjoying them. The Norwegians call this heighth Paradise Hill *(Paradies bakke)*, and who can blame them for it? No person can descend from it to the Vale of Lier without being struck with the sublimity of the prospect.

But geognosy is also a gainer when we shape our course slowly and carefully along these heights. They are highly deserving of an accurate examination. A multitude of formations, which we see spread widely over the country, are crowded together among these hills in a narrow space, and give incessant occupation to whoever has a wish to ascertain with precision and truth their different positions.

In the bottom of the Valley of Lier no rocks are to be observed projecting through the soil, but immediately on our reaching the foot of the hill great red plates of granite make their appearance, and then similar insulated and small rocks along the road. The whole of the Paradise Hill appears, at first sight, like the heights of Strömsöe, to be an entire hill of granite: but yet the granite here is somewhat different in its composition from that of the other, not sufficiently to prevent us from recognizing the same stone again, but it is much more granite, and bears less resemblance to the Zirconsyenite than that of the hills of Strömsöe and Tangen. In the lower part of the hill quartz predominates, but higher up, the other ingredients take their usual proportions, and then the granite is no longer to be mistaken. Hornblende is wanting. A single crystal appears accidentally and very rarely. There is nothing foreign in the mixture, except perhaps rarely a black metallic grain, probably of titanitic iron. The numerous little angular cavities in this rock, into which not unfrequently fine and distinct crystals of felspar, and also quartz, shoot, are highly remarkable. They are also frequent in other places, and give to the whole, even at first sight, a characteristic and decided appearance.

At nea y the greatest height of the mountain, before we reach the highest house on the hill, the red granite disappears, and small granular and clear white lime-stone, dazzling marble, makes its appearance. How often have I not left the road and investigated this remarkable separation among the bushes by the way side, before I was fully and decidedly convinced that the granite here really occupied a position below the lime-stone! So different is it to observe what appears to lie before our very eyes: yet not a doubt can possibly be entertained that the lime-stone is the more recent and superior formation. It continues for some time, and a few hundred paces from the separation it is laid quite open for a great way in large and well-known marble quarries.

We might believe it to be from mica-slate or gneiss mountains, such a deceitful resemblance does it bear to the white marble beds contained in these formations; and yet it does not belong to them. However white, and small granular it may be, it is a product of the transition formation, and is nothing more than a subordinate bed of the usual black compact lime-stone of this formation; for we have only to follow the marble quarries and the course of the road, near the adjoining post-station of Giellebeck, to see the granular changed into black compact lime-stone. The white still makes its appearance occasionally environed by the black; but at the end of about two English miles the granular no longer appears. The mountain becomes quite similar to the hills of Christiania. This is also a phenomenon which hitherto is, of all the countries yet known to us, exclusively peculiar to Christiania.

In the quarries a thin bed of white, fine, diverging fibrous tremolite lies almost every where over the lime-stone; and in the midst of the tremolite, grass green, and by no means small epidote druses, frequently make their appearance; a mixture equally singular and beautiful. The granular lime-stone itself alternates sometimes with beds of brown garnets, in which are contained violet blue, fluor spar. What a treasure of various kinds of fossils in one stone of the transition formation! As if these crystallized beds and fossils in druses, which lie over the beds of transition lime-stone and grey wacke, were preparing for an entire return to crystallized stones, such as porphyry, zirconsyenite, and newer granite.

The marble quarries of Paradiesbacken and Giellebeck lie almost in the same elevation, somewhat more than eight hundred feet above the Fiord. This is certainly by no means a plain; but the road runs for a long time at an equal elevation to the rast southern declivity from Krosgkoven The last hill of this small range is in vw a considerable way from the road: it is called Kroftekullen, lies immediately over Giellebeck and the marble quarries, and consists almost entirely of rhomb porphyry. The porphyry lies at the foot of the steep rocks of the summit, on brown flinty slate, which itself reposes on clay-slate, and then probably the lime-stone follows. To the westward below Kroftekullen, towards the Vale of Lier, I saw the siliceous strata take their course downwards to nearly the bottom of the vale near Süsdal: several white, fine granular lime-stone beds appeared between; but these were so siliceous, that they with difficulty effervesced with acids. The strata run here E. and W. (h. 6,) and dip northwards. We follow them to near the Church of Tranbye, where they are suddenly cut off by the granite of the Paradiesbacken.

It is remarkable that the strata from Giellebeck onwards, continuing along the road, stretch throughout their whole length nearly E. and W. (h. 5—6), and dip under a considerable angle towards the north. This seems also the leading stratification for an entire part of this district. We see here also frequently great and thick porphyry veins in the lime-stone as at Christiania, only they are less striking as they have been less uncovered. In the further course of the road the lime-stone entirely disappears, and siliceous strata preponderate: partly black siliceous thick-slatey clay-slate, and partly flinty slate itself. I remarked no other strata than these on the heights and in the thick woods between the Valley of Dickemarck and Asker. They appear there frequently in small rocks.

About the middle of the way between Giellebeck and Asker, I left the road to ascend a high mountain to the south, over against the Fiord, which serves as a beacon to vessels at a great distance, and which every where rises from the road high above the dark woods. It is called Asker Varde Kullen, because the signal of alarm for Asker, in time of war, is placed on this mountain. Again granite, from the bottom to the top. But I sought in vain here for the line of junction, and I do not therefore venture to decide whether this siliceous clay-slate merely surrounds the granite, or whether it is covered by it.

Great masses stand on the declivity of the mountain: immense rounded wool-packs, such as are only seen with granite. In the descent towards the Valley of Dickemarck, the precipices are frightful; and even beyond Asker, we with difficulty find our way over the cliffs and fissures. The granite itself is similar in form to these masses in both sides as well above as below.—It is small granular, and consists of both red and white felspar, of grey quartz, which is not unfrequently crystallized in small cavities, and of single, scanty, insulated, small, black, folia of mica. I saw here no where hornblende, and no foreign bed.

The mountain commands a most extensive and altogether an incomparable prospect. We see the whole of Christiania, the town, the countryhouses, the hills, and mountains around; and all the Vales of Drammen: the district of Kongsberg, Holmestrand, Dröback, and the Fiord of Christiania. Here, in the central point of this wonderfully intersected and mountainous country, we take in at one glance the relief of the whole of this remarkable district. The mountains which appear elevated from the plain, and the others which rise above the level of the range, can only have their proper altitudes assigned to them here, or be compared with one another. I first saw here that this point rose to almost as great a height as Kolaas, and higher than the porphyry mountains above Asker, Ramsaas, and Skovumaas, which look so picturesque from Christiania: but these were also higher than the rocky elevated Island of Haaöe before Dröback. The dark woody Bogstadaas, and the granite mountains between Dickemarck and Rögen, which run in a continued and almost uninterrupted chain through the whole Peninsula of Hurum, appeared, however, still higher. The height of Asker Varde Kullen itself I found by the barometer to be 1094 English feet above the Fiord.*

The mountain is continued for a little way eastwards, but soon falls abruptly, and does not reach the shores of the Fiord. I had not descended so far as a small lake at its northern base before I again saw the siliceous clay-slate make its appearance; but I was not here fortunate enough to discover the relative position of the two formations. The extension of the granite through the Peninsula of Hurum, which is only separated from the Drammen Fiord by the granite range of Strömsöe, makes it indeed probable that this granite lies also undermost and that it is followed by clay-slate and lime-stone.

In the immediate vicinity of the Fiord of Christiania, at Slaben, a black compact lime-stone again appears in thin strata above one another, and con-

---

17th Septr. h. 7 Giellebeck .... 27. I. 9. 4 Therm. S
h. 11 Asker Varde Kullen 27. 6. 1—13.
h. 10, p. m. Christiania .. 28. 7. 5—9.

tinues with clay-slate and porphyry veins without further interruption all the way to Christiania.

It is singular that the Fiord of Christiania, in its whole extent from the open sea to the town, should separate the whole of this variety of newer rocks from the older formations. All the islands even belong to the transition formation; but as soon as we set our foot on the main land on the eastern side nothing but gneiss makes its appearance. On the other hand, there is no where a trace of gneiss to be found on the western side of the Fiord, till we get many miles into the interior, when all the transition formations disappear.

This short tour was the means of throwing much light on the granite and porphyry of these countries; but with respect to the relations of the wonderful zirconsyenite, it afforded no solution. I therefore resolved to search for one to the north of Christiania, towards the sources of the Aggers Elv, where it takes its long course to the Hackedal.

M. Haussman has excellently described the zirconsyenite as a formation. This stone has in fact, at its very first appearance, when we see it scattered around in large blocks, such a singular aspect, that we cannot venture to compare it with any one known stone, but consider it even before investigation as a peculiar and independent formation.

It is strongly distinguished from every porphyry by the magnificently coarse granular, and sometimes large granular felspar, partly of a pearl grey, and partly of a red colour, which always strongly characterises the blocks by its high degree of lustre. It is equally distinct from granite, syenite, or other similar granular stones, by the preponderance of the felspar. All the other ingredients seem to be sunk in this as a basis; and they often appear only occasionally; but hornblende is never wanting, and this hornblende is generally pretty characteristic and distinct; long black crystals, which possess a double foliated fracture, by way of discrimination from mica; folia of mica also make their appearance but very rarely; and quartz shews itself in small grains, so as not to be altogether missed. It appears in general accidentally in the composition, and we search through whole hills without finding it again. Wherever the grains of the felspars meet there remains almost always a small angular cavity, into which crystals project. Among these are the crystals of zircon, which have made the stone known, and given it its celebrity. They are by no means unfrequent; and although they are not large, yet to find them it is hardly necessary to look at more than one block: they are partly brown, and partly mountain green, with every variety of intermediate colour; and their crystallization is always pretty easily distinguishable. Epidote of a grass green colour in fine needles is associated with them, but not so frequently; and in the massive felspar, grains of black titanitic iron ore also here and there make their appearance.

Blocks of this nature lie on almost all the hills around Christiania; they are known as *Kampestene* (field stones), and eagerly sought after, as they are naturally preferred for walls to the crumbling Skialleberg; but they do not appear in fixed rocks by any means so near. The Grefsen (Grefsen Nuden) lies to the north of the town, and is visible from almost all the houses; but it is at least four English miles distant from the shore of the Fiord. It closes the amphitheatre of the hills of Christiania with steep, unconnected rocks, covered at the tops with thick wood.

On the road to Drontheim, which runs along its base, there first appears on the declivity, above the usual soft clay-slate and lime-stone, several strata of compact black clay-slate resembling grey wacke, in which folia of mica are very frequently scattered: higher up, these strata become siliceous, resembling flinty slate: and further on, towards Linderud, they become perfect and distinct thick slatey flinty slate. Higher still, and almost under the signal on the mountain, there appears an indistinct porphyry; the basis white and the felspar crystals within it also white and very small. All these strata dip under an angle of 30 to the north: the mountain continues in the same direction, and if we follow its course, we advance farther from the more ancient to the more modern strata. I saw above the white porphyry of the signal a greenish black and fine granular greenstone, containing felspar crystals, in the exact manner in which it frequently cuts through the clay-slate below in the veins. Farther on, the most distinct rhomb porphyry was seen resting on the greenstone, from the height to the base of the naked and inaccessible cliffs. I went a few steps farther, and zircon syenite lay immediately above the porphyry. Red small almost coarse granular felspar, with the usual angular cavities, and with distinct hornblende in it; and this formation goes far through the woods, and ascends the higher mountains.

The entire order of the stratification is here uncommonly distinct; it lies beautiful and open like a drawn section, below the usual strata of the transition formation; and then flinty slate, porphyry, zirconsyenite.   However much this last may be crystallized, it is still the newest of these stones, and there cannot exist a doubt that it must here belong, along with the porphyry, to a principal formation.

As all the strata in the Grefsen dip towards the north, I expected to find the same order of stratification again in the wide valley in which the Aggers Elv issues out of the small Sannesöe, and is afterwards precipitated through narrow clefts, over clay-slate rocks and innumerable wheels to Christiania. The case was exactly so.   I saw lime-stone still at the Oberfoss mill, then immediately in the clefts clay-slate, like greywacke, then splintery horn-stone, then flinty slate, then several strata of porphyry of no great thickness, and at last, distinct zirconsyenite of a coarse granular red felspar, and alto-gether without quartz.   This continues to the shores of the Sannesöe; but what is very remarkable, porphyry beds lie always within it, and alternate with it; and this in such a manner, that we may frequently break off pieces, which are half porphyry and half zirconsyenite.   The porphyry is of a blackish grey colour: the base appears fine granular in the sun, and contains small imbedded crystals of red felspar and a little hornblende. Such facts are certainly highly striking, for they speak loudly with respect to the kind and nature of the affinity of porphyry and zirconsyenite; even in small pieces, which we can carry with us and preserve in cabinets.

Farther on the road, round the east end of the Sannesöe, the zirconsyenite now disappears below distinct granite.   Grey conchoidal quartz, of which not a single grain was before perceptible, now lies between the red felspar and the mica in single folia, instead of hornblende.   This is not merely an individual bed, for the stone continues to the furnaces (*Frischfeuern*) belonging to Bärum's iron-work, and through the narrow opening of Maridal.   Hence the granite lies *above* porphyry and zirconsyenite, and not *below* as at Drammen; but that does not however extend far here. In the steep rocks of the Maridal, where a dark wood covers the whole of this wild vale, the syenite becomes again dominant, and remains so for miles up the vale : every where there are crystals of zircon in it.   I frequently saw here white felspar crystals in the red felspar, as in a porphyry; and it appeared clear to me then how the porphyry may have formed itself actually in this manner.   When the concretions of the felspar diminish in size, become fine granular, at length cannot longer be distinguished as felspar, the reddish brown base of the porphyry is formed, which occasions a great change in the colour, fracture, and hardness of the other ingredients, now owing to the smallness of these grains having become invisible to the eye. New crystals of felspar form in the mass as crystals of alum do in the combination of clay and sulphuric acid. The porphyry of this country may also in fact be nothing else than a zirconsyenite, which has become very fine granular. Or this is perhaps a porphyry in which some of its ingredients have assumed a visible magnitude. We ought at least never to forget, that the basis of every porphyry is never a simple mineralogical fossil, and that its true mineralogical nature can never be discovered, for this reason alone, that our eyes cannot follow the individual parts, from their extreme minuteness ; for who would ever think of a crystallization of the compact basis of a porphyry?*

The zirconsyenite in Maridal comes into immediate contact with the great porphyry masses of Krogskoven ; or more determinately with the porphyry of Bogstadaas, and with the mountains above the Sörkedal, which are spread widely out in one plane, and which go by the general name of Nordmarcken. In the same manner as at Grefsen, this porphyry appears immediately below the syenite, but I never saw it any where distinctly ; however, I conjecture that the deep valley itself, in which there are many little lakes, Sandungenvand, Skarvvand, Mylvand, here forms the separation of   two formations to a great distance into Hadeland ; for on the eastern side of the valley the mountains are flat above, and covered with great plates of rocks like tables, as is frequent with zirconsyenite; but on the other hand, on the western side of the valley, the mountains ascend steeply and abruptly, as we expect from porphyry.

However much we may be struck with the circumstance that the porphyry under the zirconsyenite at Grefsen does not perhaps reach a hundred feet in thickness, and but four English miles from thence at Bogstaadaas, rises to a height of 1300 feet and upwards, these examples are not altogether rare in geognosy, and especially in the investigation of porphyry. It is the same however here with the sand-stone, which opposite Ringerige

---

* In Scotland we have frequent opportunities of seeing the transitions here described. In the island of Arran a distinct transition is to be observed from the uniform compact felspar basis through porphyry to granite ; in Galloway, in the mountain named Criffle, similar transitions appear; and Dr. Macknight, in his excellent description of Ben Nevis, in the first volume of the Transactions of the Wernerian Society, points out the transition from syenite through porphyry to a compact dark coloured felspathose basis.—J.

is above 800 feet thick below the porphyry; but only for a few hundred feet below Bārumsverck; and which both at Grefsen, and below Kroftekullen, at Giellebeck, altogether fails.*

Thus the country of Christiania owes to the transition formation several new and very unexpected mountain rocks *(Gebirgs-arten)*, porphyry, zirconsyenite, and granite; and if we may assume the observations hitherto made as sufficient, these formations lie below one another in the following order.

1. Zirconsyenite as the uppermost and newest rock. In one almost uninterrupted range of mountains, from the western side of the Aggers Elv to Hackedalen, on the borders of Romerige and Hadeland. The greatest height which this formation attains in the neighbourhood of Christiania is at the Wäringskullen, to the westward, above the Hakkedal, and 1629 Paris feet above the sea†.

2. Granite below and in zirconsyenite at the Sannesiö above Christiania. A proof that granite may continue till the termination of this formation. It would be singular if we were to discover petrifications in this granite. As a stratum which lies below it (the lime-stone) contains many petrifications, this is not altogether impossible, if the condition of the crystallization were not inimical to the contemperaneous existence of organic substances, and made it very improbable.‡

3. Porphyry—the widely extended height of Krogskoven; Gyrihougen in Nordrehougs Prästegieldt is the highest porphyry mountain of this district, somewhere about 1600 feet above the sea.

4. Sand-stone. Below Krogskoven, and down at the Holsfiord.

---

* The great inequality in the thickness of the beds of porphyry and sand-stone, mentioned by Von Buch, is a common appearance in similar rocks in this island. The trap sand-stone and lime-stone rocks in the river district of the Frith of Forth afford many examples of this kind.—*J.*

+ In Galloway I picked up a few fragments of a syenite containing zircon, and also grains of magnetic iron-stone. I did not find any fixed rocks of this kind, and hence cannot pretend to give its geognostic situation.—*J.*

‡ The discovery of transition granite in Norway is one of the most interesting facts as-certained by the skill and industry of Von Buch and Haussman: it adds a new member to the series of transition rocks, and renders the arrangement of Werner more complete. This granite is not unknown in Scotland. At Fassneyburn, about thirty miles from Edinburgh, both tran-sition granite, and syenite, and also porphyry, occur: similar rocks appear in other parts of Scotland; and I am informed that the granite of Cornwall belongs to the same class of rocks. Some late French writers speak of this rock as abounding in Swisserland.—*J.*

5. Flinty-slate almost wherever porphyry and clay-slate come immediately together. At the Grefsen and at Kroftekullen.

6. Compact greywacke, resembling clay-slate.

7. Clay-slate and black orthoceratite lime-stone; all low mountains, the shores, and the islands of the Fiord of Christiania, all the way down to Dröback. We hardly find these strata higher than 900 feet above the Fiord. I myself have never seen them so high in this district.

8 Granite. The extensive masses of Hurumland, of Svelvig, Strömsöe, and Eger. Its greatest elevation is probably at the Nässfieldt, between the Eger lake and Sandsvär, nearly 2300 feet high.

9. Probably clay-slate and lime-stone may also make their appearance again below the granite; but we have no determinate observations on the subject.

Here closes the list of transition formations, and next follows:

10. Lastly gneiss. The general fundamental rock in the north. Hence all the rocks which distinguish the transition mountains here from those in southern countries are the crystallized species. It is like an attempt of nature to return again to the crystallized forms of the primitive formations. It is singularly remarkable indeed, that though this attempt has not fully succeeded, yet the tumultuous secondary formation *(Flötz)* has never been able to penetrate into the north. Notwithstanding the number of the formations, not a trace of coal, of newer lime-stone, or sand-stone, has been found either in Sweden or Norway. Why may we not conclude that the convulsions which converted the formation of primitive rocks gradually into flötz formations, operated much more extensively in the equatorial regions, and only through communication extended themselves towards the north? Why should we not believe that on this account the cause of the formation of primitive rocks was of longer duration in northern countries, was inimical to the causes or movements of the secondary formation, and thereby partly introduced new productions into that formation, and partly interrupted its progress?

These are however conjectures which will be immediately overturned whenever porphyry, granite, or zirconsyenite of the transition formation shall be discovered in lower latitudes. Much of the porphyry which has hitherto been believed to be primitive porphyry may perhaps in fact belong to this formation. We have yet no sure proof that the most of the porphyries of the Silesian principality of Schweidnitz, or the porphyries of Krzeszowice at Cracow, do not belong to it.

CHAPTER IV.

JOURNEY FROM CHRISTIANIA TO DRONTHEIM.

Romerije.—Difficulty of the Journey in Spring.—Hedemarcken.—Extent of the Transition Mountains in Norway.—Gneiss and Clay-slate at Miösen.—Provost Pihl.—Fertility of the Province.—Hammer.—Greywacke at Fangsbierg.—King Oluf in Ringsager.—View of Guldbrandsdalen.—Wolves on the Ice of the Lakes.— Clay-slate of Ringebo.—Bark Bread—Colonel Sinclair in Kringelen.—Quartz at Viig —Mica-slate in Kringelen.—Gneiss on the Heights.—Valley of Lessöe.—Gneiss of Rustenberg.—DovrefieldtSneehättan —Fogstuen.—Snow-line.—J•rkin.—Straits of the Valley at Kongsvold —Drivstuen.—Opdalen, End of Dovrefieldt.—Mica-slate on the Height. —Gneiss at the Foot.--First Firs.—Ockedal.—End of the Mountain Range at Stören. —Interesting Guldal.—Arrival at Drontheim.

APRIL the 21st, 1807. The winter was now past: it could hardly indeed deserve that name. The thermometer had never been observed more than ten or twelve degrees below the freezing point;* and even this degree of cold was only experienced for a few hours on the 25th of January and the 2nd of February. All the snow upon the hills, which environ the town in the form of a gentle ascending semicircle, had been since the beginning of April gradually disappearing. The return of summer could no longer be doubtful.

The boors of the neighbouring highland districts of Romerige, Hadeland, and Oesterdalen, by no means expected this mildness around the town. They descend in long rows of sledges with deals and hides, and return laden with grain and flour: but it was in vain they now endeavoured to reach the snow in the woods over the soft and slippery clay-slate hills. The burden, which would have glided lightly along the snow tract, was now fast and im-

---

moveable on the rapidly ascending road. The snow was yet deep and firm on the ground at a distance of two English miles up the ascent, and perhaps six hundred feet above the level of the Fiord. There the warm breezes from the Fiord do not easily enter, and the thick woods prevent the warmth from penetrating the earth.

Romsaas, an inn at the entrance of the woods where I remained for the night, seemed yet entirely buried in winter. Icicles hung from the roofs, and all around was covered with snow. This place was only however five hundred and sixty-four feet above the Fiord, and was situated at the commencement of a valley, down which a stream of the name of Lo takes its rapid course, and falls into the Fiord at the old Town of Opsloe. Oos was formerly the name of an Embouchure, as for example Nidaroos (*Drontheim*) the mouth of the River Nid. Ooslo, or more recently Opsloo, is therefore the mouth of the Lo.* Such radical words are very easily lost, and yet they may frequently throw light on many combinations in kindred languages.

The following day, the 22nd of April, I was enabled to travel conveniently in a sledge: but this did not heighten the pleasure of the journey. The snow destroys every prospect. At the top of the height, about two English miles from Romsaas, there is an extensive view over the Nittedal, and the Lake of Oejeren in the bottom: but it was now one entire white with the exception of a few dried twigs here and there. In summer this may not however be one of the most uninteresting of the numerous prospects with which the country round Christiania abounds.

It was thus the whole day through: nothing could be discovered from beneath the covering of snow on which the imagination could for a moment repose; and in the disagreeable inns of Skydsmovolden (Skrimstadt), Moe†, and especially the dirty Roholt, there was nothing calculated to detain us. Two wooden bridges, the property of private individuals, are here thrown across two pretty considerable streams which fall into the great Oejeren. The last of them was sold in 1805 to four countrymen, for seven thousand dollars: a wonderful species of property and revenue, which we might perhaps have expected to find in England, but not so easily in Norway.

---

* Sevel Bloch Reise Jagttagelser fra Trondhjem til Christiania, 1808.   54.

† Moe is a name which very frequently occurs in Norway: its original signification is a small sand-hill, which can be distinguished between mountains and rocks.

Late in the evening I passed through the great *Ancker* iron-work of Eidsvold, and I was welcomed into Minde by several hundred dancing and drunken boors, who were celebrating in this jovial manner the whole night through the close of an auction. All of them had come on sledges: so that here also it was still winter.

Nothing can be truer than what we hear in Christiania or Drontheim, that for a month and a half in the year, April and the beginning of May, all communication is thoroughly destroyed by nature in Norway. I experienced this but too sensibly. Do we resolve to proceed wholly on sledges? In Romerige the sledge was perpetually immersed in water, or impeded by earth and stones. It was still worse in Hedemarcken. The snow melts much faster on the road than in the fields, partly from the trampling of horses, and partly on account of the melted snow running along the path. Shall we try the cars of the country? But a sledge is small, and a car on the other hand extremely broad. The former sledge tract now appears a ditch, and the wheels of the car cannot make their way through the snow, which is still perhaps two feet in depth at the edges. Shall we proceed on foot or on horseback? In fact, this is almost the only way to make any progress: but then who dares venture himself on horseback when the horse frequently sinks up to the belly in the woods or hollow places in the road, and cannot extricate itself without the assistance of others? We have not this danger, it is true, to fear on foot, but we proceed incessantly as in a running river; or the snow breaks, and we fall into numberless little lakes discoverable underneath. Whoever therefore wishes to enjoy pleasure and entertainment in a journey through Norway would do well carefully to avoid travelling in the months of April or May.

I had not been gone long from Minde before I again perceived that it was no time now for travelling. The sledge and horse had to proceed by a ferry over the next stream of Vormanelv, the great outlet of the Miösen. In the middle the river was quite open; but it was yet for a great way from the sides completely covered with ice. To find the place where the horse and sledge might proceed over the ice to the ferry without imminent danger required repeated and careful trials, for the ice, which was formerly firm, was now changing every hour, and falling into the stream. This Vormanelv was then as large as the Limmat at Zurich. What a stream must it be in summer, when the snow melts from the mountains for more than one hundred and thirty English miles up the country! It issues here out of the Lake of Miösen, the small inland Sea of Norway, which appears calculated to carry commerce and prosperity into the heart of the country: but it was impossible for me then to form any conception of vessels in full sail proceeding up and down the lake. For thirteen English miles upwards it was yet covered with solid ice. In the northern part, however, there was no longer any ice. This is always the case: the undermost point is much sooner covered, and every thing around wears the appearance of winter there, while the waves are power-fully agitated higher up the lake: this is probably because the warm air of the plains cannot extend over the lake. It bears a resemblance here to a very broad river, confined between high hills: for the mountains of Feygring to the westward rise to a height of above one thousand six hundred feet, and the road winds with difficulty to the eastwards over the rocks and through the dark wood (Mordskov) along the high and steep declivity of the banks of the lake. How eagerly did I wish that I could have enjoyed a few summer views at the good cleanly and comfortable inn of Morstuen, the mountain of Feygring beyond appeared so interesting, when we pictured to our imaginations green declivities, farm-houses, and woods along its sides, instead of the dismal and death-like appearance of the snow! A few ships below, on the clear and unruffled lake, would have added greatly to the pleasure of the view. But the light sledge of the boors carried me over the lake itself to Corsegaard, and as yet I saw nothing from which I could con-clude that this road over the ice would in a few days undergo any change.

I was now in Hedemarcken. Notwithstanding the snow and ice, it was easy to discover, from the general desire every where manifested to live better and more comfortably, that the road now lay through the richest province of the country. Beyond Nöckleby, the hills stretched themselves out to a table land. Clean and neatly built farm-houses appeared along all the declivities, many of them two stories high, with numerous windows, containing large and luminous panes of glass.

The fields around were inclosed with hedges, and carefully cultivated, and were pleasingly interspersed with thickets and meadows; and I now also saw, what is very unfrequent in Norway, a display of diligence in the planning and erection of the granaries. We were now in a corn country, which richly supported the inhabitants. On the fruitful clay-slate the ground yields twelve times the amount of the seed; whereas on the rocky gneiss soil it hardly repays the diligence and pains bestowed on its cultivation. The dwellings are crowded so closely together, that the whole province has the appearance of one great village. The churches approach near to one another; whereas, before they were as rare as great towns. Every place is occupied, cultivated, and cheerful; and every thing denotes the enjoyment of comfort and prosperity.

At night I reached Hiellum, in Vangs Prästegieldt, in the midst of this rich landscape, and not far from a bay of the great lake Miösen.

———

The clay-slate of the country round Christiania does not continue long on the way to Hedemarken. Three English miles above the town the zirconsyenite reposes on it; and this singular stone continues till we arrive at the bottom of the valley below Ramsaas, which follows the course of the Lo to Opslo. On the opposite declivity of the valley, gneiss rocks immediately make their appearance. The gneiss is very distinguishable, particularly from the quantity of black mica which lies in scales one above another. Its strata dip generally to the westward.

This valley of the Lo therefore throughout its whole extent divides the older and the newer formations, as is done in the bottom, and farther down to the south by the great Christianiafiord itself. We may on this account, as well as of the separation of the gneiss from clay-slate and zirconsyenite, in some measure consider this valley as a continuation of the Fiord; and it is certainly singular enough that this separation upon the whole actually preserves the direction of the Fiord almost straight north, and occasionally a little to the north-east. To the east of this line of separation there is gneiss, and to the west we find only those stones which belong to the transition formations. It traverses the Nittedal somewhat above the principal church of the valley, runs then all the way to the upper end of the lake of Hudal, above the glass-work of Hudal, reaches the Miösen below the mountains of Feygring, and is again determined by the inferior part of the lake of Miösen itself, in the same manner as it was from Christianiafiord, all the way to Corsegaard. Here this gneiss boundary again enters the main land: it forms at once the line of separation between mountains and hills, between wildness and cultivation, between fertility and want. It runs in this manner along the upper part of Rommedals Prästegieldt to the small Rollsöe. It then follows along the base of a small elevation, almost accurately the boundary between the Prästegieldt and Löiten, and Elverum. It next traverses the Glommen river between Aamodt and Elverum, about four and a half English miles below Grundset; and then it draws towards the lake of Osen and Tryssild. But its course can then no longer be known with any degree of certainty; yet the gneiss can hardly ascend much farther northwards; and I conjecture the line of separation may run from Aamodt towards the southern part of the lake of Osen, and from thence immediately towards the under part of Tryssild, where the Clara Elv crosses the boundaries of the kingdom; for from the northern part of the lake of Osen, I saw a blackish-grey very highly fine granular lime-stone of the transition formation in Christiania, consequently there was no gneiss there. We know too from the Swedish accounts that almost all the mountains of Limasocken, in Dalarne, towards the boundaries of Norway, and consequently near Tryssild Prästegieldt, are composed of sand-stones and conglomerates of the transition formation, and probably of porphyries, which rise to great heights *. No doubt, when we go farther into Sweden, a considerable way into Dalecarlia as far as Rättwick, above Falun, we find formations of this kind wholly similar to those in Norway; but they do not reach the seashore; and it looks as if these newer rocks, like the inhabitants of the upland vallies, had come over from Norway. For if we lay down a chart of the demarcations of formations, and give to the inclosed districts different colours according to their nature, we shall find that Dalecarlia, Herjeadalen, and the southern part of Jämteland, are connected with the interior part of the Norwegian Aggershuusstift in one uninterrupted province of the transition formation, and that the southern part of Norway forms in the same manner one connected gneiss region, with all the Swedish provinces which surround Dalecarlia, Herjeadalen, and Jämteland. From such a determination of directions in the spreading of different formations, to which we are guided by nature herself, we may one day expect to draw important conclusions in geology.

---

* *Hisinger Sveriges Minerographie,* 1790. In the account which Hamarin gives of these districts, he speaks always of trapp and jasper, and breccia, with angular pieces of greenish quartzy Skörlberg. And in Hemfiället and Gammel Säterfiället, on the boundaries, the rock is red stratified sand-stone. There is not therefore in this part of Dalecarlia either mica-slate or gneiss.

If it is the fact, as is proved by mineralogical charts, that almost all the newer rocks of the northern Peninsula, between the southern points of Norway and the inland parts of Sweden, spread themselves exactly from north to south, with a slight inclination towards the north-east, why should not this in reality have been the course of nature in the formation of these rocks?

The rocks disappear between the Oejeren and Miösen lake, where a stone seldom appears above the surface. The country is full of hills intersected with vallies, but not mountainous, and we pass Roholt before we come to the Mistberg above Eidsvold, the highest mountain of the country, more than seventeen hundred English feet above the sea. This mountain still consists of gneiss; but the mountains of Feygring, which inclose the Miösen to the west, do not consist of it. At the southern end of the Miösen the same granite again makes its appearance, which is so frequent in the environs of Drammen. Above this, on the heights of the small insulated mountains, we have por-phyry; and on this again we have, towards Hurdal, zirconsyenite. This is the account of M. M. Esmarck and Haussmann. Among these rocks lime-stone and clay-slate are seen to spread themselves in flat hills over the district of Toten, as in Hedemarcken; and cultivation and prosperity are also in like manner diffused through Toten.

In my way below the heights of Morstue, on the eastern bank of the Miösen, I discovered none of these rocks. As soon as I set foot on the opposite bank of the Vormän Elv, I saw black hornblende and white felspar in a fine or minute granular mixture, not merely in one insulated bed, but spread over the whole country; and about seven English miles farther on, through the Mordskov, gneiss with white felspar, and a great deal of mica, actually appears merely as a subordinate bed in this stone, and not the reverse.

I had already seen a similar formation diffused more widely and of greater height in the upper parts of Walders, below Fillefieldt on the road to Bergen. It was there evident that in age it must have followed the gneiss, and that it in general occupied the place which belonged to the mica-slate. It might also be the same here on the bank of the Miösen. The multitude of gneiss beds became gradually greater in the hornblende rock, even before I reached Morstue; and in the course of about seven English miles to Corsegaard, this multitude increased to such a degree, that the gneiss at last preponderated; and on the other hand hornblende and felspar rock appeared seldom, and in subordinate beds. The gneiss at length resumed entirely its former rights. This sort of alternation of beds, which gradually appear more and more frequently, is quite usual, however, when one stone takes the place of another in the series of formation. They are a sort of attempts made by the newer formation, which become gradually more and more successful, till at last the older sinks in the new to a mere bed, and then disappears. Mica-slate is very frequent in the north as a bed in gneiss, before it preponderates over the gneiss; and in the same manner gneiss frequently forms extensive beds in mica-slate, before the former asserts its full mastery *.

About five English miles to the north of Corsegaard, I frequently saw the white felspar of the gneiss shining through the snow on the small rocks by the road-side; but this disappeared with the external appearance of the country; and before reaching Nöckleby, clay-slate and black lime-stone had become predominant. Between Hiellum and the church of Vang the thin strata of the black and compact lime-stone make their appearance. They dip towards the north, and not unfrequently contain testaceous remains and impressions.

On Friday, the 24th of April, I went to the clergyman's house *(präste-gaard)* of Vang, which was close at hand, to pay a visit to Provost Pihl, one of the most distinguished men in Norway. I had the good fortune to find him in the midst of his amiable family; and I owe a great deal of information to the conversation I had with him that day. He possesses such a collection of astronomical instruments as is to be found in few observatories. Among others, he has a Herschel's telescope, a very beautiful Troughton's sextant, and, if I am not mistaken, two of Arnold's chronometers. These instruments are not suffered to remain unemployed: the pains he has taken in determining with accuracy by means of them the situation of his

---

* These alternations of the newer portions of one formation with the older portions of another occur in this country. In Perthshire, for example, we observe the older strata of clay-slate alternating with the newer ones of mica-slate, and the older strata of mica-slate alternating with the newer strata of gneiss. In the same county, and in other parts of Scotland, I think I have observed an alternation of the oldest transition rocks with the newest primitive rocks.—J.

country are deserving of the highest encomium; and perhaps the eastern coast of Norway, in the vicinity of Arendal, would have been still placed at random in the charts, as it was a few years ago, had not government empowered M. Pihl to ascertain its situation by actual observations. The recommendations of Zach, and the excellent work of Bohnenberger, have made him, like many others, a friend to sextants; and equipped with such instruments, and with such qualifications for observation, we may rest assured that the situation of many of the vallies in the interior of Norway will be as accurately determined as the most cultivated parts of Europe. With this knowledge, and this enthusiasm for practical astronomy, M. Pihl unites singular mechanical talents, which entitle him to a most distinguished rank in this department. His telescope is in great repute beyond the bounds of Norway: he has a three feet telescope in hand, of which the beautifully elegant an dexquisitely polished brass feet were made in his house. I did not entertain a doubt of the truth of what he affirmed with all the simplicity and confidence of truth, that after many attempts he had at last succeeded as well in the polishing of glasses of all sorts as the best English artists. His house resembles a manufactory, and his example and exortations have had the most powerful effect on the surrounding country. Pieces of mechanism are every where to be met with; and throughout the whole of Hedemarcken, in a short time, a clock or watch will not be found which were not manufactured in the province.

How excellent would it be if this were the first sort of industry which took root in the vallies of Norway! It is highly suitable to the inventive talents, the mode of living, and the climate of the Norwegians, to whom the long winter evenings afford so much time for house occupations; and it requires materials which can be much easier procured and conveyed up to them than any of those which are wrought up in great manufactories.

M. Pihl has also applied very successfully to physics: he shewed me several extremely neat and well executed thermometers of his own workmanship, besides electrifying machines, and various other articles. His meteorological observations for a number of years back are a treasure for the knowledge of the climate of the north, of the results of which it is hoped the Royal Society of Sciences in Copenhagen will not long deprive us. Vang is, as it were, the central point of the kingdom; as the climate of Christiania is modified by a degree of higher latitude, and the temperature of Bergen by a greater distance from the Western Ocean of ten degrees of longitude. This is an investigation in which the whole natural philosophy of the earth is concerned; and though we are not yet enabled to assign any mean temperature for Vang, either for years or months, it is however a subject of congratulation that the materials are actually collected, and that the determination will always be possible, whenever either M. Pihl or M. Bugge in Copenhagen have sufficient leisure to bestow their attention to the calculation of the separate observations. The mean barometrical height of Vang's Prästegaard is twenty-seven inches nine lines; consequently the mean height above the sea is nearly four hundred and thirty-four English feet. The Prästegaard does not lie however much higher than the Miösen, and we may therefore assume the height of this lake to be about four hundred and twenty-six English feet above the level of Christiania. I found nearly the same result by corresponding observations. The Miösen lies therefore higher than Rändsfiord on the western borders of Hadeland, which is almost parallel to it, and of nearly the same length, and which in the summer of 1806 I found to be only three hundred and eight English feet above the level of the sea. Is it on account of this deeper position that the gneiss makes its appearance every where on the banks of the Rändsfiord, whereas not the smallest trace of it is to be found in the upper part of the Miösen?

That it is colder in Hedemarcken by the lovely banks of the Miösen than at Christiania is by no means evinced by the appearance of the country. How can it possibly be colder? What is so well cultivated we suppose to enjoy a good climate. M. Pihl told me that almost every thing grows here which Norway produces.—Of fruits, they have apples and cherries in abundance, in Stange, in Näss, and in Helge, a large island in the lake; and they have as many kitchen stuffs and trees as they have occasion for. I sought for the oak; but it was no longer to be found. The last one I saw, and which was remarkably beautiful, grew at Skiedsmo at the Prästegaard. The region of oaks does not penetrate farther into Hedemarcken. Oak trees must here be cultivated like fruit trees, and even this only succeeds in the places which enjoy the kindly influence of the lake. Wherever the land rises in any degree they would be soon destroyed by the climate. Although cultivation is here in a more flourishing state than any where else in Norway, Vang's mean temperature, however, hardly reaches 41 degrees of Fahrenheit.

I cannot leave this valuable district without suggesting a consideration which occurs to every stranger who travels through the interior of this country. Why is there no town in such an extensive space to unite the inhabitants of the country in one general interest? Why is there no central point here, which, like a mighty spring, might every where give rise to a brisker circulation, and create life and activity in this present dead mass of produce?

Wherever we cast our eyes we perceive the urgent want of this remedy, and it is scarcely credible that at the first impulse the town should not as it were spring at once out of the soil. But formerly one existed here, and few of the present towns in Norway flourish in an equal degree, and it stood exactly at this place, on the banks of the Miösen, in the middle of the kingdom, at an equal distance from all the upland and extensive Norwegian vallies, and alike convenient for all of them from the navigation of the Miösen. The isolated ruins of the numerous churches preserve its memory, and the great palace of *Stör Hammer* its name.—Opslo had been long built, yet Hammer was always increasing in size and prosperity. The Reformation occasioned the removal of the bishop's seat from Hammer to Opslo, but still the town continued to prosper. In the year 1567 the Swedes advanced from Opslo to Hammer, highly enraged because they could not take the strong Castle of Aggershuus: they revenged themselves on Hammer, plundered the rest of the town which had been already plundered by them, set fire to the houses, and destroyed every vestige of both them and the churches. The inhabitants fled to Opslo, and never returned. Since that period agriculture has prodigiously increased; and now the soil yields even more than sufficient for the necessaries of life. There is a general endeavour every where visible to improve and ornament whatever will admit of it. But no town offers the people as yet a hand to assist them in their praise-worthy endeavours, and there is no excitation through facility of acquisition, or the collision of rival activity. We frequently hear that the Hedemarcken people give themselves up to a foolish prejudicial and ruinous luxury. The men consume their superfluity in entertainments: the women dress themselves like the merchants' daughters of Brügge. The merchants of Christiania affirm that one of their best and most important articles is silks and fine cottons for Hedemarcken. Undoubtedly they would not employ their superfluity in this manner, if, instead of going ninety English miles over mountains and vallies, there was a town in the neighbourhood which would enable them to dispose of it with facility.

We hear many people in Norway say the peasants require nothing from towns. Whatever is required for their housekeeping and labour they make themselves. They are blacksmiths, carpenters, cabinet makers, rope makers, weavers, shoemakers, and taylors, at the same time: and this they praise and consider as a great advantage for the Norwegian peasants.

It is not so however. Whoever makes so many things must make them badly, and will never be able to do with the bad what he could have done with better. The experience of these districts teaches us that the peasants are only driven by necessity to turn their hands in this manner to every thing. Whenever they have a town at hand, they soon lose their skill, and draw from it what their remoter neighbours imagine they are necessarily obliged to make for themselves.

The Hedemarckian would willingly purchase better cloth, better wheels, iron ware, tables, and chairs, in Hammer; he would willingly take the pains to manufacture a single article, that he might make it the better when he found he was always sure of disposing of it in Hammer, were such a town as Hammer to exist. What activity and refinement would it not be the means of creating in the interior of the kingdom, for this district has not nearly reached the highest degree of population of which it is capable! In the Prästegielden, where agriculture is alone carried on, and which contains no mountains, in Näss for example, there are one thousand three hundred and forty-four inhabitants to the geographical square mile, and in Vang and Stange there are about one thousand one hundred. This is unheard of for Norway, and may perhaps be the maximum of population in the present almost insulated situation of this province. But how do nearly two thousand inhabitants live on one square mile in Swedish Upland, which is certainly much less fertile? How are there two thousand four hundred and forty-two, in Schonen, where grain is never returned twenty fold, and seldom twelve times, as is the case almost universally in Hedemarcken. In Upland and Schonen there are no want of towns, communication, and collision. What fertility alone cannot do is there produced by mutual assistance and industry.

A great deal has often been said about the foundation of Hammer. The meritorious president of the Danish chancery, Chamberlain Kaas, entered zealously into the subject when he was in an official situation in Christiania,[*] and in 1802 it was believed that the project was on the point of being carried into execution ; but he was called to fill higher situations at Copenhagen, and Hammer looks yet with anxiety for the period of its revival.

On the 25th, I endeavoured to proceed to Bierke in a car (Kiärre). They are easily procured; for every peasant keeps his travelling kiär to go to church in, which is often as elegant as any thing one could have in a capital. In the lower tracts I succeeded very well with it, but not so well in the wood before coming to Bierke. The sun does not penetrate into the woods, and the air is not yet sufficiently heated. A great deal of snow was still lying in the wood, and the car proceeded with the greatest difficulty. From Bierke I was again conveyed in a sledge down to the Miösen, and across an arm of the lake to Fangsbierg. Notwithstanding the snow on the ice was all melted, the journey was passable, and a great deal more pleasant than my road over the hills from one arm of the Miösen to the other at Ringsager.

The arm of the lake at Fangsbierg was surrounded by steep rocks; and the mountains above crowned with dark spruce firs. The exhilarating prospects of the country around Vang began now to disappear. The last stages of the high mountainous range advanced all the way to the lake, and new formations made their appearance. The rocks consist of beautiful and well characterised greywacke. I saw it here for the first time with distinctness in Norway. Between Vang and Bierke single strata of greywacke made their appearance indeed through the snow; but the strata of dark smoke-grey compact limestone soon recommence, and run through all the hills which surround the Inn of Bierke, dipping gently towards the north. At Fangsbierg, however, greywacke, several hundred feet in height, alternates in coarse and fine granular strata. The pieces are often of the size of a pigeon's egg; and we distinctly recognize quartz, exhibiting various colours and other characters, and not unfrequently indigo blue, with a conchoidal fracture. We also find greenish white felspar, and scales of mica, in abundance. The pieces are however generally angular, and but seldom round. The mass which binds them together is blackish grey, and nothing else than the fine granular greywacke itself, in which the separate pieces cannot be recognized without straining the sight.

---

* Thaarup Magazin. II. B. 1. Heft, 15-½-6.

A multitude of white quartz veins traversing the strata are singularly contrasted with the dark colour of the rock ; and their unusually strong
cohesion is proved by the numerous large blocks scattered about the base
and upon the declivity of the hills. The whole of this small range of mountains extends from the higher mountains to the sides of the Bremundselv
below. If we ascend higher, the fine granular strata of the greywacke
become more frequent, and at last resemble a fine granular sand-stone. In
this way we find them about four miles and a half farther up at Narud ; and
they are then worked in great quarries, not only for grinding-stones,
but for furnace-hearths. The great iron-work of Kongsberg, and the furnace
at Moss, receive their hearths from this place. It was thought, however, that
the English ones were both better and cheaper ; but they could not be
procured on account of the war. M. Esmarck has investigated this country
still farther, and found the porphyry of the district of Christiania above the
sand-stone of Narud. This is the last porphyry after leaving the high
mountains.

That the rocks of Fangsbierg however belong only in reality to a particular small chain which advances so far out, I very soon perceived when
I descended opposite Ringsager. The black lime-stone made its appearance
again above, on the banks of the Miösen, and alternated frequently towards
Moe with thin strata of clay-slate.

These were the rocks which gave birth to such fertility in the south of
Hedemarcken ; and the lands and buildings of every description in Ringsager proclaimed the existence of the same effects here. I was in fact lost
for a few moments in astonishment, when a winding of the road brought us
immediately opposite the great and beautiful stone church of Ringsager, situated in the middle of several farm-houses, which both in size and elegance
bore testimony to the prosperity of their owners.

Names give individuality to the different points of a country, and actions
still more, when in any particular place they originated in great strength of
mind, or the fate of whole kingdoms was decided by them. Ringsager is
celebrated in the history of Norway from a stroke of presence of mind and
decision, which raised King Oluf the saint in a moment from an almost expelled monarch to be again the autocrat of all Norway. The severity and
tyranny with which he every where spread christianity among the valleys,
and persecuted the Pagans, at length roused five of the petty kings of the

country to regain their freedom and expel him. Rörek of Hedemarcken, Ring of Toten and Hadeland, Dag, the ruler of Walders, and Gudriod, a prince in Guldbrandsdalen, united their forces at Ringsager, to concert an attack against King Oluf with very superior numbers. The king learned their arrival at Minde, at the lower end of the Miösen, where he was stationed with only 400 men. He speedily manned several vessels, ascended the Miösen hastily in the dead of the night, surprised the kings in their beds at Ringsager, took them prisoners, and thus with one blow destroyed their well-concerted plan. Ring and Dag were banished from the country, Gudriod was deprived of his tongue, and Rörek of his sight. These severities did not protect King Oluf from new insurrections, which at length compelled him to leave the kingdom; and on his arrival with fresh troops from Sweden, with an intention of regaining possession of Drontheim, the capital of the country, he lost his life in the battle of Sticklestadt, on the 29th of August, 1039.

The Miösen extends still higher up the country, and I again crossed it over the ice near Freng. The prospects now became grander: the high mountains lake within narrow lim TRAVELS THROUGH NORWAY AND LAPLAND. ks of Biri, a whole chain descended in several summits to the shores of the lake, from a height of more than two thousand feet. This was higher than any of the mountains from Christiania hitherto. Steep mountains also arose above Freng on the southern bank, covered with wood to the tops, but their height did not reach one thousand feet.

It was a delightful morning as I continued my journey on the 26th from Freng, beneath the rocks along the banks of the lake. The lake, the mountains, the scattered cottages which occasionally made their appearance between the rocks, gave altogether the most enchanting variety to the prospect. Four miles and a half below Freng I entered Guldbrandsdalen, in the company of a number of cheerful peasants, who poured down from every hamlet to go to church at Lille Hammer. There was a festivity diffused throughout all nature; it every where appeared to announce the opening of spring.

At every step the views became grander. I felt that we were treading the high mountains. The boundaries of Guldbrandsdalen are very properly placed here; for a succession of high mountain views are disclosed during the whole course of the way.   How beautiful and grand is the view of the end of the lake, and up the valley of Faaberg! How wildly the great Louven Elv rushes down from the valley! The peaks of the mountains ascend boldly above one another.   Two great chains of mountains now surround one of the most extensive and remarkable vallies of the country.

I lost here at length the lake of Miösen, after having followed it for full seventy English miles, and made my entrance into Guldbrandsdalen.   Here we had no longer sledges, and cars could not by any means be used; the baggage was therefore conveyed on a horse, and I willingly travelled up a country like this on foot.   The valley did not appear to rise till we came to a narrow cleft between Moshuus and Stav; and I did not till then remark any visible difference in the climate.   The cherry trees at Moshuus giv e in this respect by no means an unfavourable impression of the country.

But at Stav the valley was full of solid ice, and I was again obliged to have recourse to a sledge to proceed over the lake of Lösness.   We gain indeed a great deal by the ice; we are neither detained by windings of the road, nor by mountains and deep abysses; and the level surface of the ice allows of a rapidity, by which distance is almost annihilated.   A journey in winter down the great Norwegian lakes, such as the Miösen, would be in fact attended with infinite pleasure, if it were not for the wolves, which are extremely dangerous in the early twilights of winter; for the wolves are no where so fierce, or assembled in such numbers, as on extensive surfaces of ice.   They avoid every thing which hangs over their heads, and therefore flee the woods: this is singular enough, but not the less true: many of the peasants can defend their possessions in no other way from the wolves than by a hedge inclosure, through which the wolf may creep, but over which he cannot spring.   He will go round the hedge before he will venture to creep through, or even to make his exit underneath.*   The wolves remain in herds of dozens together on the open ice, looking out for their prey   If a single sledge makes its appearance, they troop together from both sides, and put the traveller in well-grounded apprehension for his safety.   A rope is fastened to the hinder part of the sledge, and so long as it can be preserved, it hangs

---

* If we are unwilling to credit the accounts of the boors and many others, the same is related by Ström, who is certainly far from credulous, in Söndmors Bescrivelse..

down loosely, and is dragged along the track of the sledge.   The small inequalities of the way raise it every now and then aloft, and make it turn in
perpetual serpentine windings.   This frightens the wolves : they dare not
venture an attack, and in dread of this dancing monster, they remain at a
secure distance.

I had no longer any thing to fear from the wolves in my way over Lösness.
I was assured by the peasants that the ice was yet perfectly firm, though
nothing but water was to be seen.   The snow on the ice was here completely melted, and the surface of the ice was covered with water, at least a
foot in depth, from the streams which descended from the mountains.   We
got on at first very well, but we soon lost our way; the twilight surprised us,
and we wandered about without any certainty.   I deemed it, however,
much safer to proceed through the water on foot, than to trust myself with
the burden of a horse and a sledge on such an ice.   The guide, a boy of
eight years old, was full of courage, and dreaded nothing ; whenever I
displayed any apprehension, he assured me with confidence, that though the
way could not be seen, and the night was dark, he would be sure to reach
Lösness.   I was pleased with this early display of self-confidence, and went
silently through the water ; and he brought me in reality about midnight fortunately and without accident to Lösness, where the good-natured activity
of the people of the inn soon banished every impression of the fatigue
of the way.

This passage is no doubt made much easier by day.   I proceeded more than
four English miles in the morning still farther through straits, between mountains covered with woods, to Elstad, below Ringebo church, and I could not
therefore see the picteresque rocky landscapes on the land-road of which
Father Hell has given us some very good representations in engravings.*   I
first set foot on land at Ringebo.

Every mile onwards, the internal composition of the rocks became more
and more distinct and characteristic.   The greywacke on the steep descent
of the lake, between Freng and Lille Hammer, was beautiful ; angular pieces
of the size of a pea ; white and blue quartz grains in the dark blackish-grey
clay-slate mass, and small yellowish white shining felspar crystals between.
The stone exists in rocks by the road side.   I saw no longer any lime-stone,

---

* Ephemer Vindobon, pro. Ann, 1793,

or any slatey strata.   The greywacke is pure, and continues spread over a great extent : by this it shews that it supports the lime-stone, and perhaps greatly precedes it in the order of formation.   True black clay-slate strata first appear beyond Lille Hammer, towards the end of the Miösen, and here also but scantily alternating at first with the greywacke.   Pieces of clay-slate lie every where as usual in abundance in the granular mass ; they become more frequent higher up, and towards Moshuus, in Oeyers Prästegieldt, they are large, and considerable plates of the size of a hand and upwards.   At every step we pass from newer to older strata.   Organic remains are no where to be found in the rocks here : they disappeared along with the lime-stones.   The formation of this greywacke, notwithstanding it is composed of ruins of older rocks, does not however descend to the period of the creation and after destruction of the organic world.

In the straits between Moshuus and Stav, an opening into the valley, which puts us in mind of the vast fissures of the Alps, the greywacke becomes red, and rises into high and steep rocks.   Pieces of clay-slate lie in a heap upon it all along the way.   The primitive clay-slate does not go far here; and I should probably have seen its boundaries had I come by land, and not over the lake.   Olstad and Ringebo church lie fully in the district of the clay-slate.   That it is primitive clay-slate is proved by the beds of talc-slate between, and the uninterrupted glistening over the whole planes of the folia ; not interruptedly glimmering on insulated folia, as in the clay-slate of the strata in the greywacke.   Perhaps there is no definitely distinct boundary between the primitive clay-slate and the greywacke ; both may imperceptibly run into one another at the lake of Lösness.

The valley is very well inhabited ; much more so than we might expect in the interior of a mountainous range, in a latitude of 62 degrees.   Although some of the views were wild and savage, and appeared like a dreary wilderness, this was more however to be attributed to the nature of the clay-slate, when the torrents of Frye-Elv at Froen, and the Sul-Elv at Soedorp rush down from the mountains; and this wild appearance does not continue.   Well-built and roomy farm-houses not unfrequently are to be seen along the sides of the mountains, and small corn-fields run all the way up the valley.   When we speak of great want and necessity in Norway, we must always make an exception of vallies like this.   Not only barley and oats, but rye is cultivated here, and Guldbrandsdalen seldom stands in need of foreign assistance. From some places grain is even exported to Röraas. Whether they have ever recourse to bread made of the bark of trees *(Barke Bröd)* is very doubtful; if it does happen, however, it is at least very seldom. Yet in other countries the general belief is that the people of Norway live entirely on such bread, and that grain will not grow there. Great injustice is in this done to the productiveness of the mountain land; and in reality nothing but the greatest necessity will ever drive people to feed on bark bread; for the procuring and preparing of the bark is by no means so simple as it appears, and requires a degree of labour which such a food is by no means worthy of.

In no district of the kingdom is this bread more used than in Tryssild, and the mountainous part of Oesterdalen; but even there, where the easiest mode of preparation is understood, to how many processes must the bark be subjected before a bread can be made of it, which some evil genius must have invented in derision of the human race !

When the young and vigorous fir-trees are felled, to the great injury of the woods, the tree is stripped of its bark for its whole length : the outer-part is carefully peeled from the bark; the deeper interior covering is then shaved off, and nothing remains but the innermost rind, which is extremely soft and white. It is then hung up several days in the air to dry, and afterwards baked in an oven; it is next beat on wooden blocks, and then pounded as finely as possible in wooden vessels; but all this is not enough : the mass is yet to be carried to the mill, and ground into coarse meal like barley or oats. This meal is mixed up with *Hevel* with thrashed-out ears, or with a few moss seeds; and a bread of about an inch thickness is formed of this composition. Nature with reluctance receives the bitter and contracting food; and the boors endeavour to disguise the taste of it by washing it down with water; but in the beginning of the spring, after having lived on this bread a great part of the winter, they become weak and relaxed, and they are incessantly tormented with an oppressive shooting and burning about the chest.* If it is impossible to procure any other nourishment, such vallies ought certainly not to be inhabited.

---

* All this is related by Smith in *Tryssild's Bescrivelse Norsk Topographisk Journal* Heft XXIII. He passed twenty years as a clergyman there, and possessed sufficient experience respecting this destructive custom.

At a short distance from Froen I passed a magazine, which is kept up by the people of the place to supply their wants in time of scarcity, and by proper precaution to provide against want and poverty. The ornamental and elegant church of Froen, of an octagonal form, and built of hewn stone, was certainly no symptom of poverty.

In the evening I reached Viig. The whole family dwelt together in one room; and there was no division of any kind between them and the stable: the pigs run about between the beds. This is true laziness. Hitherto I had never seen a house of this description, and in an inn it was the more remarkable. In Lille Hammer, in Moshuus, in Lösness, and Oden, there are always tolerably well-furnished rooms set apart exclusively for travellers: the meals are served up in stone ware and silver; and though the entertainment is by no means sumptuous, for the number of travellers is not sufficiently great to admit of any considerable supply of stores, yet we almost always find Chinese tea and coffee. This is also the case further on in all the inns till we arrive at Drontheim. But Viig puts us in mind of the Polish villages. May such places be rare in the great valley! They are a proof of the greatest indolence and want of spirit in the inhabitants.

Would you not like to see Zinclar's grave? said some of the passing country people to me, as I was waiting on the road for a horse. They took me but a short way, when we came to a wooden cross on the road; upon this a tablet was placed with the following inscription.

" Here lies Colonel George Sinclair,* who with nine hundred Scotsmen was dashed to pieces like earthen pots by three hundred boors of Lessöe, Vaage and Froen. Berdon Segelstadt of Ringeboe was the leader of the boors. This tablet was destroyed in 1789 by a flood, and again restored by the boors, A. Viberg, and N. Viig." The boors with anxious expectation, and a proud feeling of self-exultation, looked to see what impression this monument would make on the stranger. I was taken by surprise, for I did not believe myself so near the scene where the action of Sinclair took place; but I felt a respect for men who could still present such a keen recollection of a noble stand against foreign invasion, and such a strong feeling of freedom and their own dignity.

---

* Jörgen Zinclar.

At mid-day I reached the narrow pass of Kringelen, where Sinclair fell. It was a true *Morgarten* conflict: the road was narrow, and cut out of the solid rock, and overhung the steep and precipitous banks of the river which rushed along at the bottom. Sinclair had no where met with any opposition; for almost all the youth of the country had been drawn to the Swedish war in the south of Norway. He had no suspicion of any attack here, and carelessly pursued his way: the boors with great address proceeded unperceived over the rocks, and dexterously detached a small division to the other side of the river, which made its appearance over against the Scots on a large meadow, and with considerable irregularity kept firing on their enemy below. The Scots despised this ineffectual attack, and passed on ; but their attention was however directed to the meadow on the opposite side of the river. The boors suddenly made their appearance on the rocks in every direction : they closed up every avenue of advance: they prevented every means of retreat. Sinclair fell in the foremost ranks, and the rest were *dashed to pieces like earthen pots.** This is again repeated to passengers on a table here. *And thus let the enemy and the world learn*, they add, *what Norwegian valour, firmness, and fidelity, are capable of in their native rocks.* About sixty of the Scots interceded for life, and were taken prisoners. They divided them among the hamlets; but they forgot that prisoners are no longer enemies. They grew soon tired of feeding an enemy, and the defenceless Scots were collected together in a large meadow and murdered in cold blood. Only one escaped.

This fact is not told in the monuments, but they have not destroyed its reality, and may it continue to be handed down as a frightful warning with the recollection of this heroic action.

But how came the Scots into Norway, and to penetrate so far into the Norwegian mountains ? In consequence of a plan, which, as experience has shewn, was of too bold a conception. King Gustavus Adolphus, in his first unsuccessful war with Christian the Fourth, dispatched Colonel Munckhaven in the spring of 1612, to enlist men in the Netherlands and in Scotland. As the colonel was endeavouring to return in the end of summer with two thousand three hundred fresh troops, he found the fortress

---

* Nihundert *Skotter.*
Bley knuset som *Leer potter :* this expression is partly for the sake of the rhime.

of Elvsborg at Gottenberg in the possession of Christian, and the whole coast in consequence, from Norway to beyond Calmar, shut to the Swedes. Necessity compelled him to break through Norway. The greatest part entered the Fiord of Drontheim, landed in Stördalen, and found no Guld-brandsdalians to oppose them. They were thus enabled to proceed over the mountains to Jemteland and Herjeadalen, and by their arrival preserved the capital of Stockholm, which was threatened by the Danish fleet; but Colonel Sinclair landed in Romsdalen. He had already proceeded many miles through Romsdalen, Lessöe, and down the valley below Dovrefieldt, and might well believe the Swedish frontiers at hand, when he was destroyed by the circumspect and daring attack of the boors in Kringelen.

" Sinclair came over the salt sea,
To storm the cliffs of Norway,"

is a ballad which we hear in all the Norwegian towns; and it will long hand down to posterity the memory of Sinclair and the Guldbrandsdalians.

The views become now wilder, and the valley ascends more rapidly. The rivers (*Elven*) fall into one another with greater noise and impetuosity; and the mountains around are covered from top to bottom with dazzling snow, and appear to enclose the valley in every direction. We can scarcely perceive where the two great vallies of Vaage and Lessöe unite, notwith-standing both are the principal vallies into which the upper part of Guld-brandsdalen is divided. The Ota-Elv, which comes from Vaage, dashes down through a narrow gullet, and we have no suspicion that there is a large valley above of thirty-five English miles in length, and containing seven thousand inhabitants.

I durst not venture to go through the pass of Rusten among the mountains in the dark of the evening, and I remained during the night in the great farm-house of Formo, where, without, I found the men occupied with their small patches of cultivated ground between the rocks, and within, all the order, the cleanliness, and the prosperity of an unconnected and remote situation. I was here 1269 English feet above the ocean.

Through the whole of the lower part of Guldbrandsdalen, the clay-slate predominates to the tops of the highest mountains. In several places it might be used for roofs of houses, if these places were somewhat more accessible, as for instance, at the *Säter* of Froen.　It is strongly distinguished by beds of a talcy nature, which frequently make their appearance in it, and which bear a resemblance to the pot-stone (Grydesten).　Small fine granular folia of talc are mixed with quartz; and the whole mass of the bed becomes from that circumstance greyish white, resinous, and glimmering, coarse splintery in the fracture, and at the same time very fine granular and semihard.　When the quartz is lost in the mixture, and the talc folia increase somewhat in magnitude, the rock becomes sectile, cohesive, and soft, and it may then be made into furnaces, pots, and kettles; but such pure beds of grydesten are rare in the clay-slate, and much more frequently subordinate to the mica-slate.　Towards Froen, quartz beds become frequent in the clay-slate, and we are often surprised with the appearance of small druses of rock crystals in the white beds, which shine out of the velvet green chlorite earth of the roof.

The mountains rise from the valley almost without interruption to an altitude which at their tops reduces the firs to a very stunted appearance. They are almost every where about three thousand two hundred English feet in height; yet we see from many of the views up the great valley, and the vallies which run into it, that these mountains are but the first elevations, and that the mountains beyond rise to a much greater height.

The high rocks along which the road from Viig to the lake of Breiden runs are pure quartz rocks, having a striped appearance, owing to intermixed scales of mica, and a variety of colours in the quartz; and in the fissures and druses with numerous very small grass green acicular epidote crystals, which are often closely crowded together, give a colour to the quartz, and can no longer be recognised as distinct fossils.　This quartz divides the clay-slate from the mica-slate; it bears however probably more affinity to the latter, at least we must conclude so, from the appearance of the mica and other crystals in it.

But the pure mica slate soon also makes its appearance.　At the upper end of the lake of Breiden, and through the pass of Kringelen onwards, it appears uncovered for a considerable length of way with the greatest distinctness, and shews how far we are here removed from the clay-slate. The strata dip towards the north-west, as is the case the whole way up the valley, and as in the Alps, in the inverse order of the formations according to their age.

At some distance from Kringelen, the fissured and drusy quartz again makes its appearance, wholly as a mountain rock; but it only appears below in the valley; for all the streams which descend from the highmountains on the east side tumble down monstrous masses of gneiss, which are remarkable and beautiful on account of the great number of considerable white felspar crystals which the gneiss incloses in the manner of porphyry. The mica of the gneiss is scaley foliated, and not forming a continuous surface as in mica-slate. Hence both formations are from this circumstance characteristically different, even more almost than from the presence or absence of felspar; for fine granular felspar is in the northern mountains as in the Alps by no means unfrequent in mica-slate.

I had scarcely proceeded half a German mile beyond Form on the 29th of April, when I crossed the foaming Louven Elv by a bridge. On the opposite side we stood beneath huge rocks. The valley was now at an end. The road winds steeply up the height; and then the views open through the wood into wild and dark glens. The tops of the mountains, illumined with snow and the rays of the sun, now appear of an insurmountable heighth: and the stream is lost in the dark fir-wood and in rents, and the foam and the noise of the falls ascends only occasionally from the deep below. The valley of Lessöe lies above.

This is the last stage of the great valley from the Miösen upwards. The spring, which had commenced in the under parts of Guldbrandsdalen, was here entirely changed into winter. The snow had so far left a few of the slopes on the northern side, that they might just begin to think of the cultivation of their fields; but the ground of the valley, and all the other declivities, lay yet deeply buried under snow.

The snow was several feet in depth, and no where firm. The horse could no longer extricate himself from it, and trembled in every joint through fear. It was not much better with the foot passenger, for the water run in streams beneath the soft snow. One foot fell in while the other remained on the outside of the snow; and frequently the snow sunk beneath both feet at once, and then we rode astride upon it with both feet dangling in the water underneath. It is almost impossible to travel through the valley in this time of warfare between spring and winter. I was therefore extremely glad, when, in the afternoon, on passing the church of Dovre, I reached the high lying, large, and delightful Toffle.

I should have very much liked to see the valley of Lessöe in summer. However incredibly affected the inhabitants may appear in their stiff dresses with inflexible parallel folds, the coats of the men with monstrous flaps over their pockets, of which the pointed stiff lappets stick out a great way, and strike against the back on both sides; they are notwithstanding industrious people. There is scarcely in any of the other Norwegian vallies so many institutions and regulations for the purpose of assisting rude nature, and diminishing her hostile influence, as in these districts. Before the snow leaves the declivities, they make small hedges all the way down: the snow hangs on these hedges, and is not very easily carried down. The ground would be otherwise too early uncovered, and the immense streams of melted snow from the woods would occasion great ravines, and carry down all the soil with them. When the dry month of June arrives, in which in all the high Norwegian vallies the heavens are never obscured, the grain would be parched up before it had well appeared above the earth, if water were not brought from a distance of several miles, and every where distributed among the fields where necessary for the sake of producing fertility and abundance. In harvest they place new hurdles between the ears, to prevent the stalks from being laid by the heavy winds, and thereby rotting. Such diligence and precaution is highly deserving of being rewarded every year with a good harvest; but a premature frost in the end of summer or in autumn puts frequently an end to their best hopes, and the grain is destroyed by the frost on the field. This is the great plague of all the vallies which are in high situations. But is it wholly impossible to diminish these evils?

It is not the intensity of the cold which destroys the corn; it appears rather to be the too rapid increase of this cold that shrivels the husks of the delicate grains on the ear, by which means they immediately burst in the heat that succeeds to this cold. Cloudy nights never injure the grain, neither do clear nights; but in the latter the ground and the plants are covered with rime. The moment of alarm is in the morning when the sun breaks out. If a small drop of water falls from the ear it is irretrievably lost.* It is evident from this that the ear is only frozen by the immediate withdrawing of the great quantity of caloric which the rime requires in order to be converted into water. If

* Ahveatz Bescrivelse over Söndfiord. Norsk Topograp. Journ. XXIX.

the rime is shaken off before sun-rise, and before it begins to melt, the grain will not be frozen. This might be effected by drawing tight ropes over the fields, or still perhaps in several places more powerfully and securely by means of syringes kept in readiness for that purpose, and with which the rime might be washed away by water, and the too rapid heat from the sun thereby prevented; and this could hardly fail to succeed, as the water itself communicates a higher temperature to the ear than it possesses so long as it is still surrounded by the rime.

Lessöe is the only valley in all Norway which descends from the east side to the Western Ocean without our being under the necessity of previously ascending high mountains. This is a singular phenomenon. The chains of mountains which run through the whole length of Norway are here intersected by a great valley, and completely separated from one another. *Lessöe värks vand*, a small lake about nine English miles above Dovre, and on the banks of which there is an iron-work, communicates at the same time with both oceans, and is not certainly more than 2343 English feet above the sea. There is a beautiful spruce fir-wood from Lessöe over the greatest height all the way down to Romsdal; whereas, in the other passes, even in the south of Norway, bushes will hardly grow at top.

The interior of the rocks is here no less remarkable. At first the quartz continues from Formo onwards. It then frequently resembles porphyry; for dark quartz crystals lie scattered in the pale base, and the rock is almost every where divided by *drusy* fissures. At length, about half a German mile from Form, the gneiss also makes its appearance in the valley, and the quartz rocks disappear. The gneiss rises immediately to a great thickness. The Rostenberg commences immediately after its appearance, as well as the ravine towards Lessöe : and in these straits it becomes very remarkable. It generally abounds in mica; the mica is not scaley foliated, but in considerable folia, which are continuous, and it abounds with beds of quartz. But there is also every where considerable pieces of gneiss scattered in it, in which the felspar predominates; the mica only appears in separate isolated folia, and the quartz very sparingly. The mica in these pieces forms more straight and parallel running streaks than slate; while on the other hand the slatey composition in the surrounding gneiss is more strongly marked and distinct. These pieces are all angular, and the most of them are even quadrangular, of so considerable a magnitude as a foot and upwards, and they appear in fact very thickly heaped together, but still in such a manner that we always distinguish the connecting gneiss mass between. The streaks of different pieces lying near one an ther are often parallel, but they also frequently take completely different direct ons. They do not consequently follow the direction of the slatey structure of the gneiss which constitutes the basis. This wonderful rock is not a conglomerate, the pieces being too small. The basis is too distinct and too strongly characterised as gneiss. But it must be owned that this appearance bears some resemblance to the manner in which the pudding-stone is found in gneiss at Valorsine, and in the lower Valais, according to the account of Saussure :—an older gneiss which was destroyed in the period of the formation of the newer. At Toffle there are even large plates of mica-slate with pieces of gneiss full of felspar enclosed. But the gneiss of the Rostenberg does not descend so low in the order of formation.*

————

I was here at the immediate foot of the great and celebrated *Dovrefieldt,* and close under the highest mountains of the north. I began the ascent in the evening with considerable apprehension at first, partly on account of the softness of the snow, and partly from the exaggerated descriptions of the cold in the higher regions of the mountains ; but my fear was unnecessary. The road from Toffle rises so very rapidly that we soon left the spring climate behind us, and got to firm snow.

The road continued in an unbroken undulating line all the way up the mountain, somewhat like that at Airolo. In a short time the tops of the trees scarcely made their appearance through the deep snow, and in the course of a quarter of an hour we saw nothing but the row of high poles gradually disappearing before us which serve to point out the course of the road through the snow.

In the course of two hours we gained the top of the height. An interminable plain of dazzling snow lay stretched out around us. The hills on it appeared only like soft waves, and the great valley of Lessöe like a slight cut. The mountains over against the valley were nearly of the same heighth,

————

* The conglomerated appearance of the gneiss and mica-slate mentioned in the text is probably an original formation, not an instance of gneiss and mica-slate containing fragments.—*J.*

and there also no summits of any consequence appeared above the plane. Nothing but the repose of a boundless waste was to be seen.

I proceeded onwards with difficulty, having the furious north-west wind in my face, which whistled over the snowy desert, and which had collected a number of dark, rainy clouds above the Fiord of Romsdal; but it did not bring the clouds up the vallies, and notwithstanding its boisterousness, the air was pure and clear.

The road now ascends the last inconsiderable hill of the *Harebacken.* This is the greatest elevation of the road, 4573 English feet above the sea, and 2130 English feet above the valley of Lessöe. At last, the high pyramidal form of the Sneehättan makes its appearance, as in a cloud, several miles to the north. Montblanc rises in a similar manner from Breven, above the surface of the ice. It does not resemble a mountain, but an assemblage of mountains above another range;— an immense and elevated mass, rising above every thing in the wilderness.

The nearest hills were scarcely 300 feet higher than the road. The wind had frequently blown the snow from the small rocks, and then I distinctly perceived mica-slate, and not gneiss. Small mica folia lay scattered over the continued fine slatey mica and quartz, between the slates. The strata dip towards the north-west.

The whole chain sinks gently from the height into a level valley, which I followed, and in which I reached Fogstuen in the dark.

It seemed as if I had reached the Cloister of St. Bernard. Fogstuen, like the cloister, is one of the highest habitations in the country, and buried in a similar manner in almost perpetual winter. They are here accustomed to strangers suffering from the severity of the frost. They conducted me in a very friendly manner into a clean and well-constructed room, exclusively destined to travellers; and the landlord contrived, with admirable dexterity, to kindle such a blazing fire of birch-boughs and flaming twigs, that I soon forgot both ice and snow, and the raving storm without, and from the very bottom of my heart blessed the memory of good King Eystein, who built in the year 1120, on the Dovrefieldt, the four " Fieldt-stuer," for the good and welfare of travellers.

They equipped me next morning with a pair of large gloves for my journey, such as are here used in winter in travelling over the Fieldt: they were truly characteristic of the country, being made of sheepskin, and they reached over the elbows almost to the shoulders: they were tied together behind the shoulders with a thong: add to this a sheepskin cap, tied with lappets under the chin, and under the nose, and by which the brow and the eyes are covered: a great wolfskin over the body, and sheepskin boots, and the traveller has no longer any thing of a human appearance.

We proceeded in light sledges down the level valley, and over three or four frozen lakes, which are connected together as in a chain, and the banks of which are surrounded with birch and alder. From these lakes the Folda Elv takes its rise, and proceeds to the eastward, through the copper-works of Foldat, and at last enters the sea with the Glommen. The road only continues for about nine English miles down the vale, when it leaves it, and turns to the north, and ascends through birchen thickets, up a ravine, to Jerkin. A dwarfish, miserable, crooked, and branchless Scotch fir stood there, wholly isolated; the first we met with again in these mountains; but it was sufficiently evident from its figure that it had been driven by mere accident into a climate in which it could only draw out a miserable existence. How can pines thrive in a latitude of sixty-two degrees, on elevations of three thousand six hundred and seventy-two English feet above the sea! M. Esmarck found them in a thriving state, for the first time, in the Foldal, at an elevation of two thousand nine hundred and eighty-two English feet.

The mountains rise again with great rapidity on passing Jerkin, and the road soon here reaches its highest elevation, four thousand five hundred and sixty-three English feet above the sea. This is properly the head of the principal chain of the Dovrefieldt, which connects the great Kiölen mountains between Sweden and Norway with the Langfielden mountains on the western coast of Norway: it is as it were the middle point, from which these chains of mountains part; and it is by far the greatest elevation of the northern Peninsula. The pass at Jerkin exceeds in heighth almost all the known passes over the northern mountains. The road over Fillefieldt, between Christiania and Bergen, which is generally accounted very high, only rose at the marble pillar, on the boundaries between Bergen and Aggershuusstift, to three thousand nine hundred and seventy-three English feet; and it is therefore much inferior to the road over the Dovrefieldt. But then the mountains which overtop the pass! From the heighth of Jerkin, the Sneehättan rises like a mighty giant above the plane: its immense form is lost in the clouds

o

over the fields of snow, and we look up to its summit as from a deep valley. It was never heard of before that its summit had been reached by any man, till this was accomplished by M. Esmarck nine years ago: and they still speak of this undertaking at Jerkin with a sort of astonishment.   M. Esmarck carried a barometer with him, and determined its heighth to be three thousand nine hundred and forty-four Danish ells, or eight thousand one hundred and fifteen English feet.   The Langfielde mountains are now sufficiently known, as well as the mountains of Kiölen, since the memorable journey of Wahlenberg to the top of Sulitjelma, to enable us to pronounce with certainty Sneehättan to be the highest summit of the whole north.   It is a mountain worthy of standing by the side of Montrose, to which, as seen from the Valais, it bears a considerable resemblance of form.   We continued only two miles above on the mountain.   We then proceeded through a wooden gate in the midst of the level, which serves to mark the boundaries between Aggershuus and Drontheim, and also between what are commonly called Söndenfields and Nordenfields.   We next descended a ravine, down which the Driva precipitated itself from the Sneehättan towards the north; and we soon saw ourselves between immense perpendicular rocks, in a rent which hardly afforded room for the water of the stream.   Great fragments, like pyramids and towers, have in some places fallen down and completely choaked up the valley, and round which we had to seek a way for the sledges, which we found with the greatest difficulty.   If the stream had not been entirely covered with ice, we should never have been able to extricate ourselves from this perilous situation.   In the midst of this strait, and surrounded with monstrous rocks, lay Kongsvold, the third of the "Field-Stuer," erected on the mountains, which was provided with the same conveniences as Fogstuen and Jerkin, and which were equally acceptable to us.

But from Kongsvold the road was dangerous and painful in the highest degree. We had to cross the stream in the straits times out of number. The water now flowed under the ice, and the ice was gradually consuming underneath. It broke with us; the horse sunk deep in the opening, and the foresight and dexterity of the guide alone prevented the sledge from being also drawn to the insecure part, and plunged into the deep. I could hardly have believed that it was possible for us to preserve our horse in these painful crossings, which were repeated for a hundred times at least.

I was here again compelled to feel that in spring no person travels in Norway. The ravine of the Driva is the winter road from the mountains. In winter the ice is firm and the road easy, but in summer people are compelled at Kongsvold to climb the mountains again to a height almost equal to that of Jerkin, and then to descend precipitately the well-known and frightful *Vaarstie* before reaching Drivstuen. It is conceived utterly impossible to proceed in summer through the straits; and to regain the mountain road requires a re-ascent to nearly its greatest height.

The valley is in truth surrounded by steep and savage rocks of a most alarming height. We can no longer measure their heighth from below, and the huge blocks at the bottom seem inconsiderable pieces, when compared with the surrounding masses. It is such another fissure as the Schöllenen at St. Gothardt, or the abyss of the Hongrin above Chateau D'Oex. It is not a valley in which the mountains incline gently towards the plain, but a rent which divides the mountains throughout the whole extent of their breadth. When we reach Drivstuen the rocks begin to give way for the first time, the valley becomes somewhat enlarged, and the bottom frequently bears a resemblance to a plain.

I passed the night at Drivstuen. The wind drove the snow in such quantities up the village, that at last the rocks were no longer discernible. Several boors came over the mountains late in the night, who affirmed that there had been no fall of snow on the heights, and but very little at Kongsvold. The weather had been constantly clear at Jerkin. The wind in the valley was entirely local, and could not penetrate to the upper regions of the mountains.

Drivstuen is a considerable hamlet. Grain does not indeed grow here, for the place is two thousand four hundred and fifty-seven English feet high; but the mountains, the *Säter*, or Alps, and the valley itself, are excellently adapted for grazing, and the inhabitants avail themselves of this advantage. They keep about thirty milch cows, send a number of cattle for sale to Drontheim, and breed besides not a few strong and useful horses, highly prized for their docility and hardiness. Drivstuen is something more than a welcome asylum for travellers in these dreary deserts.

The prospect changed on leaving Drivstuen the first of May. The valley of the Driva still continues it is true to resemble a mountain vale for several miles downwards; but it becomes gradually broader, larger, and of more importance. On the west side a chain of mountains, among the most striking of all those belonging to this range, follows the direction of the valley. The whole chain rises at once from a rocky base to the naked summits. We can view the whole height from below; and the pine-woods at the foot, followed by birches, the fields covered with snow, and lastly the naked rocks, above all, afford us a scale for measuring such an astonishing magnitude. The chain follows the valley for three German miles down to Opdalen. There is nothing on the Dovrefieldt which can be compared with it for heighth, and Sneehättan alone raises his summit still higher.

The valley becomes now covered with wood: but only Scotch firs *(Fichten, Pinus Silvestris)*, and no where are spruce firs *(Tannen, Pinus abies)* to be seen.* They appear first as a wood some hundred feet below Drivstuen, or two thousand three hundred and forty-three English feet above the sea. This can by no means be considered to determine the greatest heighth at which they will grow, for it differs too much from the heighth of the fir at Jerkin; but it may be deemed upon the whole the medium heighth up to which the declining temperature will allow the firs to grow vigorously. It appears also as if the fir-wood rose much higher on the beautiful declivity from the north towards the church of Opdal.

The Dovrefieldt ends in Opdalen, nearly in the same manner as St. Bernhardt at Martigny, and St. Gothardt at Altdorf; for there great vallies meet together here, all of them different in their direction, and each with a peculiar and distinct character. The Driva, instead of running northwards from the mountains, now takes a western course towards the Fiorde of Romsdal; and from the church of Opdal towards the east a valley opens

---

* Without the Linnean names we might here labour under some mis-conception. *Fichte* it is believed does not correspond with p. sylvestris but with p. abies, and *tanne* with the p. piceà. The vulgar name for p. sylvestris is nearly the same in all the Teutonic languages: In German *Föhre*, Danish *Furr, Fyrre*, Norwegian *Furu*, Icelandic *Fura*, Swedish Furu, and English *Fir.—T.*

The German word *fichte* is applied to all the species of the genus pinus: the pinus sylvestris, the Scotch fir is the gemeine fichte, the *fichte* of Von Buch. The pinus piceà, is not *tanne*; it is the *edel fichte* of the Germans. The *tanne* of Von Buch and the Germans is the pinus abies, the Norway spruce fir.—*J.*

more than four English miles in breadth, and quite flat and even at bottom. A determinate boundary is thus placed to the mountains of Dovrefieldt. New chains and new ranges of mountains begin here which do not belong to the great chain.

If we reckon the breadth of the Dovrefieldt from Opdalen to Toffle, we shall find it to correspond accurately with the breadth of St. Gothardt from Altorf to Airolo. It is also possible to find some resemblance in the declivities between both passes, for as St. Gothardt gently rises from the north for thirty-six English miles to the hospital, and then falls suddenly and abruptly towards Airolo, in the same manner there is here an uninterrupted rise for thirty four English miles to the heights of Jerkin, and a sudden descent from Harebacken down to Guldbrandsdalen. It is true there are five thousand one hundred and twelve English feet from Altorf to the hospital, and two thousand seven hundred and sixty-nine English feet from the hospital to Airolo. On the other hand, Oune, the inn at the church of Opdal, lies two thousand one hundred and ninety-two English feet above the sea, and three hundred and twenty English feet above the valley; consequently from the valley to the height of the mountain is only two thousand seven hundred and sixty-nine feet; but then there are two thousand one hundred and thirty feet from Harebacken to Toffle. If it were possible in mountains so rich in divergent lateral chains to determine the slopes with any degree of accuracy, we might pronounce that the southern declivity is here, as well as in the Alps, by far the most steep.

But who would think of comparing the Dovrefieldt with St. Gothardt with respect to the variety, alternations, and richness of the landscape?

Sneehättan is a mica-slate mountain, says Esmarck. In the rocks immediately before entering the straits of Kongsvold, mica-slate was visible, and strata of black carbonaceous mica, such as abound so much in Graubündten, and at the Huffener in Valais. But Kongsvold is surrounded by strata of gneiss. The gneiss is fine slatey with detached folia of mica, which lie parallel behind one another; but this may only happen in the beds of gneiss which are so frequently enclosed by the mica-slate in these mountains; for down to Drivstuen the mica-slate appears almost always with that splendour and alternation which have made Airolo at St. Gothardt so well-known in mineralogy. There are not unfrequently beds of hornblende also here; then garnets in the mica-slate; then on these slates the beautiful fascicular and divergent hornblende crystals. The same stamp is impressed on nature from the Alps to the north pole.

Deeper down in the straits of Drivstuen a multitude of blocks and rocks of a most beautiful gneiss are to be seen. The great and generally twin crystals of felspar, nearly round, and of a white colour, shine forth in the midst of the thick scaly mica, which surrounds the crystals like a frame. The felspars are very much heaped together, and at least the size of a hand, and the mica scales are shining and easily separated. All the ingredients in the gneiss are distinct and determinate and form a remarkable contrast.

Soon afterwards the mica-slate again appears. But who in such a ravine as this can pretend to determine what is in its original situation, and what may have fallen from above ? Notwithstanding external appearances, it is possible that this gneiss may actually be the fundamental rock of the mica-slate, and may be connected with it in the farther progress of the universally diffused gneiss formations. How much is concealed by the snow, and withdrawn from our observation !

The gneiss actually appears very distinctly towards Rüse, about half way to Opdal; it is straight slatey with white granular felspar; and the former gneiss, with the great felspar crystals, imbedded in the manner of porphyry, soon afterwards also makes its appearance; whence it is evident that this beautiful transformation is not subordinate to the mica-slate, but actually belongs to the independent gneiss.

———

The plain which connects Drivdalen and Oerkedalen beneath the steep and high declivities of the mountains is an extraordinary appearance in this alpine country. There is hardly any thing here to equal it, for length, size, and evenness. It is nearly nine English miles of a continued dense fir-wood, with a gentle and almost imperceptible inclination towards the middle. To see only snow and trees, with rocks in the distance, for such a length of way, is a sight to which, after leaving Christiania, we are by no means accustomed; but these plains lie in a latitude of 63 degrees, and upwards of two thousand feet above the level of the sea, and are consequently of little utility to the country. Even then, in the beginning of May, every thing wore the appearance of the deepest winter, without the smallest symptom of change.   The thermometer at mid-day did not rise above the freezing point.   We may in this way account for the scanty population of this wood.   The hamlets are scattered along the northern declivity, but even there very thinly.   From this elevated situation, the unfavourable climate, and the neighbourhood of such high mountains, we may easily conceive why the population of Opdals Prästegieldt amounts only to two thousand seven hundred and seventy-two souls, notwithstanding it contains a space of more than eight hundred and forty square English miles, equal to many a large province.   There are here only three inhabitants to the square mile.

At Sundset the first spruce firs *(Tannen)* make their appearance among the Scotch firs *(Fichten)*; and shortly afterwards the latter entirely disappear, and the wood is wholly composed of the spruce firs.   Sundset is one thousand five hundred and seventy-eight English feet in heighth.   Can spruce firs actually rise no higher? It appears evident however that in this latitude a height of more than two thousand one hundred English feet will not admit of their growth; and therefore we do not probably exceed the truth when we assume six or seven hundred feet as the difference between the two boundaries beyond which the spruce and Scotch firs do not ascend; rather more than two degrees Fahr. of difference in the mean temperature.

It was now evening, and we continued to descend lower and lower for more than one thousand feet to the bottom of the great Oerkedal.   The day had entirely disappeared, but a clear twilight did not forsake us the whole night through, and we saw our way before us as in a faint moonshine.   What a change is effected in the climate by a cleft like this.   Above we had snow of a great depth, but below the lands were in cultivation, and we had to drag our sledge over the newly ploughed land—with what labour may be easily conceived.   The road crosses the valley, and at the opposite extremity again ascends.   We began then to have ice again and winter; and this continued till we came to Bierkages about midnight.

We waked the people, and entered the great room.   One cried out Fire, and I felt alarmed.   I imagined myself in the midst of savages.   The men were entirely naked.   They also were frightened, and sought for their cloaths, and they then began to look like Norwegians.   I now learned for the first time that almost all the boors here, young and old, sleep the whole summer through naked in their beds like the Italians; and this also arises from the very same cause, to save their bed-linen and shirts.   Perhaps this is the only feature which the boors of Drontheim have in common with the Italians. It shows however the wants of the country.   They conducted us into a room, and I soon found that this was by no means the worst place of our journey.

The valley of Sockne begins above Bierkages.   I proceeded over a frozen lake, and then for about nine English miles down a level valley, where I saw nothing but snow and dry twigs, and here and there the roof of a house above the high waves of snow which surround all the hamlets like a fortification.   At Hoff, however, in the middle of the valley, I found the snow beginning to melt, and at last, at the church of Sognedal, I again saw the ground under the snow.   At Sogness, opposite to the church of Stören, the winter had entirely disappeared.   The clear and grand prospects which now opened to me into the extensive and beautiful *Guldal,* and the large spots of fresh green on the slopes, with the labourers in the snowless fields, loudly proclaimed the joyful tidings that a better season was at no great distance.   Hoff was only one thousand and five feet, and Sogness only four hundred and eighty-nine feet above the sea.   I had now therefore completely descended from the mountains.

It was *Ting* or court-day in Sogness.   The boors of the whole Bailliwick *(Vogtey)* were collected together: the *Sorenscriver* (judge) had settled their differences; and the *Foged* had levied from them their duties and taxes. I was pleased to see the kindness, humanity, and patience with which the boors were treated by the royal functionaries, and which they seemed to repay with the most hearty and cordial confidence.   This state of things is pretty general throughout Norway, and hence, even yet the Fögde would have little difficulty to stir up the brave Norwegians to deeds like those in the Kringelen, or against Charles the Twelfth at Krogskoven and Frederickshall.

The Guldal is a beautiful valley: it is long and broad, delightfully environed, and well peopled.   The views down the valley, over numerous and considerable hamlets and churches, with the broad and glittering stream in the middle, are altogether enchanting.   Fertility and cultivation smile upon us from every hill.   The whole antiquity of the nation is crowded together in this valley: it is the cradle of the land.   Here Norr came first over from

Sweden. Here dwelt the mighty Hakon Yarl. In this valley he was found. out and conquered by the valiant, noble, and wise adventurer Oluf Tryggvasön. Here many of the heroes of the country dwelt in their courts, and those kings who bloodily contested the dominion of the land never imagined they had made any considerable progress in it, till they had conquered Drontheim and its vallies. Now we every where see wealthy boors, and no Hakon Yarl, no Linar Thambaskielver, no Duke Skule. Their peace and repose have sometimes been disturbed by the tempests of Swedish wars; but the inhabitants continue to advance in an easy yet perceptible progress in all the arts of peace towards their higher destiny.

I passed the night in Fosse, and entered Meelhuus next day about noon. I saw so many hamlets every where around me, that I did not remember to have ever seen an equal number at once in any district of Norway, with the exception of Hedemarcken. *Meelhuus Prastegieldt* is 165 square English miles in extent, and contains many small mountains; but yet it is inhabited by three thousand nine hundred souls, or twenty-three to the square English mile. The best cultivated and most populous vallies of the south of Norway, the counties of Laurvig and Jarlsberg, do not amount to this. The undermost level of the valley is very little elevated above the surface of the sea. Leir at the church of Flaa is not above fifty feet in heighth; and Fosse, situated on a high hill above the valley, did not rise more than three hundred and twenty English feet above the Fiord. We see here at length several large and extensive and almost marshy levels. As on the road to Oust we were ascending the declivity, the sun shone down the valley on the silvery surface of the Fiord. Pure and hospitable valley! Here I again took courage, and began to live. I had escaped from winter, and was once more among men.

Drontheim was now at no great distance. We have only to cross a small range of hills about six hundred feet in heighth, beneath the black forest of the Byenäsfieldt, little more than four English miles, and the enchanting view of the town opens upon us from the height of the Steinberg. How the Nid-Elv from the valley winds round the large town! Considerable buildings rise above the regular streets, and the old cathedral, the last remnant of northern magnificence, raises aloft its grave and venerable form. The view of the broad Fiord and vessels in the harbour is magnificent; the appearance of the small island of Munckholm in the water, and above all, the distant views of the Stördalsberge, Frosten, and Strandt, are highly delightful. These views might well be compared with that from the Egeberg down on Christiania; if Drontheim had such a distance as Christiania, and the prospects of the distinguished and characteristic mountains of Bärum and Asker.

———————

With the Guldal and the termination of the higher mountains newer rocks again make their appearance. Rocks of a black clay-slate rose on both sides of the valley, even before reaching Sögness. The strata were almost perpendicular, at most, a little inclined towards the south. About two English miles down the Guldal, the clay-slate gives way to greywacke, which is as distinct and beautiful as in the lower part of Guldbrandsdalen. Numbers of small white quartz grains are connected with a mass of blackish grey clay-slate; there is little felspar, and still less mica. All were in a very small granular mixture, with many quadrangular pieces of clay-slate between. This appears to be the fundamental rock of the Guldal all the way to the sea bay.

But to what formation does the rock belong which predominates between the Guldal and Drontheim? Are we to consider it as mica-slate or as clay-slate? On the Steinberge towards Drontheim downwards, it appears at first sight completely to resemble clay-slate. The rocks are very fine slatey; the slates are not shining, but there are black and pinchbeck-shining mica folia every where scattered over the surface of the slates; and these betray the true nature of the rock; for such mica folia are not frequent in primitive clay-slate. Small crystals of hornblende not unfrequently also appear, but no quartz, except in isolated and rare beds. The strata stretch h. 3. and dip towards the south-east.

Nearer to the valley of the Nid, in the neighbourhood of the estate of Munckvold, the mica appears continuous, and shining in the manner of mica slate, but always fine and straight slatey, and without quartz, which we should rather expect to be the case in clay-slate. In the neighbourhood of Küstad, about two miles from Drontheim, the rocks appear again in other forms. The mica as chief ingredient can no longer be mistaken. The folia surround a kernel, and form large balls of two and three feet in diameter. This interior kernel is extremely compact and hard; blueish grey, fine splintery or fine granular in the fracture, probably a fine granular mixture of much compact felspar, a little quartz, and fine mica folia. The surrounding mica is also blueish-grey, glistening and continuous, and every where covered with a multitude of beautiful pinchbeck-brown shining scales of mica. These balls lie close together, and whole rocks consist of them. We might be tempted frequently to believe them to be a conglomeration of large blocks, which they certainly are not.* The nature, the multitude, and the distinctness of the mica, and the position of this rock betwixt other strata of mica-slate, sufficiently prove that we ought not to separate it from the mica-slate formation; but this indistinctness of formation, this constant change and restlessness in the ingredients, shew on the other hand that this mica-slate approaches very nearly the transition to clay-slate. Perhaps in the mountains between Meelhuus and Kläbo, which rise to a heighth of three thousand two hundred English feet, the clay-slate might be again found.

---

* The clay-slate of Thorand in Saxony exhibits the globular structure described in the text as occurring in mica-slate; and I have observed the same appearance in the transition slate of the Pentlands near Edinburgh.—*J*.

CHAPTER V.

DRONTHEIM.

Reception in Drontheim.—Spirit of the Inhabitants.—Branches of Subsistence.— Röraas.—Manufactures.—Architecture, and Appearance of the Town.—Fate of Wooden Towns.—Society of Sciences in Drontheim.—Provost Wille.

In the south of Norway, and in Denmark, it is generally understood that no traveller returns from Drontheim without feeling a sort of enthusiasm for the reception he there met with. From this number I must certainly not be excluded ; for who could be insensible to repeated acts of the most hearty kindness, to a politeness that anticipates every want, that is always affecting and never oppressive ? Who would not be filled with gratitude at seeing so many worthy men anxiously labouring to make the time you spend in Drontheim a time of gladness ? This warmth of heart, this conviviality and sympathy, appear to be characteristic of the inhabitants of this town. They are in fact by no means foreign to the character of the whole nation, and are here displayed as we might expect to find them among men of higher refinement and cultivation.

But how are we to account for the refined tone which prevails in the societies of this place, for the graceful and attractive manners, and the taste, which greatly exceed any thing we meet with in Christiania? This is more than we expect, and more than we have a right to expect ; for Drontheim lies in fact very high northwards, and is separated by numerous obstacles, great distance, and high and impassable mountains, from the other parts of the world. The general prosperity of the place may have greatly contributed to this, and perhaps also the fortunate.circumstance that almost all the generals and superintendents *(Stiftsamtmännern)* sent here from Denmark have been distinguished for their worth and superior politeness. We must own, however, that this circumstance is not in general of such universal influence.

Probably this superiority of disposition may have arisen from some accidental circumstance, the consequences of which still continue to operate. May Heaven grant that this character of the noble inhabitants of Drontheim remain long uncorrupted! It is indisputably true, that in no district of Norway is there such an attachment to their country, such true patriotism, and public spirit as in Drontheim; no where are the people capable of making greater sacrifices, or more easily united in the accomplishment of any object beneficial to the country. The causes of this however are not difficult to discover. The patriotism of Drontheim is more concentrated in the country and less diffused. Christiania sends boards and planks to England, from whence it draws the means of living with comfort and even splendour, and therefore it naturally wishes the prosperity of England, with which its business has been always successfully carried on. Trade has thus given an extension to the country, and enlarged the sphere of interest. Bergen sends fish to Holland, and expects garden stuffs in return. In Bergen therefore the people cannot be indifferent to what passes in Holland, and they have no cause to wish more for the injury than the advantage of Holland. But in Drontheim these foreign relations are not so determinate: their view is alone fixed on the country, in which they live in security and repose; and every attempt to disturb that repose awakes most powerfully in them the spirit of self-defence, and repulsion of foreign attacks by which their peace may be endangered. Whatever outcry may be made in Christiania against the plunder of the Danish navy by the English, it is in the nature of man that every appearance which has the least tendency to justify the English by whom they expect to be benefited should be anxiously laid hold of. Let inquiry be made at Christiania respecting the number of those who in their hearts blame the English for this unheard of deed, and he will find that the prevailing sentiment is, that it was not done by the English, but their ministry. If an armed vessel appear on the coast, it is sent out by the hostile ministry; but if an English letter, an Englishman himself, or the news of an English action arrives, this is attributed to the friendly nation, and not to the ministry. Hence a vigorous preparation for defence can never be so universally excited there as when an invading neighbour insults the inhabitants with the promise of restoring to them their lost honour. Drontheim possesses the patriotism and public spirit of a solitary republic; and Christiania the spirit of a trading town, with extensive connexions, in a large monarchial state.

At the last enumeration, the inhabitants of Drontheim amounted to eight thousand three hundred and forty souls. This is a considerable number for a town situated so far north ! There are few towns in Denmark equal to it. These inhabitants are also in general brought together by commerce, but not so much by foreign commerce as the internal communication between numerous valleys and districts, to which this place forms a central point of union. The boards which are exported from hence to Ireland are of small importance compared with what is exported from the south of Norway. The exports of cod, herring, train-oil and hides, are more considerable, and especially, the copper from the mines of Roraas. The two hundred thousand cwt. and upwards which for centuries have been procured from Röraas, not only enrich numbers of families in Drontheim, but give life, population, and cultivation, to what would otherwise be waste and dreary mountains, keep the whole valley between Drontheim and Röraas in perpetual activity, and create a brisk circulation through the very heart of the country. If it were not for Röraas, Drontheim would at least lose the fourth part of its inhabitants, and a considerable share of its prosperity. At present, an immense number of horses are kept in perpetual employment between the two places. In winter the copper is brought down in long rows of sledges, which return with provisions and other necessaries. In summer, also, there are always horses and cars on the road, employed in carrying and drawing what in winter is much easier conveyed in sledges over the snow.

I know not whether the quantity of fodder consumed by so many horses, which deprives the other cattle of their proper share, has been the means of inducing the inhabitants to avail themselves of their horses in feeding their cattle. But we find this custom, which appears so very singular to natives of the south, only prevalent in Röraas, and a few of the vallies which surround Drontheim ; for in the whole of the rest of Norway, so far as I know, nothing similar is observable. They carefully collect the horse dung, and give it to their cows, who eat it with great eagerness. It is also frequently boiled in great kettles, and a little meal mixed up with it; and then, not only cows become thriving and fat upon it, but also sheep and geese, hens and ducks. Even horses themselves are fond of this mess. It is also the usual mode of fattening pigs. The horses eat scarcely any thing but Norwegian herbs. Perhaps the other domestic animals might not be so fond of digested barley and *hexel*. At present, however, this stuff seems of such necessity to the Norwegian boors for the support of their cattle in winter, that the want of it would expose them to great embarrassment.

The inhabitants of Drontheim are also employed in a few manufactures, which in time, perhaps, may be of some importance. The commander of the town *(Stadthauptmann)*, M. Lysholm, has made a successful attempt to prepare colours out of the extraordinary richness of the Norwegian lichens, and the collection of these lichens seems to give employment to a number of boors in Opdalen. It is by no means an unimportant branch of trade, as is easily proved by the immense quantity of lichens which the English annually drew from the small harbour of Christiansand. M. Lysholm also carries on a saltpetre manufactory, and another for converting impure sea-salt into white kitchen and table salt.

Cloth linen and carpets are manufactured in the great house of correction; and the poor-house also gives out manufactured linen. This is, however, of no great importance.

Every time we proceed through the streets of Drontheim, we are struck with the beauty of the town, and yet it is altogether built of wood. I do not believe there are more than four stone houses in the whole circumference of the town, and these are miserable and inconsiderable buildings. But the wooden houses have an uncommonly agreeable appearance here, as in every one we see the endeavours of the possessor to ornament the exterior as much as possible is strongly visible, and the endeavour is frequently crowned with success; for the delicacy of feeling and taste of the inhabitants is not confined to their mode of living, but extends to every thing around them. At least, I was impressed with the idea that there was a greater air of ornament, neatness, and beauty in this place, than in Christiania; something more in the Dutch, or rather more in the English taste, than we perceive in any of the other Norwegian towns. It would have been better, however, to have gradually built houses of stone; for Drontheim has not only more than once experienced a total destruction from fire, but wood is also a material which can never be converted into a good, durable, and ornamental edifice. In the Munkegade, for example, the principal street in the town, a large palace, such as Copenhagen perhaps cannot match, rises above the other buildings, and is conspicuous at a distance of more than two English miles. It is built in a simple and noble style, and produces a striking effect; but it is composed of wood. The boards, through sun and moisture, are in a perpetual motion: in the side exposed to the sun they are quite dried, and draw the building down. Whatever ought to be uniform and regular, becomes distorted, and all the little ornaments which should aid the general impression, in the course of time grow disfigured, and serve only to excite an unpleasant idea of disorder and decay. The evil cannot be remedied without pulling the house entirely down, and building it anew. This great town-house *(Stiftsamthause)* does not yet, it is true, exhibit such a ruinous appearance; but it is the inevitable fate of all wooden edifices. I never passed this immense palace without experiencing a strong feeling of regret that it was not built of materials worthy of its simple grandeur. It will long remain a monument of the good taste and sublimity of idea of the respectable general, Von Krogh, who constructed it; but if it were of stone, it would serve as a perpetual monument, and a model to preserve a feeling for good taste alive in Drontheim. The general was not at full liberty in his erection. The building has been sold to the king, and is now the residence of the chief magistrate, and the public bodies of the district.

The remains of the old and highly celebrated cathedral, to which the whole of the north formerly went in pilgrimage for the remission of their sins on the grave of St. Oluf, stand at the end of the same street. The great and extensive ruins yet remain to bear witness to their former state, notwithstanding the town has been seven times burnt to the ground, and that Swedish plundering parties have also contributed their share to the general devastation. It is still evident that there is no edifice in Norway to be compared with it, and that even yet it is the largest in the whole country. Should the downfall of Drontheim be decreed by fate, and its revival be transferred to another situation, these ruins would still keep alive the recollection of the place, the people, and their actions. It will never be completely annihilated like the wooden towns of Hammer and Intin, or the eastern cities of Babylon, Ctesiphon, and Ninevah, built of brick, and cemented with bitumen. From the remains, which point out very distinctly the extent of the building, this cathedral appeared to me much larger than even the cathedral of Magdeburg: the choir alone is at present the principal church of the town. There is more external ornament, however, about the cathedral of Magdeburg; but if we dare trust the description, the inside of the church of St. Oluf exceeded every thing of the kind which was known.

This *Munkegade* is a noble street, such as few towns can boast of. It runs through the whole breadth of the town to the shores of the Fiord, and the buildings on both sides of the street are very respectable. The charming island of Munkholm, with the castle, rises in the back ground in beautiful perspective above the bright and clear Fiord, and the prospect is closed by mountains covered with snow, which rise above the water wholly in the distance. Nothing can be conceived more attractive. We should scarcely credit a drawing, however faithfully it might represent nature; but no drawing could convey the perpetual fluctuations of light on the works and towers of the island, and the deep ground which disappears in the blue ætherial mountains, the tops of which are illumined by snow.

On proceeding down the Munkegade, we perceive a large, simple, and beautiful stone edifice, which was erected a few years ago: the first and only building of the kind in the northern part of Norway. This house is occupied by the Drontheim society of sciences, and the high school. The school-rooms are below: the society occupies the first story, and the teachers of the school live in the second story. The society is an institution well adapted for the extending and advancing of science in these northern latitudes; for they have ample means at command, and amidst all the rubbish heaped up in these rooms, there are at the same time good materials for excellent collections. It possesses the libraries of two famous historians, the Rector Dass, and the learned Schiönning, both of them of considerable extent, and a great number of manuscripts for the most part connected with the topography of the country. A very vain collector of curiosities, Counsellor *(Justizrath)* Hammer in Hadeland, who died about six years ago, bequeathed all his collections to this institution, with a very considerable sum of money, which might be very usefully applied, were there not an oppressive and almost impracticable stipulation tacked to the testament, that this money should in the first place be applied to the printing of all the manuscripts of the deceased. That the manuscripts of such a man as Hammer should remain unpublished, is a circumstance which the world has certainly no great cause to regret.

It is consolatory to observe that all these means are in existence, and cannot easily perish, and that they only require the presence of an active mind to watch over the institution, and enrich the country and community.

at large with the scientific advantages which may be drawn from it.   But
every thing vital in the society, as at present constituted, is limited to a few
sparks, which are hardly visible through the gloom ; and it bears scarcely any
resemblance now to what it was in the times when its founder, Bishop
Gunnerus, Suhm, and Schiönning, gave it so high a celebrity, and when
its writings might contest the palm of superiority with those of the most
distinguished societies of Europe.   The demon of popular utility has extended its workings to them, as well as many other institutions, and, as has
always been, and ever will be the case, completely destroyed every beneficial result.   New statutes have, it is true, been lately enacted ; but these
only increase the dreariness of the prospect.   Writings and instruction
cannot soon be expected; and the hope that the union of the sciences
should find a respectable asylum in such high latitudes, has only been accomplished for a short period.   But, perhaps, this is but a shock, and the flames
may yet burst out again with increased brightness.

I was acquainted with Provost Wille in Drontheim, who is since dead.
He was also a collector; but a collector with more discernment than generally falls to the lot of such people.   He was in possession of very various
acquirements : the activity of Ström had in part devolved to him, and his
zeal for the collection of books, manuscripts, maps, and materials of all sorts,
connected with the investigating and clearing up of the geography of Norway,
had in him become a real passion.   At his death he left such a valuable collection behind him as no person had ever before assembled together, and
from which, by a careful selection, many remarkable articles respecting the
country might have been brought to light, if the materials had not been scattered and dispersed since his decease.   He had, in fact, formed too extensive
an idea of a description of Norway ; and alarmed at the boundless project,
he never possessed courage to commence the work ; but he acquired lasting
merit by his excellent description of Sillejords Prästegieldt in Oevre
Tellemarken, one of the most remarkable districts of the country.   At a
later period, in a journey through Tellemarken, he had given a complete
description of the whole province ; it was in the press in Copenhagen, and
burnt along with the drawings and maps in 1794, in the general conflagration
of the town.   He possessed a considerable and well-chosen library, a neat
physical apparatus, and many good specimens of remarkable natural objects.

I saw a small, but singular sheet, in his possession : it was the plan of the ineffectual investment of Drontheim by the Swedes, if I mistake not, under General Armfeldt, in October, 1718. To compare such an excellent speaking representation with nature was not without pleasure. The Swedes concluded that Drontheim was without any defence, and had not calculated on any effective resistance. In the vallies and roads leading to the town they certainly met with very little opposition ; but the town itself, however, they were unable to take. The Swedish accounts say it was for want of cannon ; but the small plan in Provost Wille's possession represents cannon playing on the town, both from the Steenbeerge, and from the citadel of Christiansteen on the other side of the river, which is now wholly demolished. The Swedes at last withdrew to Röraas; but as they were afraid of being enclosed there by the Norwegians advancing from the south, they were conducted by General Armfeldt to Tydalen. He and the whole corps were frozen to death on the mountains towards Jämteland, between Handöl and Tydal. The Swedes have always anxiously attempted to obtain possession of Drontheim, when they were the most powerful, and Charles Gustavus even dismembered the whole province of Drontheim *(Drontheim Stift)* from Norway in the peace of Roskild. They were in the right; for Drontheim might easily have become of the same importance to the north of Sweden that Gottenberg is of to the south; and Gustavus the Third, who never forgot Norway, would hardly have founded the new town of Oesterby in Jämteland without some particular view. But these dangers to Drontheim are now past, at least from the part of Sweden.

CHAPTER VI.

JOURNEY FROM DRONTHEIM TO FINMARK.

Country-Houses at Drontheim.—Stördalen.—Cultivation at Levanger.— Värdalsöre Gloves.—Tellegrod.—Clay-Slate in Stördal, Gneiss in Värdal —Shell Clay at the Figa-Elv. — Steenkiär. — Bestadfiord. — Ship-Ellide. — Järn-Nätter at Eilden. — Embarkation at Lyngenfiord —Appelvär.--Näröens Market.—Riisoe.—Difficulty of living here.—Egg Islands. The Waves smoothed by Hail.—Lecko.—Battle b tween Eagles and Oxen.—Marble in Gneiss.— Helgeland.—The Night diminishes the Strength of the Wind.—Climate of Helgeland.—Counsellor Brodtkorb.—Granite of Bevelstad. — Tiölöe.—Alstahoug, a Bishop's Seat. —Sörherröe.— Phœnicians in the North. — Stratification of the Islands. — Catching of Land Birds. — Luröe.— Kunnen.—Glacier at Gaasvär.—Boundaries of the Spruce Fir Region.— Vigtil.— Hundholm. — Bergens Trade to Nordland. — Bodöe. — Lazaretto. — Expence of Wood in this country.—Grydöe.—Stegen.—Prästekonentind.—Mica-Slate.—Laws of Stratification.—Simon Kildal, a Preacher.—Schools.—Westfiord.—Mälström.— Saltenström.—Lödingen.—Climate.—Gneiss.—Sun at Midnight in Midsummer's Day.— Northern Lights in Winter.—Does the Climate of the North grow worse?—Fisheries at Vaage.—Their Importance.—Arrival of the Fish.—Method of catching them with Nets.—With Lines.—Angling.—Form of the Hooks.—Laplanders in Lödingen.— Intermixture with Finns. — Arrival in Finmark's Amt. — Cassness. — Faxefieldt.— Fishing and Agriculture cannot be united.—Klöwen at Senjen.—Strata of Tremolite.— Contempt of the Norwegians for the Laplanders.—Lenvig.—Tremolite.—Colony of Bardon-jord. — Bensjord. — The whole of Tromsöe private property. — Saw-mills.— Arrival in Tromsöe.—Difficulty of the Town's advancement.—Shell-Beds.—Impression of perpetual Day.—Culture of Grain in Lyngenfiord.—The Laplanders are dangerous Neighbours to the Norwegians. — Lyngensklub, Lodde, Salmo arcticus.— Lyngen's Glacier. — Maursund.—Alt-Eids Fiord. — Appearance and Climate.— Glacier of the Jöckulfieldts.—Smaragdite and Felspar at Alt Eid.—Similar Rock in the Neighbourhood of Bergen. — Passage over Alt-Eid. — Langfiord. — Arrival in Alten.

STÖRDALSHALSEN, 20th of May, 1807. The neighbourhood of Drontheim is distinguished by the numerous and beautiful country-houses which surround the town. On the road to Houan, towards the Fiord, several considerable ones lie very pleasantly along the shore of the bay, or on the declivities of the hills. This is a strong proof of the prosperity of the town which can afford such environs; for citizens occupied with the mere concern for a livelihood would not think of enjoying the delights of summer in this way beyond the town. These country spots here, it is true, want the charm of the fruit gardens of Christiania; for neither cherries, plumbs, nor pears, ripen here, and apples ripen with difficulty. Touteröe, an island about fourteen English miles to the north, and in view of Drontheim, possesses peculiar advantages in this respect; for there cherries, and even good cherries, are to be had in abundance, and at no great distance there is a wood of oak, lime, and ash. But at Drontheim the oak does not grow easily; it will live there, but never grows any larger. Fruit-trees and oaks thrive in nearly the same temperature; and where oaks cease to grow, fruit-trees are as great rarities as a palm at Rome, or a chesnut at Lund. This temperature is nearly 40 degrees of Fahrenheit, according to observations made in Sweden and the southern part of Norway. The observations of Berlin in Drontheim for several years give the same result.

The hills which fall gently towards the Fiord, end nearly nine English miles from the town, where the mountains become higher and more abrupt, and the intermediate vallies narrower and deeper. This gives rise to several wonderfully enchanting situations in the bays round which the road winds; and sometimes, particularly before descending into the Stördal, the road ascends rapidly through a thick wood for several hundred feet, and then descends with equal rapidity. These heights may be considered as the last branches of the chain which runs between Sälbo and Stördalen, and which is again connected with the great Kiölen mountains on the borders of the country. The rocks appear to be somewhat different in the extreme points. The formations at the Fiord resemble, however, the mica-slate at Drontheim, which approximates so much to the nature of clay-slate. We no where see the characteristic lustre of the continuous mica, by which elsewhere the mica-slate is so much characterized; neither did I see on the roads either garnets or crystals of hornblende, or white lime-stone beds in the formation. Fragments of calcareous spar, however, without number, appear at the top of the height of Stördalen. The formation is there undulating, slatey, and altogether singular. The undulations may fairly be compared with the waves of a stormy sea; they are so large and long, and rise and fall in such a manner, that it becomes extremely difficult to discover from these slates

the inclination of the strata; besides, the surface of the waves is so wonder
fully indented, that the slates appear to be fixed in like swallow tails, and this
throughout the whole length of the mountain. Yellow quartz, and small
veins of white calcareous spar, intersect every where, and on all sides these
singular forms. These mountains certainly deserve an accurate investigation of their composition, for they would instruct us in an admirable manner
how clay-slate is produced from mica-slate.

In general we perceive distinctly enough that the strata, every where from
Houan to Stördalen, dip pretty strongly towards the east. The small lateral
vallies run parallel with the direction of the strata.

LEVANGER, 21st of May. The manufacture of earthenware at Stördalshalsen
is the greatest and almost the only one in Norway; and it may well be of
some importance in a country where clay is every where a rarity. Hitherto
it only occupies seven turners. There are works for burning tiles and lime
connected with it, the first almost the only thing of the kind, and the other
at least the nearest to Drontheim. The lime-stone is procured in the neighbourhood, from the hills at the foot of the steep mountains of Stördalsfieldt,
which rise in so prominent a manner from Drontheim; for they are isolated
like Saleve, and at least two thousand one hundred and thirty English feet in
heighth. The lime-stone is blackish grey, and very fine granular. It does
not belong to the mica-slate, but is wholly subordinate to the clay-slate.
Would it be possible to find greywacke in these mountains?

The clay-slate becomes first distinct and conspicuous in Langstenen, a
small rockey valley towards the Fiord, about nine miles from the Stördal.
The strata dip towards the east, and occasion great perpendicular precipices
towards the west. From these precipices two charming cataracts descend
over the rocks to the bushes beneath, and the stream runs between green
meadows to the bay. This is a view which seldom occurs in the high
mountain on this road; and this is also nearly the last valley of this nature;
for, in the rich and fertile country towards Levanger, from the appearance
of cultivation, and the size and cleanliness of the farm-houses, we might
easily be disposed to forget that we are in a mountainous region, and in a
northern latitude. This fertility is again to be ascribed to the clay-slate.
The boors are deserving of such a country; for their zeal and activity in the
improvement of their agriculture has acquired the greatest celebrity to
Skongnens Prästegieldt; and prosperity every where follows in the train of

industry. The greater skill here displayed in the mode of managing the ground, in which this quarter so much excels the southern part of Norway, is attributed by many to the connexion with the Swedes, who come annually in numbers from Jämteland to the market of Levanger, and who have long enjoyed the reputation of being excellent cultivators. Most of the farm-houses are surrounded with gardens, in which hops, turnips, and roots, grow in great abundance. They are conveyed from Levanger to Drontheim, and render the importation from Holland, which was formerly so general, almost wholly unnecessary: an alteration equally for the advantage of both country and town. In Bergen the case is very different; for there the whole summer through vessels are weekly dispatched for cabbage, turnips, onions, and a number of other kitchen herbs, from the gardens at Dordrecht. These productions might in all probability be easily and advantageously raised from the beautiful clay-slate hills of Vossevangen.

Levanger resembles a small town. About fifty families are collected together here, and we again see pavements and streets, a most unusual appearance in this country. This is the consequence of the great fair which is held here yearly in the beginning of the month of March. The inhabitants of Jämteland exchange iron and copper wares, and hides, for provisions, grain, dried fish, and herrings. The road over the mountains through Vär-dalen is much less laborious to them than the great distance to Sundvall, the nearest town in Sweden, and which is not in a situation to afford them the wished-for productions in such a manner as Levanger receives them from Drontheim. These connexions will continue to prevent the inhabitants of Herjeadalen and Jämteland from easily forgetting that they were formerly Norwegian colonies, who, in the tenth century, took possession of the meadows and woods on the opposite side of the mountains, to avoid the persecution of the tyrant Harald Haarfager. They became Swedish provinces for the first time by the peace of Brömsebro in 1645.

STEENKIAR, the 22nd of May. About two o'clock in the afternoon we were conveyed with our carriages over the great Värdals-Elv at a ferry; and we found ourselves on reaching the opposite side in the streets of a considerable village, called Värdalsore. We staid at this place an hour at least, as a mark of respect due to the great reputation it possesses for its excellent gloves. In Christiania even I frequently heard of the gloves of Värdal: they were praised not only for their superior softness, but also, and chiefly perhaps, on account of their highly agreeable smell. How often have I heard a wish for Värdal's gloves, and commissions for them given to every one who was bound for Drontheim. But when we came to the place itself we found only prepared leather, and very little even of this; and no gloves whatever. We were told by the people that they were in want of orders, and that they could only work at the express desire of an employer. It is a pity that they are unacquainted with the demand for their article, and that they do not seek after it. It is asserted that the preparation, which they endeavour to keep a secret here, is made of the bark of the elm-tree *(Ulmus campestris)*, and hence it receives its colour and agreeable odour. However, no person has yet been fortunate enough to produce gloves or leather calculated in the minds of the ladies to supply in point of agreeable quality the place of the gloves of Värdal.

Our ladies proceeded in a large four-wheeled chaise: a true phenomenon for the inhabitants. This succeeded very well till the vicinity of the Figa-Elv; but the road became there frequently extremely narrow, and the snow lay often on one side, so that it became necessary for several men to support the chaise. We were more than once in great alarm as the chaise proceeded over Tellegröd. This was the first time I ever saw this frightful Tellegröd; but I was soon convinced that it was dreaded with great reason. The earth freezes in winter for a depth of several yards; and when the thaw comes on in spring, it is long before the warmth can finally expel the frost from the ground. Hence the winter has been long past on the surface, and the earth become dry and firm, while the undermost rind remains still frozen. The middle thawed part remains beneath the dry surface like a morass, and cannot penetrate deeper into the earth. Such places on the road cannot be known; and horses and coach sink at once like a vessel at sea. The firm rind shakes for a great way around, and rises and falls in continual undulations. The carriage dances, the horses take fright; the crust immediately gives way, and the horses and carriage sink for a number of feet into the abyss. The common expression at parting to all strangers who travel in spring, is throughout Norway—" Heaven preserve you from Tellegröd!" and truly they are in the right. It is frightful to see the carriage and horses rolling over the firm ground as if convulsed by an earthquake; and every moment we expect to see the ground open and swallow them up. However, our ladies were more fortunate: the covering of the Tellegröd held firm, and we came all fortunately over it.

But in Steenkiär nobody would advise us to proceed farther with the great carriage, which they said would be completely impeded by the road and the snow. We left the carriage therefore behind us; and a land-proprietor, who was travelling to Drontheim, was to take it back. He had no want of inclination for this; but his courage failed him. It was too bold an undertaking he thought to proceed in such a vehicle. He stared at the chaise, considered, and again considered, and then viewed the chaise over again. It seemed to appear to him as if we had yoked horses to a tower to proceed over the mountains in. He took the carriage to Drontheim, however; but he only sat in it in Steenkiär before the horses were yoked, and in Drontheim after they were unyoked. Half a century may elapse before another four-wheeled chaise comes to Steenkiär.

For the rest of the road we only used Swedish travelling Kjärs; a sort of cars, with one seat and two wheels, and drawn by one or two horses.

The clay-slate frequently appears in the hills towards Värdalsore. We must account it primitive clay-slate; for we frequently find beds in it foreign to the transition clay-slate. In this way several miles from Levanger there lay in the valley numerous large blocks of actynolite, with felspar and jasper in the mixture, and epidote in small round masses. The epidote is easily to be distinguished from the actynolite by its darker colour and its pointed crystals, which are never, or very seldom, united in one whole, and then through the want of the foliated fracture. For the folia with difficulty divide themselves into such small parts. Even in the crystals of Arendal, of a foot in length, this foliated fracture is never in great perfection. But the actynolite is soon discovered by its constant inclination to the radicated fracture; even in single fibres we may find rhombs which point out the double cleavage; and the fibres seldom cross one another in such a reticular manner as the needle crystals of the epidote. Värdals-Elv appears to determine the boundaries of the clay-slate, and all the new formations which connect gneiss with greywacke; for the nearest rocks on the other side of the stream are small and long granular *hornblende* with a little felspar, but with many large silver-white mica folia between, and containing white quartz

R

beds: a variety of the gneiss; for gneiss frequently alternates with the hornblende. Farther on, towards Berge, garnets frequently lie in the white mica of the hornblende; and almost in the same manner as we see them so frequently in the pieces of Philipstad and other mines in Wermeland. We might almost believe that nature intended to show here only the manner in which gneiss and greywacke are connected together by intermediate formations; for all the newer stones which make their appearance at the Fiord of Drontheim, and in general on the north side of the Dovrefieldt (*Norden fieldts)* are merely cabinet masses compared with the gneiss which every where preponderates here, and excludes every other formation. These masses appear in the vallies as if inserted and driven into the ground. On both sides the mica-slate soon again makes its appearance in the height, even in the gneiss itself.

Hopes have frequently been entertained in Drontheim of discovering coal; for we always hope for what we wish. But, alas! the series of formations goes only down to greywacke, and consequently many members of the transition formation, before we can expect sand-stone and coal, are wanting. If greywacke and clay-slate have hardly room to stretch themselves out, how should there be room for the coal formation, which every where seems to dislike narrow vallies, and to require wide and extensive spaces. In the neighbouring district of Jämteland, the black lime-stone, which is rich in orthoceratites, is at least spread over the greywacke; but there is no trace of this in the neighbourhood of Drontheim; and Drontheim is consequently much farther from coal than Hedemarcken and Christiania.

About four English miles before coming to Steenkiär, the road descends from the height into a deep valley, in which the Figa-Elv, a considerable stream, runs to the Fiord. On the declivities before reaching the bottom, thick beds of blue marly-clay (*mergelthon*) make their appearance, in which a great number of shells are every where scattered. The most are pieces: some are bivalve, and few can be distinctly recognized. They are not impressions in the clay, but the natural calcareous shells. This is an uncommonly singular phenomenon, and worthy of notice in a country where marine productions are almost altogether unknown on the dry land. The valley becomes narrower in the neighbourhood of the Fiord, when the road descends completely to the bottom. This clay there terminates, and the gneiss again makes its appearance. The extent of this small formation is therefore very limited; but it rises to four hundred or five hundred feet above the surface of the bay. It is however very local, and amongst the newest of which we know any thing among mountain rocks; but when we see that it is an appearance which extends nearly over all Norway, it will be deserving of the highest attention. This shell-clay has in reality been frequently found in the southern part of the country. At Hafslund, on the Glommen, and immediately above the Sarpenfoss, one of the most considerable waterfalls in the north, it is successfully used as marle for the lands. Oysters lie in it in great abundance with their natural shells. In Rachestad Sogn, near Frederickshall, there are two hills in which anomia pecten is so frequent in the clay, that the whole mass might be applied to the burning of lime (Wilse Spydbergs Bescrivelse S. 154). At the Drammen-river, above the town of Drammen, we again find this clay on both sides of the stream, for about fourteen English miles beyond Eger, and quite full of shells and corals. That excellent naturalist Ström (Egers Bescrivelse) accurately investigated the shells, and recognized them to be the same with those he had often observed on the Nordenfield sea-coast.[*] Nothing of this is visible even on the greatest heights when we penetrate far into the country. The beds of the Südenfieldts scarcely ascend so high as those on the Figa-Elv. It appears as if the vicinity of the great ocean were unfavourable to the deposition of this formation; for wherever it has yet been found, it is deep in the interior of the Fiord, and protected by many miles of interposed gneiss mountains from the agitation of the open sea. The whole phenomenon may perhaps be a consequence of what has been proved so clearly and undisputably in the north, the decrease of the ocean's surface towards the surrounding land. In this case these beds would acquire a new geological interest; for they would excite a hope in us of being perhaps able to trace that decrease to its first beginnings.

All these beds repose on gneiss. In the straits of the Figa valley it appears in great rocks on the road side. The mica in it is scaley, and the felspar very frequent. Some of the beds, which appear of a coal black colour, and which seem in fact to have been taken for coal, are exceedingly singular:

---

[*] Mytilus edulis et barbatus—Anomia patelliformis—Cardium edule et echinatum—Mya truncata—Venus islandica et cassina—Ostrea maxima—Nerita marina—Turbo littoreus—Buccinum undatum.

they appear to be immature coal, which await in the earth the period of their maturity; for the beds have evidently been worked. They were entirely pure dark-black, thick scaley, and shining mica. In isolated masses of this nature the mica is well-known to be not unfrequent in the gneiss; but whole beds of it are certainly a striking circumstance.

———

The situation of Steenkiär, at the uttermost extremity of the long and extensive Fiord of Drontheim, appears somewhat dismal and monotonous. The place itself, a small village of low and crowded houses, has not the agreeable appearance of Värdalsöre, or the life and stir of Levanger. There was formerly a town here, it is said, which King Oluf the Saint destroyed and transferred to Drontheim. But how could a town possibly exist in the neighbourhood of Drontheim, which King Oluf Trygvasön wisely founded, where four of the principal vallies of the country, Stördalen, Sälbodalen, Guldalen, and Orkedalen, meet together as in one common centre? The flourishing state in which Drontheim has remained for so many centuries, is a sufficient proof that this excellent situation fully compensates for the difficulty of the entrance from sea, and the inconvenience of its harbour.

BEITSTAD, the 23rd of May. The mountains at the end of the Fiord were now several hundred feet in heighth; but we had great difficulty in proceeding over the wet and loose snow. The cultivation of the fields was hardly thought of here. But this delay was somewhat extraordinary, and general in the whole of the north; for the labours of the field are usually begun in the middle or first half, and at Drontheim in the beginning of May. The country here is by no means agreeable; it is destitute of character, and possesses no grand or striking prospects. This is also the case with Soelberg, where the principal church of Beitstad is situated; the narrow Fiord of Bertstadsund resemble a river or a stagnant land-lake, and makes very little impression. This extreme arm of the Drontheim Fiord, which penetrates so deep into the country, is scarcely a musket shot in breadth; the banks rise flat and without rocks, and the dark wood every where stretches to the tops of the hills. The country however is not without animation, for eight or nine saw-mills cut a great quantity of deals from the firs of the surrounding woods, which are conveyed by water down to Drontheim. In harvest the herrings not unfrequently seek shelter in this remote corner, and afford the inhabitants an easy and frequently considerable supply.

The white gneiss, with scaly mica, has now taken the place of all the other rocks, and we see it in considerable purity by the road. At the Fiord, below the church of Beitstad, we discover thick and beautiful beds of hornblende in it.

EILDEN, the 24th of May. After a passage of about two English miles across the Fiord, we reached the extremity of the bay, Gielleaas, a sort of harbour for the shipping of deals to Drontheim. From thence we have hardly to ascend two hundred feet to Eilden; and at this place we enter the province of Nummedalen. This great valley forms an isthmus between the Fiords of Beitstad and Lyngen, of scarcely fourteen English miles in length, and an elevation very little above the surface of the lake. If the surface of the sea were a hundred feet higher, the greatest part of the province of Fosen would be an island, inclosed towards the north by the Namsenfiord, and towards the south by the great Fiord of Drontheim, which winds in such a singularly tortuous manner, for the most part parallel with the principal chains of mountains. This isthmus is consequently well known, and as the greatest part of it belongs to Nummedalen, it is generally called Nummedals Eid *. Eilden owes also its name to it, and that in a singular enough manner. When Norr came by land to Norway, from what is now called Finland, through Lapland, and Gorr arrived there in the ship Ellide, round the southernmost point of the country, in quest of their sister, who had been carried off, they determined, after the conquest of the country, that Gorr should reign over all that a vessel could circumnavigate, and that Norr, on the other hand, should reign over the continent. Beitr, Gorr's son, proceeded up Drontheimsfiord to where Beitstad yet preserves his name, and at Gilleaas ordered the ship Ellide to be drawn upon land. He then placed himself at the helm, unfurled all the sails, and ordered the vessel to be dragged by the whole crew; and with great labour, the ship, with Beitr in it, was conveyed on a sledge over Nummedals-Eid to the Lyngenfiord. He had gone round the whole of Fosen in his vessel, and taken possession of the country as a regular succession. In this way they played upon words, even in those times, to which in more recent ages we would attribute superior virtue. Such instances will never be wanting, so long as we treat the laws of morals as a sort of positive insti-

---

* *Eid* is the Norwegian expression for every isthmus; Viig for every bay that usually ends with an isthmus. The German word *wick* is derived from Viig.

tution, and do not endeavour to unfold them from the nature of man, and the order of the world; and in this we shall never succeed so long as we struggle against nature, and consider man as the object and end of the world.

Elliden, or Eilden, on the greatest height of Nummedals-Eid, bears the name of the ship Ellide, which, after this event, was for a whole century employed in numerous other similarly heroic actions.

This valley, between two chains of mountains, is long and broad, and almost every where covered with thick woods, nearly in the same manner as the plain between Opdalen and Oerkedalen below Dovrefieldt. The few cottages in it lie all scattered along the eastern declivity; but the grain, which the boors obtain here with great labour, is seldom conveyed uninjured from the fields: early night frosts destroy it for the most part nearly every year. The same complaint is also made in Opdalen; but agriculture is there carried on beyond the region of spruce firs. In Nummedals-Eid almost the whole wood consists of spruce fir; and Scotch firs are even very rarely interspersed. Where spruce firs grow to such a size, and thrive so well, the industry of the husbandman generally overcomes the difficulties of the climate. Why does this circumstance not also take place here? The general temperature of this district is probably not unfavourable to agriculture; and it is likely that the destructive period may be limited to a few nights; such noxious nights as are well known in Sweden by the name of Järn-nätter (Iron nights). They are only local, and seldom extend over great districts. When we look from the church of Eilden down upon the valley, and see every thing dark and black, everywhere wood, and bogs, and morasses intervening, we can easily conceive what influence all this must have on the climate. The wood prevents the ground from being heated; and the evaporation from the bogs also produces a similar effect. In other places, when in harvest the power of the sun no longer heats the atmosphere, throughout the whole duration of the night, the ground returns the warmth which it received from the sun during the day, and the nights are not dangerous. But what warmth can be returned by this fir wood? The sun does not reach the ground, and the cold air remains as if in cellars between the trees. Were there less wood in the valley, and more upon the height, the järn-nätter would not be more oppressive to the inhabitants of Eilden, than to those of Beitstad and Inderoen.

APPELVAR, the 25th of May. Lightsome, and glad of heart, next morning I went along by the Sledge of the Ladies, down the valley to Aargaard. Rocks, and those of a picturesque description, again made their appearance on the east side : the river flowed through meadows, and its banks were lined with alder ; and there was every where more animation.   To the westwards Oyskavelenfieldt rises to a great heighth, and far above the region of trees. This is the highest range of mountains of these latitudes near to the sea. We were assured that the snow is never perpetual at top, but disappears every summer.   This heighth of the mountains can, therefore, be no where above four thousand six hundred and eighty-six English feet, which is nearly the limit of perpetual snow here ; but we should not be warranted in attributing to them more than a heighth of three thousand two hundred English feet.   The Oy-elv comes mostly from these mountains, and unites at Aaargaard with the smaller stream of Eilden Elv.   A large open boat, with sails and six men to row it, had been expecting us at this place several days. We left the dry land here, with the intention of proceeding the whole way to the north cape by water.   The Elv quickly drove the boat into the Lyngenfiord, which goes up the country like a river.   Close and green thickets along the banks, and beautiful distances before us, made our passage entertaining and easy.   In an hour's time we were carried by a gentle south-east wind across the great Namsenfiord, a name well-known throughout the whole north ; for an immense quantity of logs and deals proceed down the Namsen Elv, one of the most considerable streams in Norway, and are by means of the Fiord dispersed all over the north, even to the boundaries of Russia.   All the way to Vadsö and Kola, there is scarcely a house of any consequence, or a church which is not built of logs from Namsen ; and the woods of Malanger and Balsfiorden have been yet very little used, nay, remain almost untouched.   We passed with rapidity to Sörvigen through the Sörsund, and then to Seyerstadt.   We wished to remain here, but the fresh south-east wind was so favourable for us, that the boat's crew intreated of us to cross the same evening the Foldenfiord, a passage of about nine English miles, very generally dreaded.   We crossed over without any accident, and landed at about eleven o'clock in the evening, at the hospitable and elegant house of Appelvär, where we met with a very friendly reception.   The inland part of Foldenfiord, which begins at Appelvär, and goes deep into the country, is much less dangerous ; but in the outward part, several of the northern yachts are annually lost in their long and difficult passages to Bergen.   Exposed to all the storms and swells of the ocean, the vessels are driven upon the cliffs and dashed to pieces.   Hence Foldenfiord and Statland at Söndmör have always not a little contributed to impede the rising prosperity of the north of Norway.

RIJSOE, the 26th of May. Appelvär is merely a small, and very low island, and like all islands on the ocean without a single bush on it.   The strata of gneiss are every where visible ; great reddish felspar crystals, undularly surrounded by their scaley mica; and all these strata dip towards the west at a considerable angle.   There are innumerable islands of this description on the coast here, and particularly in Naröens Prästegieldt, round the great islands of Vigten.   In each of these cliffs the strata have a different direction, or, at least, a completely opposite inclination, and in each of them, the inclination of the strata seems to owe its origin to a different cause ; yet on more narrow inspection, we find a correspondence between several of the Skiärs in their direction and dipping, which extends for several miles ; and the direction of the ridges of mountains generally determines the direction of the strata. We find horizontal strata no where.

About mid-day we passed by the church of Näroen on the dismal, bare, and desert island, and shortly after we entered the narrow Näroensund, between a number of store-houses supported by the rocks, or built in the water, and closely crowded together.   In these uncomfortable and crowded places, a considerable fair, held on the 24th of July, brings annually a number of people together from the surrounding country.   The mountaineers of Nummedalen and Helgeland here barter their produce for goods brought by the merchants of Drontheim: the bustle becomes then so great, that the number of boats form nearly a bridge from one side of the Sound to the other; and in these boats a bottom is obtained which nature denies to the situation.

Hornblende beds are here evident on the rocks between the gneiss, and small veins of white felspar stretch in various directions through the stone.

A favourable south wind drove us still farther from Näroen.   But it rained heavily, and we were thoroughly drenched.   The wind then changed to the west, and it became a complete calm, but the rain continued to increase.   We proceeded slowly onwards, and were compelled in the evening to seek shelter from the merchant on the small island.

RIJSOE, the 29th of May. The wind changed in the night to the north-west, and increased in strength ; and we were kept prisoners on the island by the inclemency of the weather.   It is not more than the size of a garden like

Isola Madre. In such a small space it is impossible to live comfortably. Scarcely any of the most common necessaries of life are to be found on the spot, not even water; for the springs are dry, except in the rainy season, which extends only to a few months in the year. In winter they are obliged to bring the water they drink from the continent, and this not only for themselves, but for all the vessels, which in considerable numbers lie at anchor here among the cliffs for security from the storms and high waves of the sea. These are vessels belonging to the north of Norway, and also vessels going or returning from Archangel. The distance to the stream on the continent is at least a league. There is still less wood on the island; not a single twig: the soil is however excellent and fertile, and would well repay the pains bestowed on its cultivation. There is, in fact, a sort of agriculture carried on, by which barley, oats, and potatoes, are raised; but that they raise them at all is a great wonder. The black ground between the cliffs has no outlet; the water keeps it as wet as a sponge, and renders it a constant morass; and horses, sheep, cows and pigs, wander about every where at will, tread out the fruit, and eat the leaves and ears. The whole of their attention is directed to fishing, and all the time that is withdrawn from fishing is considered as lost. It must be owned that it requires much more determination and courage, constancy, and durability, much higher powers of mind, to live by fishing than by the quiet labours of agriculture. The first requires incessant agitation, perpetual life and activity, a constant endeavour to save themselves from unceasing dangers, and to attain their object during the very existence of that danger; in the other we have complaints upon complaints against the weather and the climate, and hopes of things turning out better, merely from the wish that it might be so; for neither processions with relics in Lombardy, nor church-prayers in Norway, have power to change the grand progress of nature in the accomplishment of her general objects. The difficulties which those employed in fisheries have to contend with we had an opportunity of learning from the fate of our landlord, M. Ravaldson. He ventured himself, with all the other people on the coast, in the storms of February, with his yacht and twenty-two men, in the long passage to Lofodden, the great rendezvous of the fish and their enemies. The currents drove the yacht from one sound to another; and with wind and sea against them, before they could reach their place of destination, they were wrecked on an unlucky cliff. M. Ravaldson returned with expedition to procure another yacht for the prosecution of the fishery at Lofodden. His masts and ropes were broken by the wind; the time was completely lost; the fish had left Lofodden before the arrival of the yacht, and he had all his labour for nothing. He was obliged to return with his expence and his toil, his sufferings and his hopes, unrecompensed. This was not his case alone; for the same thing happened to almost all the yachts which are annually sent from Helgeland and Salten to the fishery at Lofodden. Few of them arrived there; and fewer still, when they did arrive, from the violence of the storms, could think of fishing. This was an unusual misfortune; but the difficulties and labours are every year the same, and yet every year we see thousands of men repair to Lofodden in the stormy sea of February. Surely such men deserve to be named among those who inhabit the earth.

Beyond Riisoe, there are a number of small uninhabited islands, which form various small groups. They are called *Holme* when they are high and rocky; Vär when they are entirely flat—a defence and protection, as it were, against the raging of the great ocean. They have often endeavoured to keep cattle, pigs, sheep, or goats, on these islands, but they are not allowed to remain there. In a single day they are stolen. These *Värs* are much more profitable for the eggs of the numberless sea-fowl which brood here. An aegge vär *(egg vär)* is reckoned among the advantages of a possession. The fowl do not easily leave a place which they have once chosen to lay in. When the possessors of the Vär come in quest of the eggs, the bird knows them and remains quiet; for he knows from experience that the superfluous eggs only are taken away, and that one is always left in the nest. When they come to its nest, it flies to a short distance, looks quietly on the spoliation, and then returns when they are gone. But foreign vessels and strangers frequently fall upon the Vär, and rob the place of the whole of the eggs, when all the flock, to the number of many thousands, rise at once, and fill the air with the most frightful aad lamentable cries. In their despair they fall down again on the nest, and for a long time are inconsolable. If this robbery is repeated they lose all patience, and leave in a body the ungrateful vär for another, whose situation seems to promise them better protection and quiet. The most of them are sea-mews *(Meven Maasfugl)* Maage in Denmark : their eggs are large, and of by no means a bad taste.

In the afternoon of the 29th, the storm changed with great violence to the south-west. It began to hail, and afterwards to snow. The black sea seemed to become all at once calm, yet the wind continued to rage with the same violence. I then heard, for the first time, that it is well-known to sea-faring people that a fall of hail and snow lays the most furious commotion of the waves. This appeared to me very extraordinary. The phenomenon is probably to be ascribed to the same causes as the allaying the motion of the waves by pouring oil on them; a consequence of the dissimilarity between the two undular agitated substances.

————

Forvig at Vevelstad in Helgeland, 1st of June. Notwithstanding the boisterous weather, we found an interval on the 30th of May about mid-day to proceed onward with a faint south-west wind. We soon reached Leckö. The island in the southern part is higher than any of those we had hitherto seen, and is at least upwards of one thousand feet in heighth. The northern part, on the other hand, is flat, and only covered with hills. We landed in that part at Skey not far from the church of Leckö. For several years it has been a filial or annexed church to Kolvereid; but formerly the whole of this circle, and a great way into the main land, belonged to Näroens Prästegieldt, which was one of the most extensive and laborious incumbencies in Norway. At present the islands are all that remain to Näroen, with only the single annexed church, or parish of Mittel Vigten. Kolvereid became the head of a new Prästegieldt, and includes under it the remaining annexations, viz. Foldereid in the innermost part of the Foldenfiord, and Leckö.

Skey is a considerable property. The cultivation of the farm requires four horses and thirty cows; and a number of sheep, geese, and ducks, find sufficient sustenance. These may in fact be considered as a sort of luxury here; for the boors never allow themselves to think of keeping geese and ducks. The fishing expedition to Lofodden is here, as well as every where else, the principal object, and in harvest, when the season is favourable, the herring fishery on the coast of Helgeland. We learned with astonishment that eagles were very much dreaded on these islands; for they are not contented with lambs and smaller animals, but even attack oxen, and not unfrequently master them. The manner of their attack is so singular, that we should have doubted the truth of the account, if we had not heard it so circumstantially and distinctly confirmed to us in the same terms at places a great distance from one another. The eagle plunges itself into the waves, and after being completely drenched, rolls itself among the sand on the shore till its wings are quite covered with sand. It then rises into the air and hovers over its unfortunate victim. When it is close to it, it shakes its wings, and throws stones and sand into the eyes of the ox, and completes the terror of the animal by blows with its powerful wings. The blinded oxen run about quite raving, and at length fall down completely exhausted, or dash themselves to death from some cliff. The eagle then mangles undisturbed the fruits of his victory. In this way the nearest neighbours of Skey lost an ox a short time ago.*

On the shore of the island I saw for the first time on this passage wedges of white fine granular marble ; small beds, which gradually in their continuation decline in thickness, and are at length lost between the slates of the rock. They are two or three feet in heighth, and about twenty feet in length. Other smaller wedges are often to be found in the same direction, five or six feet in length. They lie in the mica-slate with continuous mica, in which probably there is no want of garnets, though in the haste I could not find any. The marble beds at Hope, in the neighbourhood of Bergen, and on the islands of Moster or Salthellen at the outlet of the great Hardangerfiord, have exactly the same appearance. This enables us to see somewhat more distinctly the true position of these masses; for otherwise we should be very much embarrassed by this perpetual change; at one time gneiss rocks, at another mica-slate, and again gneiss, and that as frequently close to the land as far out at sea—a great want of determination apparently in the order of both formations ; and we might find it extremely difficult to discover the true connexion in such small islands. But at Bergen it is evident that the mica-slate first follows the great gneiss formation, which constitutes the fundamental part of the mountain, with all the beds which are so remarkably peculiar to it. Then comes a new formation of gneiss sufficient for the creation of whole islands. Beds of mica-slate are very frequent in it, by which it may

---

* Pontoppedan Nat. Hist. II. 143.

The bird mentioned in the text is in all probability the falco ossifragus or sea-eagle. Dr. Edmonstone, in his interesting description of the Zetland islands, relates several curious particulars in regard to the address of this bird in obtaining its food. Dr. Edmonstone's statement, although less remarkable than that of Von Buch, has been called in question by observers of skill and experience.—J.

well be distinguished ; and then comes mica-slate of a nature which approximates very much to the clay-slate. The series of islands on the coasts is, however, generally closed with the newer gneiss, and clay-slate and newer rocks are reserved for the interior of the country. Hence we can only expect gneiss on all extreme islands: the middle ones are mica-slate, and on the larger islands or the main land, the gneiss again rises to considerable heighths without being here interrupted to any considerable extent by mica-slate.*

Gneiss is the fundamental formation, we may almost say the only formation, of the north ; for almost all the other rocks are surrounded by it, and then, even when they possess a peculiar and distinct character of their own, from their being surrounded in this manner, and from their small extent, are in a manner subordinate to the gneiss, and overpowered by it. Nature in the higher latitudes is so accustomed to the gneiss formation, that she always returns to it ; and even when mica-slate, lime-stone, and clay-slate, make their appearance, they merely resemble a series of movements, which have spread towards the north pole without having their origin here, movements which have merely been able to hide from us the cause of the gneiss formation without destroying it as in lower latitudes.

Towards evening we sailed with a faint breeze from Leckö over the Bindalsfiord, which for the space of more than nine English miles lies quite open to the great ocean, and is therefore much dreaded by boats. We were now opposite Torgehatten's singular figure, which points out the boundaries of Helgeland to a great distance. This mountain rises like a pyramid to a heighth of more than two thousand feet. It is seen many miles out at sea, and serves as a signal to mariners. At three o'clock in the morning we reached Saalhuus in Brönoen, at no great distance from it, completely wet and wearied. For a long time we had had no nights. In such dull days as those we then had the day was not to be distinguished in brightness from midnight.

At the shore here, thick beds of white marble run through the rocks, and can be traced to a great distance. As all the strata dip very strongly to the east, they become on that account the more remarkable. About an English

* This alternation of gneiss and mica-slate is not uncommon in Scotland. I first noticed it in the Zetland islands.—J.

mile into the country, we find beds of continuous mica-slate resembling lime-stone, with innumerable little garnets in them. All these protrude themselves out of the rock, and are not surrounded by mica. These beds may therefore be used for mill-stones like Sälbostein above Drontheim. About forty pieces of mill-stone are annually exported from this place. The coast belongs to the continent, but it is low, compared with the other coasts of the north.

The wind became again strong as we reached Saalhuus. An hour afterwards we had a complete storm from the south, accompanied with rain as yesterday, which continued beyond mid day. As the day advanced, the violent gusts of wind followed at greater intervals, and about nine in the evening we had so gentle a south wind, that we again ventured ourselves in our boat over the Fiord into the open sea. This is a singular phenomenon, and by no means accidental. Universally along this coast, after such storms, we expect a calm in the evening, and for several hours of the night; and in this expectation we are as seldom deceived as in the hope of a clear day in summer after a clear sun-set the previous evening. Our own experience sufficiently proved to us how well-founded this expectation is. But this is only in summer, in the months in which there is so little difference between evening and morning. In winter, on the contrary, in the storms which in November blow along the whole coast, the wind rages much more by night than by day; and when trees are torn up by the roots, or houses blown down, it is generally in the middle of the night.

What can produce so opposite an effect in summer? the difference of temperature no doubt. But how can temperature operate so powerfully in such dull weather? At the bottom of the atmosphere on land there is no difference of any consequence. The rays of the sun do not penetrate through the clouds, which are spread over every place. The storm, which with the different air brings along with it also its southern temperature, very soon diminishes the greater heat of mid-day, and increases on the other hand the lesser heat of the night. The sun no doubt during various hours produces some influence on the upper limits of the clouds; but this cause appears to bear very little relation to the effect, and cannot clear up the phenomenon. The greater heat of the air in the higher regions compels the clouds and the vapour to rise still higher, by which the storm obtains a greater and higher course to move in; the mass of air spreads itself out, and flows with less rapidity farther; and from this cause we might also expect a result quite different from the reality. The true cause probably lies hid in the vicinity of the continent; for during the day the storm consists of shocks or gusts, which follow one another in a continued succession ; and when its rage begins to decline, it is not from the diminished violence of these gusts, but from their gradually taking place at greater and greater intervals, till they at length wholly cease. But wind and storms at sea are always equally violent, and do not consist of gusts: when the storm ceases it is gradually ; and then it either does not appear again, or a new storm immediately succeeds from an opposite quarter. We may therefore conjecture that this phenomenon of a night calm in summer on the Norwegian coast is confined to the sea in the vicinity of the coast, and does not extend to the open sea. The ceasing by shocks is besides sufficient to shew that the removing cause is not of gradual operation, as a decrease of the temperature of the stormy air would be, but a cause which also produces its effect by shocks. It may therefore be the land wind, which in summer nights on all the coasts of the world flows from the colder land to the warmer sea, and which must here necessarily weaken, diminish, or completely remove the storm from the south.

The fog lay deep on the water, and concealed from us the view of the mountains. That the northern chain of mountains of the Wellenfiord are high and steep, we did not perceive till we had crossed the Fiord, and almost reached the foot of the rocks. We were shortly afterwards agreeably surprised with the sight of the large, clean, and elegant farm-house of Vevelstadt, and then of the church of Vevelstadt; and about midnight we landed at the beautiful, large, and distinguished farm-house of Forvig, which is in fact a palace in these latitudes, and seems a kingly residence for the princes of Helgeland.

Forvig, the 31st of May. The spring had accompanied us on our northern voyage to this place. The weather was clear and warm; and the snow was disappearing in the woods and the declivities of the mountains. The gooseberry bushes were shooting out vigorously, and the birches were putting forth their buds, and we were in hourly expectation of the leaves. The sun did not disappear till ten o clock, and was again seen in the heavens before two. The temperature only fell from $59°$ of Fahrenheit at mid-day to $49°$ at night, and the following day it rose as high as $70°$; a warmth sufficient to create a movement throughout all nature. This warmth had been long expected; for it was unusual to see the trees so late in putting forth their leaves. This was owing to the snow, which fell in such immense quantities in March and April, of which considerable remains were yet to be seen in the wood close to the sea. So long as the ground is covered with snow, the genial warmth is lost in the melting of the snow as in an abyss; and the temperature of the ground is kept at a depth which does not allow the trees to shoot out. However desirable and necessary it is that the ground should be covered with snow in winter, it is equally prejudicial in spring. Both are however grievous plagues to this province; for the winter along the sea is not severe enough in proportion to the latitude, and it frequently rains in the drizzling manner of a fog when there is a deep fall of snow farther in the country. Hence the frost in January finds the land almost without any covering, and the cold of the surface reaches and destroys the roots of the plants. In harvest also the corn frequently suffers from night frosts. This is a considerable change compared with Drontheim. There fruit-trees are not altogether forgotten; but in Helgeland the climate will not allow them to think of any thing but the berries of the country. In Drontheim, although they have not oaks, they have limes, ashes, and sycamores; but here they have none of these trees. However much the climate of the western coast excels that of the Bothnian Gulph in the same latitude, yet this difference seems almost to disappear under the polar circle; for when the trees begin to assume a green appearance here in the end of May, the same thing takes place at nearly the same time at Luleå or Raneo, both which places are at an equal heighth with Vefsen or Vevelstadt. Although the winter is somewhat less cold in Helgeland than it is in Westbothnia, the summer on the other hand is not so clear, and consequently not so warm. We therefore allow a great deal when we assign 36° of Fahrenheit as the mean temperature of these coasts, a temperature which we can hardly reconcile with observations and experience. This may well excite astonishment; for from 60°, the latitude of Bergen, to $63\frac{1}{2}$°, the latitude of Drontheim, the temperature only falls from 44° to 40° of Fahrenheit, or about 1° 28 for the degree of latitude; whereas the decrease at Helgeland amounts to 1° 8' for the degree of latitude. This is no doubt a very unfavourable climate, and many of the enjoyments of life must be dispensed with; but the little which nature offers may by ability and understanding be turned to such an account as to leave them small regret for the want of a more advantageous country. Beyond

Forvig and Vevelstad there is a beautiful wood of Scotch firs, with Norway spruce firs now and then intermixed, which runs up a broad and level valley at the foot of the higher mountains. The waters are precipitated from the rocks in small resplendent and noisy cascades, and are collected below into placid streams, which in their various windings give a wonderful animation to these woods. Frydenlund, a neat country-house here, is situated along the side of a hill, between majestic birches and alders, and with a view of Vevelstad and of the Sound. Above, in the wood, a number of pleasant walks lead up to the mountains, and to similar embellishments among the high and dark trees; and close by them there is a passable road of more than four English miles in length, which is an unheard of phenomenon in these rocky coasts. All these improvements owe their origin to the excellent proprietor of Forvig, M. Brodkorb, a counsellor of justice (justiz-rath), who has distinguished himself by the most praise-worthy and enviable exertions for the province which he inhabits. He has several times saved it from perishing by famine by the activity and vigour of mind which he displayed in freighting vessels with grain from Archangel at his own cost and risk. In former centuries the Russians had made their appearance as robbers on the coasts; but they never before appeared in the transactions of friendly commercial intercourse. From these transactions, M. Brodkorb derived much more respect and confidence than personal advantage.

FORVIG, the 4th of June. The white lime-stone beds are again not unfrequent along the shore of the Sound. They are enveloped in mica-slate, in which no felspar appears. The hills near to Forvig, about two hundred English feet in heighth, are composed exactly in the same manner all the way up to the flat valley above, beneath the higher mountains. The slates here assume a singular zigzag form, and run in serpentine lines above the rocks. The strata rise considerably here as well as elsewhere, and dip towards the country and into the mountains. This is the general mode of stratification on these coasts; we can recognize it at great distances, through declivities and more firm protruding strata, of which the lines of inclination are distinctly visible for a distance of several leagues. But this composition of the rocks terminates whenever we reach the foot of the higher and connected range of mountains. Felspar immediately makes its appearance then in every quarter, in large shining crystals, which press closely on one another, and almost drive out all the other ingredients. There is also nothing more of a slatey nature in the blocks and rocks.   It is exactly as if we were ascending the rocks of the Kynast, or were among the blocks in the neighbourhood of Hirschberg and Schmiedeberg.   This unexpected formation stretches out in exactly the same manner, not in a single or thick bed, like the gneiss with the large felspar crystals at Kongsvold and Drivstuen; for from the foot to the top of the mountain of Bevelstadsfieldt, which is certainly more than one thousand six hundred English feet in heighth, its continuation is scarcely interrupted by a small bed of a little granite abounding in felspar.   The large felspars every where shine on the large disentangled blocks down the declivity. However, when such large surfaces are uncovered, we cannot disguise the circumstance that by far the greatest number of these crystals lie behind one another, and in places at a considerable distance from one another are exactly in the same direction.   Hence we are enabled to ascertain the general dipping of the rock which is 60 degrees towards the east, and which stretches about h. 11. (N. N. W. and S. S. E.); exactly similar to the mica-slate below at the Sound.   We do not find such a parallelism in the rocks at Hirschberg, nor in the granite at Feurs on the Loire.   The mica between the felspar crystals is black scaley, and never continuous as in the mica-slate, and the quartz is not abundant.   The small chain stretches for more than nine English miles between Vetsen and Velfiord; and probably at the last named Fiord penetrates into the country.   Its greatest heighth is a point above Högholm, which is thence called Högholmstind, probably upwards of two thousand one hundred English feet in heighth.   How far however does the granite extend into the interior of the mountains?   And what date shall we assign to it? Shall we represent it as actually older than gneiss, and of equal age with the granite of France, Silesia, and the Hartz? Or shall we believe that this old granite only makes its appearance for once in this single spot of Norway exactly at the extremity of the great Kiölen mountains, where there is neither the commencement of a plain, nor a termination of the newer formations? I took a good deal of trouble to find at the foot of the mountain the immediate line of separation between the granite and the mica-slate.   Both formations are here characteristic and determinate, and do not alternate with one another; and however difficult it is to arrive with accuracy at the point of separation, yet it was evident that the mica-slate not only displayed itself at the foot of the mountain, but actually went up the mountains in a considerably greater degree.   If we add to this the stratification; the dipping of the strata towards the east into the mountains, it becomes a matter of certainty that the mica-slate runs *beneath* the granite, and that the latter is consequently newer, and lies above the other. We may compare it in respect of position with the granite of the summit of St. Gotthardt, which also lies above remarkably fine slatey mica-slate, and which does not yield to this granite of Vevelstadt either in extent or heighth.* From its first appearance below the Hospital to the tops of Proza and Fieudo, there is a heighth of more than two thousand one hundred English feet. But where shall we again find it in Norway or Sweden? Hitherto no points are known where we may search for it with any hopes of success. The whole of the newer gneiss formation, which lies on the mica-slate, is in general more rare in Sweden. Are we to include in it the granite which Cronstadt describes in the mineralogical account of Jämteland? *(Svensk. Akad. Handl. 1763)*. To the south of the road, he says, between Jämteland and Värdalen, the greatest part of the mountains consists of mica-slate, which again makes its appearance at Röraas. But northwards, and especially in the two parishes of Sörle and Nordli, which belong to the Prästegieldt of Sneaasen in Inderven, the mica-slate changes more and more into granite, and the latter appears at last of a very coarse grain and of a red colour. Towards the boundaries of Nummedal it again disappears, and in place of it we see at Portfield a quartzey conglomerate, which is not unfrequent in these mountains, and which probably belongs to the transition formation. We might well conclude from this position that the granite also was newer than mica-slate; but were there in it as many large and beautiful felspar crystals as distinguish in so remarkable a manner the rocks of Vevelstadt, Cronstadt would certainly not have forgotten to mention it. In the granite at Forvig tourmaline is not altogether rare; even close to the house there are great blocks lying about, containing large beautiful and black crystals of tourmaline surrounded by

---

* It is worthy of remark that several of the most striking granite mountains in Europe are composed of newer granite: this is the case with Saint Gothard as mentioned by Von Buch. Mont Blanc, the highest mountain in Europe, is composed of newer granite: and the magnificent granite mountains in the Island of Arran in the Frith of Clyde rest on strata of clay-slate, &c. and therefore are composed of a newer granite. Many of the Zetland islands nearly agree with Norway in structure and composition, and I am of opinion that the granite of the Mainland, the largest of the Zetland islands, belongs to the newer formation.—J.

large shining folia of mica of a silver white colour.   But I could see no hornblende ; a few black beds, especially at the top of the mountain, consist of
fine scaley mica, with a little felspar between.   We find there also a few pure
and thick beds of quartz.

The prospect from the *Fieldt* of the Archipelago of islands around is not
without interest.   We are astonished at the diversity of shapes they exhibit.
Some of them are rocks, which are hardly accessible ; others again are completely flat, and scarcely elevated above the surface of the water; some
exhibit the appearance of mountains ; while others look like a collection of
several high islands, between which the water has forsaken the bottom.
There is none of them more striking, even from below, than the elevated
island of Alsten, with its high mountains, the Seven Sisters *(Syv Söstern)*,
seven points which ascend far beyond the region of snow, and the base of which,
consisting of black rocks, rises almost perpendicularly out of the sea.   They
are by far the greatest heights as far as the eye can reach, and certainly
exceed four thousand two hundred and sixty English feet.   In the whole
way northwards, to the farthermost cape, we see but few islands which equal
this in heighth.   If it were not for it, the singular Vegöe would appear high,
whose two points rise with great abruptness, but do not exceed two thousand
one hundred and thirty English feet.   It is one of the outermost islands
towards the sea, and exactly opposite Vevelstadtfieldt.   It is so much the
more remarkable as we can recognize from this Fieldt with great distinctness
the manner in which the strata of both points dip, not as in the Mainland
towards the east, but in an exactly opposite direction towards the west.
May not the greatest number of these islands, points, and rocks, only rise
above the surface of the water, because this diversity in the inclination and
dipping of the strata in their opposite extremity universally raised several
strata above this surface?

Sör-Herröe, the 7th of June.   We changed our boats at Forvig.   The
Sorenscriver Holst of Tiötöe, gave up to us what was called a *Venge-boat*,
that we might proceed with it to Lödingen.   It was a prettily painted boat,
well ornamented, with a roomy and comfortable cabin or *Venge* behind,
almost resembling that of a ship.   Our ladies and children found some
consolation in this for the hardships of the journey, and when the weather
was calm, it would have resembled a pleasure excursion, if every boat and ship
were not a prison.   Six men rowed during calms.

The north wind was against us. We rowed tediously over Vestenfiord towards Tiötöe, expecting, but in vain, that the wind would fall, and allow us again to make use of our oars. It is singular enough, that a single small island, even a rock, is sufficient to put a stop to the waves of an unfavourable wind not too powerful. Defended in this manner, we can make leagues of way; and when we pass the rock, we scarcely proceed a few hundred yards in the hour. We were compelled to remain at Rösöen opposite to Tiötöe in expectation of a different wind. Vestenfiord is the farthest point at which oysters have ever been found. They have never been discovered farther north, which is singular enough. Oysters do not live at the surface of the water. Is the colder temperature communicated to the water at oyster depth? If not, why are they not found still farther north?

The stone of Rösöen is white gneiss abounding in felspar, on which beds of mica-slate frequently repose, consequently it is not gneiss of the older formation. The strata here also dip towards the west. Notwithstanding the island is not above an English mile in length, we can see rocks on it of two and three hundred feet in heighth, separated by flat and green vallies, which run from the one end of the island to the other, and the greatest heighth of which seldom rises more than forty or fifty feet above the surface of the sea: a group of *skiärs* in which the water between the rocks is wanting. The vallies assume the place of the sounds. On the other hand, Tiötöe, which lies at the distance of not more than an English mile, is throughout flat and without rocks, an entire plain. It may easily be conceived how much we are struck by the appearance of such a plain almost close to the foot of the great Giants of Alstahoug, and in the midst of so many singular rocks all around. It was from the earliest times accounted one of the most fertile plains in the *northern lands*, and the vassal sovereigns of the Norwegian kings, or their governors of these provinces, used to dwell here.* The ground is yet excellent, and is said to be excellently cultivated. The proprietor, John Brodkorb, a cousin of the counsellor in Forvig, to whom the greatest part of the island belongs, is universally spoken of as one of the best and most active agriculturists of the north. He endeavoured

---

* *Heimskringla—Kong Olafi hinom Helgasaga.* 162. *Hareck, Evind Skalduspiller's son* drove all the boors from the island, and became rich partly by cultivating it, and partly by trading to Finmark.

to grow wheat on this island; but notwithstanding the grain ripened, the profit did not repay the trouble. The large and beautiful house of the proprietor, and the church, close adjoining, have a very agreeable effect when viewed from Rösöen.

It grew completely calm in the night, and early in the morning we rowed pretty quickly past by the bishop's seat of Alstahoug, and about mid-day reached the flat island of Sör-Herröe, two leagues from Rösöen, and one of the farthermost islands towards the sea. But we were stopt from proceeding farther by the winds rising again. It appeared as if by an accident we had been at once transported from the northern mountains to the plains of Denmark. The view from the house extends over an endless plain, where no rock, no cliff in the direction of the sea, recalls the mountains of the coast to our recollection. The island is more than two English miles in circumference, and yet it is scarcely more than forty feet above the sea. The water on that account does not easily find an outlet, and a great part of the land for the most part of the year remains a complete morass. What a pity! for the ground is here very good. Even in its present condition the island maintains forty cows, four horses, which are necessary for the tillage, and a number of sheep, among which there are many Spanish. It is the property of the chaplain of the Prästegieldt of Alstahoug, who lives on this island. The whole parish was given to the bishop for his maintenance, when about three years ago Nordland and Finmark were separated from the bishoprick of Drontheim, and both provinces received their separate bishop's seat. This was very necessary. It has been sufficiently shown by experience how difficult, and frequently even impossible, it was for the bishops of Drontheim to perform their visitations. These provinces remained less known on that account, and the country could not avoid being subject to many a small grievance, which a bishop in districts less remote could easily have remedied. The Laplanders, or Finns as they are here called, since their conversion to christianity, were never subjected to the bishop, but only to the college of missionaries. The college is now no more, and the new bishop exercises spiritual authority over all the Norwegian subjects in Nordland and Finmark:

It was an interesting spectacle to us to see on a Sunday several hundreds of these amphibious men flocking to church. They are all dressed alike, as if in uniform, in earth-brown frocks, like those of miners, only close on every side except the breast, in the opening of which small blue trimmings run down on both sides. They had large white trowsers on above boots, and a red woollen cap on their heads, with a felt hat above it. This is the characteristic dress of the fishermen of Nordland; those of Drontheim, Bergen, and Christiansand do not wear it. Their figure is still more striking. The flat faces, fair hair, which are generally believed to be universal among the inhabitants of Nordland, are not often seen here. On the contrary, I saw with astonishment several true Turkish physiognomies, and this was by no means a rarity: noses and bones extremely prominent, black dazzling eyes, and no trace of the fair physiognomy of the Danes. The muscles do not here appear swoln out at the expence of the bones; all the features are sharp and determinate. It is true all are not of this description; but we should in vain look for such forms in Upland at Geffle, in Westmannland, in Wermeland, Westgothland, or in other provinces of the interior of Sweden. This brought to my mind the recollection of the endeavour of the worthy Schöning, in a highly learned and instructive treatise, to prove that the Phœnicians in all probability visited these remote regions, and that Carthaginian seamen repaired to the fishery at Lofodden, and conveyed their fish to Africa.* For Thule lies here and no where else, if there was ever such a place: it is certainly not in Iceland, nor between the Orkney and Shetland islands. In Iceland, which the polar circle scarcely touches, we cannot speak of day for several months in summer, and night for several months in winter. Agriculture was never in a flourishing state in Iceland, which Strabo states to be the case with regard to Thule, where the grain was preserved in large granaries and thrashed out and kept. Iceland was an uninhabited and desert island, when the Norwegians, under the government of Harald Haarfager, sought and found freedom there; but in Thule, which reached to the Frozen Ocean, the inhabitants were by no means in a low state of civilization. Mela, who follows in this the Greek writers, relates expressly that Thule is to the north of Sarmatia, and opposite the Belgian coast. However, Thule cannot be one of the British islands; for Thule was six days journey distant from Britain; it was large, and lay far beyond Ireland; and it was the

---

* *Kiöbenhavns Videnskabs Selbsk. Sckrft.* Tom. IX. Four days journey to the south-east, a great fishery was formerly carried on in the Straits, and the fish salted and taken to Carthage.—p. 162.

farthermost land towards the north.   How can we suppose that the Phœnicians could be so well acquainted with Britain and Ireland, that they
could accurately determine their extent without ever visiting the neighbouring country of Norway, or receiving any information respecting it in
these islands. The coast of Scotland and Bergen are so near that the
passage from the one to the other is frequently performed in one day ; and
Scotch vessels laden with butter have sometimes been freighted by Scotch
farmers direct for the town, which is at so short a distance, instead of disposing
of their commodity for general exportation.   The currents run from Scotland
and Ireland towards the northern coasts, and not towards Iceland.   Thule
included the whole coast to the northernmost point; and the oldest writers
do not say that it was an island, though even if they did so, it was very
easy for them to mistake Norway, or even parts of the country, for large
islands. When we find that the memory of the punic residence is still
preserved in the language and manners of Ireland, how easily might words
and individual expressions peculiar to the inhabitants of Nordland display
their foreign Phœnician or Carthaginian relation in the same manner as
their prominent features.   In this point of view a collection of the most
remarkable peculiarities in the language of this country would be of the
utmost importance; and it were to be wished that M. Schnabel, a clergyman
in Sör-Herröe, who is extremely well versed in ancient and modern literature,
could possess leisure and inclination to think of such an undertaking.*

Sör-Herröe, the 8th of June. We overlook here from the north the seven
high rocks of Alstahoug.   From Vevelstad we had hitherto only seen them
from the south. The view is sublime beyond all description, especially when
the nightly sun gilds their tops, and leaves the immense mass at their base in
mysterious uncertainty.   In the steep vallies, bushes reach upwards along the
sides to a considerable heighth, after which the rocks appear steep and entirely naked, and the snow lies fast and immoveable at top, and shines in the
rays of the sun like ice.   The chain runs parallel with the great Vefsenfiord ;
and it is evident that it has formed this fiord. The strata dip towards the
north-west, and the greatest and most frightful declivities are towards the

---

* M. Schnabel has since been removed to Drontheim. He is a brother of the late Dean
Marcus Schnabel in Eger ; a very good fragment of a description of Hardanger, written by
whom, has been given to the world by Provost Ström—Kiobenh. 1781.

Fiord. These strata have been elevated from below, and hang in a threatening position over the abyss which they have opened.   Vefsenfiord changes its direction immediately on their account, and enters the land in the crooked shape of a half moon, when the range of rocks terminates; but lest we should be tempted to imagine from the appearance of these rocks the phenomenon to be more universal, and the same stratification to belong at least to the whole province, we have only to cast our eyes at the same time on the large and high mountain of Dunnöe, which withdraws the sun from Sör-Herröe for several hours together, notwithstanding it is at a distance of almost five English miles.   On this mountain there are two tops in the shape of small mountains, rugged, bare, and rising abruptly upwards; the one westwards towards the sea, and the other eastwards, each upwards of three thousand two hundred English feet in heighth. All the strata may be distinctly seen to the top; but they no longer dip, as at Alsten, towards the north-west, but distinctly inland towards the east; and the outgoings of the strata are in the direction of the open sea towards the west.   It is undoubtedly difficult, perhaps it is even impossible, to discover any general law of stratification here; but it would be deciding precipitately, were we to ascribe the raising and formation of these islands to secret powers, to *earthquakes*, which have generally so much imputed to them.   Have earthquakes then heaped together all mountains in exactly the same manner? There are no mountain-chains throughout the earth's surface with *horizontal* strata: the cause is not sufficiently powerful to effect this.   What are all these rocks of Nordland? What are these islands of upwards of four thousand feet in heighth? What are the seven thousand five hundred feet heighths of Dovrefieldt, in comparison with a power capable of agitating at the same time the greatest part of the earth's surface? Could it throw up the whole of Helgeland; mountains many miles in heighth, and not rocks, whose summits we can ascend in a few hours? Others again say, if it was not such an universally operating cause, we may, however, reasonably ascribe this throwing up of the strata to elastic fluids, inclosed in cavities beneath the mountains, which have been produced by the dissimilar deposition of the mountain-masses.   It is true we may comprehend in this way the constant fluctuation of the direction, and the origin of many a singular form in the mountains: but this cause would destroy every remains of regularity or law in the phenomena of

U

stratification, and consequently would be granting a great deal too much. For we find not only stratifications which continue for a great way through countries, and perhaps too on the coast here; but it also seems as if at a certain distance from the main mountains the strata always incline towards the central chain. Besides, such fluids would operate throughout their whole extent. When they experience the slightest opposition, there they break out: the pressure upon all the other points is diminished, and ceases with the breaking out, and consequently no effect is produced on the more firm and solid places. Such causes may throw up rocks, small islands, a *Montenuovo*, horns and peaks; but could hardly be supposed to produce a chain running for many miles, like the rocks of Alstahoug, or many of those on the coast of Bergen.

But these elastic fluids are merely conjectural causes, and we know how dangerous it is to admit conjecture into geognosy. By such means we make a bold plunge into the midst of the slow course of geological experience, and if the adventure does not succeed (and how can that be determined), we lose the beautiful clue which the progress of this experience promised would connect us more nearly with the living world, which allowed us to hope that we should have a nearer prospect of comprehending the general plan of nature in the formation of organic and inorganic bodies, if we could only again find the same path: for the great course of nature is but one and the same, from the coagulation of granite, to the career of man.*

LURÖE, the 10th of June. A fresh south-east wind soon carried us from Sör-Herröe, along by the bottom of the high rocks of Dunnöe, in a few hours into the open sea. We were struck with the singular appearance of the rock Lovunnen; a cliff wholly insulated in the open sea and separated from all the other islands. We passed it, and in the afternoon we reached the group of islands of Luröe. In the evening we lay at anchor in the small harbour of Luröe.

---

* The hypothesis of the connexion of organic with inorganic bodies alluded to in the text is at present much in vogue in Germany. The supporters of that speculation maintain that minerals, plants, and animals, have been called into existence by the agency of the same general powers; that minerals were first created, plants and animals last. M. Von Buch, in the year 1806, published a memoir on this subject, which he read before the Royal Academy of Sciences in Berlin. It is entitled, " Ueber das Fortschreiten der Bildungen in der Natur."—*J.*

Lovunnen is the resort of numerous birds *(Lundfugl, alca arctica)* called puffins, which are very much caught on account of their feathers. It is by no means difficult to catch them. They sit collected together in small clefts of the rocks. The fowler lays hold of the first with iron hooks ; or if the opening is deep, he sends a dog trained for that purpose into it, who returns with one of the birds in his mouth. The next bird follows the first, another succeeds, and so on in their order, till the last of the republic. In this manner the fowler extracts the whole train of them at once, and in a short time collects a great booty. (This manner of catching birds was known also to Pontoppedan. Nat. Hist. II. 141). The bird does not live on flat islands, or those which are not very rocky ; and it also shuns the appearance of men. Its constant residence is high and difficult cliffs, which stretch wide out into the sea. It does not enter the Fiord, and scarcely comes between the islands which surround the mouth of the Fiord.

The residence of the proprietor of Luröe lies an English mile from the harbour, in the interior of the island, in a pleasant situation, between green meadows, along a running stream. The most frightful rocks rise perpendicularly over it, which inspire a feeling of grandeur, and contrast in a powerful manner with the calm sensations excited by the contemplation of the lovely green below. The numerous and amiable family of M. Dass, a merchant who lives here, is altogether in harmony with this beautiful scene. We ascended the mountains even in the night. The birches above were putting forth their leaves, and the birches at the foot of the rocks were already entirely green. We found here again the spring of Vevelstad. After ascending the first stage of the mountain, we entered a valley of perhaps four hundred feet in heighth : there the leaves disappeared, and in the distance we saw snow along the declivity. The *Fieldt* is steep, even from this place, but not rocky ; we soon reached the precipice above the Gaard, and scarcely saw the buildings and human beings in the giddy depth beneath us. The birches at last left us, and the places covered with snow became more frequent and large. The summit appeared to us like a sort of town formed of stones, with a signal on it. We were there on a small ridge between two abysses. The snow here resembled a mound, which hung over the depth to the east for a great extent. We were obliged to advance with great caution towards the edge, where we saw the rock fall steep and perpendicularly. The heighth appeared very great. The barometer gave it two thousand one

hundred and eighty-seven English feet above the Gaard.* This is a great deal
for rocks which rise with such abruptness. We commanded an extensive
prospect: the islands in the distance were concealed in a fog, which joined
sea and sky together. Towards the land, snow mountains of a much greater
heighth than Luröe towered up before us in continued rows. Close at hand
we saw the mountain of Hermanöe, equally high, steep, and small; and far
out at sea appeared close to one another the four high points of Tranöe,
between which the polar circle runs. It is astonishing to see the manner in
which the sea and mountains here contend for dominion, and in which the
immensity of the ocean at last prevails. It was a melancholy consideration
for us that we could not advance and penetrate into this immensity. We
hastened back, and rapidly descended the mountain; and we were softly
and gently received below by the limpid streams, the thickets, and the
friendly habitation amongst the rocks.

What could be wanting to this island, if the fishery of Lofodden allowed
the inhabitants to think of any other mode of industry than fishing ? The
soil is excellent in the extensive hanging vallies, in which the water may be
either carried off, or preserved as is found necessary. Sand and calcareous
earth from shells are to be found in the neighbourhood in sufficient quantity to
admit of giving the greatest possible fertilization to the soil. There is as
much wood on the mountains and the Fieldt as is necessary for household
purposes. The property is separated from other places in such a manner, that
all entrance to bears, wolves, and lynxes is denied; for they cannot pass from
the continent across the Sound, and therefore they never visit these islands.
Lastly, the climate, like that of Tiötöe, is not altogether unfavourable to
agriculture. Luröe is universally accounted warmer than any of the surrounding islands; and we were willingly induced to give credit to this; for
72°. 5 of Fahrenheit for four or five hours in the middle of the day, which we
had here, is a degree of heat in spring which is not given to every island.
The flat part of the island lies like a hothouse before the Fieldt, which defends
it from the north and east. In such a situation it appeared surprising to us
that the Gaard only maintained thirty cows, fifty sheep, and one hundred
goats. What the few boors maintain, who also live on the island, is hardly

---

* Luröe Gaard Barom. 28. 0. 2. Therm. $\frac{-}{1}$ 10 R. h. 8. p. m.
  Luröe fieldt.......... 25. 11. 2...——.. 5.——h. 10. p. m.

worth mentioning.   M. Dass also keeps for his pleasure two rein deer on the
Fieldt, which, when they multiply, will be banished from it ; for rein deer
do no good in the neighbourhood of cultivated fields.   This place may have
lost in comparison of what it was in former times, for it appears to have been
of old in high estimation.   Three or four immensely large tumuli of earth
(Kiämpehoie) preserve the remembrance of the old inhabitants, their wealth,
and prosperity.   For tumuli, such as they are, could only be thrown up for
great and powerful individuals.   But history, which is full of the mention of
the heroes of Nordland, and often commemorates those who dwelt in Helgeland, is altogether silent with respect to Luröe.   The fear of the inhabitants prevents the tumuli from being examined ; for they imagine that they
would waken spirits, who would then take up their perpetual residence on
the island.   In a half-opened mound there were a good many remains of
antiquity found.

Gneiss only appears here, and no trace of mica-slate or lime-stone.   The
whole way up the mountain we are dazzled by the felspar, which is not
concealed between the slates, as is so often the case in the gneiss of the other
islands.   Were we only to have pieces and small blocks before us, like those
of which several walls are constructed, we should believe in the existence
of granite in this mountain.   But the mixture is too indefinite and too
various to allow of such a supposition ; and even in the rocks which have
fallen down, and lie scattered over the meadows, the slatey structure of the
gneiss is betrayed by the flat position of the scaley mica.   We can hardly,
however, find felspar of a larger grain than what we see above in the middle,
and at the summit of the rocks ; for there we not unfrequently meet with
pieces, single crystals, of at least the length of a foot and upwards ; but the
felspar which lies between, though of an admirable flesh-red colour, and
but seldom white, is not small granular.   Quartz frequently appears also in
this gigantic form, but mica does not. Mica, which also makes its appearance
in small masses, consists of small black scales over one another, and not large
folia.   Deeper down, at the foot of the rocks, the grain of the mixture is
certainly less striking and large ; but it is so much the richer in foreign ingredients, which appear here almost essential.   None of them deserves to be
named before the black titanium, which is exactly similar to the titanic iron
ore of Arendal; for none of the ingredients is more frequent, or more striking,

from its dazzling and almost metallic blackness, between the red of the felspar. The pieces are of the size of small eggs, like the *augite* in the beautiful mixtures in the Saualpe in Carinthia; almost conchoidal in the fracture, and in large blocks, granular, mixed with felspar and quartz, and nearly always with hornblende. This is singular enough, for hornblende never appears here as a bed, but in a mixture in this manner it frequently preponderates over the mica. A beautiful green covering of chlorite in druses and cavities of the quartz also occurs, similar to what we observe at *Mont Blanc*, and the south side of the Alps. Should not tourmaline also be found in these mixtures ? Or is its place occupied by the pieces of titanium.

The strata of the mountain dip towards the north-east; and hence the abrupt precipices of the south side and towards the Gaard.

The flat part of the island is covered with later productions. In the middle, between the Gaard and the harbour, there is a church; and at no great distance from it the ground is uncovered. Under the black moorish mould there appears a stratum two feet in thickness, of a dazzling snow white colour, of a loose texture, and to appearance of fine shining sand. Sand however it is not. In the whole stratum there are nothing but yellowish broken shells, and small bivalves in infinite number; several large ones lie between, but there is no earth or marly clay, as at Steenkiers. The stratum runs in this manner in a horizontal direction, throughout all the plains of the island, covered with moss and peat earth. Its greatest heighth is not perhaps more than twenty or thirty feet above the surface of the sea. Is this also an effect of the gradual sinking of the surface of the ocean ? The phenomenon is at least more remarkable than all the marle beds containing bivalve shells at Südenfields and Steenkiers; for it does not bear the appearance here of an inundation as in those places, but of gradual and calm deposition on the earth. We find something similar in the neighbourhood of Drontheim, on the entirely flat and level peninsula of Oereland. We meet there with bivalve shells, lying scattered under the great peat moss, and covered with marine plants. The whole moor in the undermost beds consists there of marine plants, but in the uppermost, of meadow and marsh plants. Fabricius, 264. Where is there any thing similar in the Sudenfields ?*

---

* The following is the passage on Fabricius alluded to by M. Von Buch. " We obtained a fresh proof of the decrease of the water of the ocean by examining some masses of peat newly dug

VIIGTIL, the 13th of June. On the morning of the 11th, we left Luröe with clear and charming weather ; and immediately after we cut through the polar circle nearly below the high and rugged Hestmano, the strata of which appear to be a continuation of Luröe, and dip in a similar manner towards the north-east. Before mid-day we saw Rodöe before us, a celebrated and consi- derable island, and now the central point of the extensive Prästegieldt of Rodöe. The form of these rocks, which rise like a high Gothic ruin, high above the green surface towards the north, presents us with one of the most beautiful and striking objects in the north. We passed it, and reached in a complete calm the small island of Svinvär in the evening, where M. Hvit gave us what we almost every where met with, a most friendly reception. The sounds which run between them are sufficiently deep and clear to allow the vessels of Nordland a passage through them : and on this account these vessels frequently wait here for a favourable wind, or the end of a storm ; for the waves in such long and narrow sounds are by no means dangerous. These islands are of gneiss, with white felspar, very small granular, and extended lengthwise between the scaley mica. Mica-slate beds are not strikingly visible, but are probably contained in them.

The wind on the following morning was again favourable to us, and brought us quickly through the Stött Sound. Here, as well as at Svinvär, there is a place of trade and a merchant's seat, which is situated under the frightful rocks of the Kunnen. This is a broad cape, which stretches out for a long way. It is a peculiar mountain, which divides the winds like Stattland at Söndmör. We had scarcely approached the point, when we were deserted by the south wind, and the boat's crew were obliged to have re- course to their oars. We proceeded close under the rocks for more than an English mile. They appear indescribably high and striking. The precipices are completely perpendicular and naked, and certainly a thousand feet in heighth ; and there are but very few places at the base where they are washed by the sea where there is any thing like a footing. Some

---

from the earth. The lower beds were almost entirely composed of putrid marine plants. We observed distinct examples of the zostera, with its long and narrow leaves ; the upper bed was formed of sphagnum. In general not a trace of wood, nor roots of any kind, so frequent in other bogs, was to be seen in this vast morass : but we found abundance of shells and sand, which are evident proofs of an anterior inundation." *Fabricius, Voyage en Norwege.—J*

of the rocks advance before the rest with majestic shapes, like colossal Addystones in the sea. Every thing around appears of a gigantic shape, and the small and audacious boat on the waves dwindles down among them to a point on the surface of the water. The strata run horizontally in clear and dark stripes, but in a similar gigantic proportion; for these stripes may probably be forty, nay even sixty feet in thickness. We follow them for a long way, as far as this steep cape extends, through all its windings, and over all the projecting rocks. We were at last enabled with great difficulty to force our way through the deep waves into a bay, where they were driven back from the rocks with double fury; and after a great deal of labour we steered between a group of rocks and low islands, which formed a secure harbour, and landed at the friendly farm-house of Viigtil, situated beneath a green declivity.

Kunnen divides Nordland into two parts, distinctly separated from one another. It is in fact a considerable mountain, a large and insulated moun-tain-mass, like an island which does not join the principal range of mountains. The perpendicular rocks at the cape rise immediately from the water to a heighth of more than one thousand feet, but they are, at least, another thousand feet higher farther on, and in the course of four or five English miles they may be about four thousand feet high. Perpetual snow lies here, and what is still more, this snow has generated glaciers. About four or five English miles south from the cape, opposite the trading station of Haasvär, a glacier descends from the height, and the ice comes into immediate contact with the sea, a circumstance perhaps peculiar to this glacier. Even then, the warmth of the summer had merely driven it a few steps from the shore, but it would probably regain its former space in a short time. This is the first glacier in the north from the glaciers on the mountains of the Nord Fiord and the Söndmör in the latitude of 62°; for we look in vain for them on the Kiölen mountains to the south of Salten.

These masses of ice give us a pretty clear idea of the extent of Kunnen inland. For to produce glaciers it is not enough that mountains enter the region of perpetual snow; they must be preserved there by means of a con-siderable space; for an insulated high mountain like Sneehättan, and a small chain of mountains like the rocks of Alstahoug, can never collect so much ice in one place as is necessary to drive forth a glacier from the upper regions to warm vallies. The glaciers of Alstahoug resemble icicles, which melt in a warmer temperature: but the glaciers of Folgefonden, Justedal, Salten, and Kunnen, impel new masses of ice downwards from their inexhaustible stores, when the undermost half must yield to the higher temperature.

Helgeland ends here, and Salten begins. The boundaries of the former province are about four English miles beyond Kunnen. This no doubt is not the ancient Helgeland; for Halogaland was the general name for all the country inhabited by the Norwegians in the north as far as Finmark; or from Senjen to far beyond Lofodden. Of this nothing has been preserved by tradition; for Salten and Helgeland are now considered as two very distinct provinces; and in this they are not only justified by political institutions, but by nature herself. Helgeland's boundaries are also the limits of the Norway spruce region. These trees still continue to grow in the Fiords round Kunnen; and in the woods of Beyernfiord they may also be found; but they do not extend farther north, nor are they to be seen deep in the interior of the Fiord. From Saltdalen upwards towards the mountains, and in the deep and narrow Saltvattudal between Salten and Foldenfiord, nothing but Scotch firs are to be seen, the spruce firs having altogether disappeared. In this remarkable way the climate becomes more severe from Helgeland upwards. In the latitude of Kunnen 67° at Känger, on the river Torneä, in Sweden, there are excellent woods of spruce firs; and in that country these trees do not disappear till we reach the latitude of 68°, where they are still to be found on plains about eight hundred and fifty English feet above the level of the sea. Hence we were led to expect that we should find the spruce firs on the sea-coast, or in the interior of the Fiords, for at least a hundred and eighty miles higher, and till we passed Tromsöe; but the long winter of the northern latitudes has here obtained the ascendancy over the cloudy summer of the sea-coast; and the brighter sun of Swedish Lapland is more powerful in its effects than the excessively severe winter there. We can hardly assign to Saltenfiord a higher mean temperature than 34° of Fahrenheit; nor a higher temperature for the warm month of July than 57°. 8. In Sweden, July under the polar circle rises far beyond 59°, perhaps close on 61°; and when Salten's August does not reach 52°, at Pello and Känjis that month rises to 55°. 6. On this account Helgeland is held in the north to be a highly favoured province; and it yields in population to few provinces in Norway. In the enumeration of 1801 it was found to possess twenty-four thousand three hundred and sixty-four inhabitants, of whom four thousand nine hundred and ninety-three were contained in the Prästegieldt of Alstahoug.

This is nearly fourteen men to the square English mile, which is more than the general average of Drontheimstift, and almost equal to that of Bergen and Christiansandstift, though much inferior to the populous districts of Aggershuusstift. In the extensive vallies which ascend from the Fiords towards the mountains, in Vefsen and in Ranen, agriculture is successfully followed, more so perhaps than in the middle provinces of the country; and in these situations there is also no want of woods. But on the sea-coasts, and on the innumerable islands, the inhabitants are principally subsisted by the herring fishery. For of the quantity of herrings exported from Bergen, which are generally known by the name of Bergen herrings, the half is nearly supplied by the single Prästegieldts of Alstahoug, and Rodöe. No doubt in more recent times, the herring has here, as well as every where else, removed to a greater distance from the coast; but they still appear in sufficient abundance to occupy with their fishery many thousand hands every season, from August to the end of the year.

Our stay at Viigtil was not without its pleasures for us. Here we first saw the sun at midnight in the greatest clearness, and even giving forth heat. The birch bushes on the hills were now green, and the flowers appeared in multitudes along the declivities. A noble waterfall, not far from the house, precipitates itself between blocks of rock of a most picturesque form, and the stream winds with a loud murmur through green meadows to the Fiord. The prospect over the numerous cliffs is singular, and especially the view of the high, steep, and rocky Fuglö, from which two cataracts, like shining silver bands, fall from a height of about a thousand feet. In the distance the island of Landegade rises to a similar height, with equal steepness. All the other islands are low, and scarcely raised above the level of the sea, compared with these powerful masses. They look like two gigantic teeth, and serve to point out to mariners at sea the entrance of the Saltenfiord, and the passage to Hundholm; for they are seen in the open sea at a distance from land of many leagues. Fuglö is exactly opposite to the Kunnen, and appears like a piece torn from it, for the rock is equally high and abrupt; but it is considerably more than four English miles distant from the main land.

The rocks of Viigtil consist of mica-slate, and not of gneiss. This is not only proved by the continuous mica of the masses, and the want of felspar, but also by the numerous garnets which are here incorporated in the

mountain rock, as in the mill-stones of Sälbo at Drontheim. The strata dip
here every where towards the south-east, as also at Kunnen, as we may distinctly see in the distance. Hence these strata appear horizontal at the sea
side beneath the cape, for their outgoings are at this side. This coincidence,
and their external appearance, warrant us in conjecturing that the whole
mountains of Kunnen are of mica-slate and not of gneiss; and, even if otherwise, we should still have obtained several further results to enable us to
determinate which of the adjacent islands belong to the newer gneiss,
and which to the older. We should thereby have found a place again
for the greater and independent mica-slate formations between the two
gneiss formations.

BODÖE, the 15th of June. In the afternoon we passed in calm weather
the clergyman's house of Gilleskaal, and about five o'clock reached the
trading station of Arend. The wind entirely ceased as usual towards
evening, and the rowers brought us quickly over the Saltenfiord, more than
nine English miles in breadth, through flat islands, to the harbour of
Hundholm, which we reached at one o'clock. We lay close to a large brig,
and in sight of a newly-built wooden quay, here called Brygge, large storehouses, and a considerable and new dwelling-house at a short distance.
However indifferent objects these may be elsewhere, here they are by no
means so. A vessel waiting for a lading is a sight which does not occur
for centuries. A quay, or Brygge, worth more than ten thousand dollars in
a country where every log must be brought from a considerable distance
southwards, gives rise to well-founded hopes; and these boats and houses
prove that the owners neither want spirit nor means to carry their views
into effect. Hundholm is in fact one of the most remarkable places in
Nordland; for since its foundation the dawning of the prosperity of that
province has brightened up. The object is to carry on immediately from
this place the trade of Bergen; and for that purpose several of the richest
merchants in Drontheim have joined together. We can hardly doubt that
with their means, zeal, and activity, they will accomplish their objects; for
nothing appears more simple and agreeable to the nature of things. The fish
is caught in the north, and sold in Spain and Italy. What obstacle can there
be to the saving of toil and labour, of time, gold, and men, and to the sending
the fish from hence directly for Spain and Italy? Nothing but custom. In

ancient times, before the existence of Bergen, or any other town on the coast, while Lofodden abounded with fish as at present, vessels were dispatched immediately from the north to England and Flanders. Thoralf, King Harald Haarfagers Jarl in Helgeland, in the year 890, pursued the fishery the whole winter through, below Vaage in Lofodden, and exchanged the produce in England for a rich cargo of wheat, wine, and honey.* But when Oluf Kyrre founded Bergen in 1069, at the time when the Hanseatic League was formed, the Germans soon settled there, and got into their hands the whole trade and exportation. It was more convenient for the inhabitants to dispose of their fish directly to the counting-house, than to convey them to foreign countries through unknown tracts. The same thing took place with the inhabitants of Nordland. The unceasing demands of the German factory excited in the people of Bergen a great activity in bringing fish from every quarter to this general magazine; and this eagerness brought them to Nordland and to Finmark. In this way the fishermen of Nordland disposed of their produce immediately to the vessels of Bergen, which imposed on themselves the necessity of a very short carriage; for the fishermen did not seek for the merchant, but the merchant the fish. As the latter was unwilling to have a fruitless journey, he brought along with him to the fishermen all that they stood in need of, to catch as many as possible. These were the times when the town exercised a beneficial guardianship over the country. But the town itself suffered under a much harder oppression. The German factory long remained in Bergen like an independent state, not amenable to the laws of the land, and frequently set order and justice at defiance. The courage and wisdom of a Walckendorf was required to heal this dangerous wound in the heart of the state. The freedom and licence of the Germans were at length curtailed. The Hans Towns, however, to revenge themselves perhaps for their lost superiority, and the Pomeranian towns in particular, and Wismar and Rostock, appeared in 1539 before Bergen with a numerous fleet, and plundered this town in so cruel a manner, that the impoverished townspeople were rendered incapable of carrying on their usual traffic with Nordland.† The population of Nordland was too great to maintain itself; and if they wished to preserve the country from destruction, it was necessary

---

* Rigla Schöning Norske Hist. II. 455.　　　　† Holberg Bergens Bescriv. 110.

for the inhabitants to perform themselves the long and dangerous passage to Bergen. In this way the famous Bergen craft had their origin, which after a lapse of 270 years were considered in the country, and still more in Bergen, as a sort of law of nature. That Bergen's Vog should be closely crowded in the month of June with innumerable Nordland yachts, and again in winter, was a matter of as great certainty as the succession of summer to winter, and winter to summer. The purchaser puts his price on the fish, and not the vendor; and the latter has an increased gain in the productions, grain, and articles of luxury which the inhabitants of Nordland derive from the town. In the year 1807, there came at once from Nordland, Tromsöe, and Senjen, one hundred and twenty-six yachts to Bergen, and that twice in the same year. What a distance! It is almost half the passage to Spain, and by much the most dangerous and difficult; in a zone in which the winds are daily changing, and frequently change in the same day, and blow with equal fury from opposite points of the compass; and on a coast which on account of the numerous cliffs and Skiärs is well known for one of the most dangerous on the globe. Notwithstanding of the immense distance (one hundred and eighty miles of Nordland for many), these yachts are by no means constructed in such a manner as to allow them to expose themselves to the open sea without danger. But there are too many places, great fiords, where they must renounce the protection of sheltering islands, and where not unfrequently the heavy laden yacht is driven by the winds and waves on the cliffs. There scarcely passes a year in which some of the yachts of Nordland are not lost in the Foldenfiord or at Statland. Such a loss is easily borne by the merchant but not by the boor; and several villages have been completely ruined by a calamity of the kind. The fishermen or boors, called here Bönderne, * generally freight the yacht, which is provided by the

---

* That fishermen should be here called boors can only appear singular to a German; for the derivation of boor (*Bauer*), from *bauen*, to cultivate the land, proves that boor can only be applied to the inhabitants of the land. But *Bonde*, which is translated boor in all the Danish dictionaries, signifies an inhabitant, and comes from *Boe*, to dwell. *Bönder* signifies nothing more than *dwéller*, and may therefore be equally applied to landsmen and seamen. In the Icelandic it is still nearer the original word *Boanda*. The soldiers under their generals are called in the same manner; for example—*Erlings Skalgsöns armies*—*Heimskringla*, II. 305. olafi hinom helga Saga—Several European languages have corresponding expressions—*Paysan* from *pays*, *villagiano* from *villagio*—*Agricola* however is closely connected with the soil.

retail merchant in their neighbourhood. If the vessel is lost, the boors lose along with it a great part of their property; and if the same accident happens again after an interval of a few years, the whole island remains waste and ruined. The heart sinks, and the poor wretches must begin the world again as if they had newly come into it. A short time ago collections were made throughout the whole country for two communities in Vestvaage, who, through the repeated loss of their yacht in two successive years, were suddenly reduced from a state of prosperity to the greatest poverty. The want of men from these expeditions is equally prejudicial to a province which in the short summer of that part of the world has the greatest need of them. A yacht, with mast and sails *(Raaseil)*, like all those of Nordland, cannot be navigated with fewer than from eight to twelve strong and active men; and more than a thousand men are thus taken from the land, not for the purpose of advancing its produce, but of diminishing it considerably by the expence of the passage. Two months are thus completely lost for them; and this is severely felt. For not only the fishery but agriculture stands in the greatest want of them; and the cultivation of the meadows is essentially necessary to the preservation of the cattle during winter. To think of circulation in these passages is altogether vain. They bring back with them only the most necessary articles, corn and meal for their families; they receive it from the merchant for fish delivered. All the families live near one another in an insulated manner: the desire of thriving can only be excited among a very few, and then can produce no beneficial consequences; for there are too few means of gratifying it when men are not collected together from a general interest, and impelled to their performance of reciprocal services. What is gained beyond the mere necessaries of life is spent in the simplest sensual enjoyments, the consequences of which are as perishable as the enjoyments themselves, and have never here enobled the people. This seems the unavoidable destiny of fishermen.

It seems to have been the beneficent intention of the government by raising Tromsöe to a town to awaken new life in Nordland, to render the ruinous voyages of the fishermen to Bergen unnecessary, and by means of a circulation among the inhabitants to generate activity and prosperity among them. The consequences have not answered the expectations which were formed. Tromsöe probably lies too far from Lofodden, the central point of the fishery. The passage to Barcelona was for the merchants of Tromsöe such another voyage of discovery as that of Gama to Calcutta. In twelve years they have never ventured on it. But what is not to be expected from Hundholm, which is situated in almost the middle of a province carrying on the fishery, and which is in the way of every vessel from Lofodden or farther north? It is also near to Helgeland, and affords an excellent and secure harbour, with entrance and outlet to the south and north, and a very easy access from sea. The brig which we found there had anchored with the assistance of a chart, and without a pilot, in the harbour of Hundholm, a thing which they durst hardly have ventured on at Tromsöe. Besides this, there are men at the head of the undertaking at Hundholm well acquainted with the nature of the trade, who know the markets of Barcelona, Leghorn, and Naples, without the necessity of previous instruction in Bergen; and who, from what they have already done, have shown that they look beyond the moment, and have in view the duration and stability of the undertaking. It was with pleasure we contemplated the ten or twelve small wooden houses which on the plain have the appearance of a sort of street, where the labourers live who are employed in the works. Such was the origin of Carthage, New York, and Boston. Hundholm, it is true, is not yet a town, nor has the government as yet conferred on this place the privileges of a town; but how easily might it be possible, and how desirable would it be, by means of this kindled spark, to transport Bergen six degrees farther northwards. Would Bergen be thereby destroyed? Would it be doomed to the fate of Pisà or Wisby? This could scarcely happen, for Bergen would always be a central point, and the soul of the circulation of very considerable and partly very fruitful provinces; and from its situation it would never cease to be a staple for the north. Hence Hundholm can never attain the heighth of Bergen, even if the six hundred thousand dollars, which the trade of Nordland is at present equivalent to, were all exported from it. But when reasons are alleged against the rise of towns in Nordland, such as the low price of grain in Bergen compared with all the other Norwegian towns, because the corn-ships are sure of disposing of their cargoes there on account of the great competition, it amounts to the same thing as if it was said that Bergen must supply all Norway with all foreign commodities, because these commodities on account of the certainty of disposal must always be cheapest at Bergen. The inhabitants of Nordland would not pay higher for their grain should a town rise up amongst them; but they would be much more secure from famine than formerly, and this consequence alone would be of no small advantage to them.

But Hundholm is yet a great way removed from all this. Every thing in the undertaking is new. The passage from hence to Barcelona has not yet been performed, notwithstanding it has been attempted with several vessels for two years back ; and the fishers have not yet convinced themselves that they may derive here all the advantages of a voyage to Bergen, without the necessity of sacrificing their time, means, and health, in Bergen. With peace at sea, and courage and constancy on land, the Nordlandian town will have existence in the course of a few years.

The country of Hundholm is flat, and the whole of the north side from the outlet of the Saltenfiord is almost mountainless. Towards Bodöe, a good English mile in distance, the road runs over a flat morass, containing perhaps more than ten square English miles. What a pity that this land should be wholly lost! The cattle run about in it, and prevent the grass from coming up by breaking the surface of the ground. This level resembles Oereland, and still more the flat vallies of Luröe; for a stratum of white small vivalves and broken shells lies also every where here under the black peat earth ; and this stratum is also at most not more than thirty feet elevated above the surface of the sea. This phenomenon is certainly deserving of attention, as it seems to be common to the whole of the coast.*

Our arrival at night in Hundholm was mentioned at Bodöe in the morning. M. Kegge, the magistrate *(Amtmann)*, dispatched his active and vigorous sons with coaches for us, which conveyed our ladies rapidly along a good and beaten road over the plain. We had almost forgotten that there were such things as country roads, having so seldom found them. We followed slowly on foot.

At the end of the plain lay Bodöe, on a green declivity, rising gently up from the Fiord, and surrounded by hills covered with birches to the very tops. Between the houses at the bottom there is a beautiful and clear running stream. The large and well-built houses of the magistrate and the clergyman are here situated quite close to one another, and on the opposite side of the stream the church appears from among the trees. The

---

* Beds of shells along with sand occur in different parts of the river district of the Frith of Forth, and in some places many yards above the present level of the sea.—*J.*

other royal functionaries of the province *(Sorenscriver—Judges,* and *Foged— Tax Gatherers)* live at distances of about an English mile; and hence it is possible to procure a select and polished society in every respect in this remote situation in a very short time.   This gives Bodöe a great superiority over every other place in Nordland.   The inhabitants have also, for a great number of years, been distinguished individuals.   The late clergyman, Professor Schytte, was long celebrated and esteemed as a learned man, and a scientific physician.   His acquaintance with natural history was very great, as is demonstrated both by the writings of Bishop Gunnerus and by his own treatises in the collection of the Scientific Society of Drontheim.   His labours as a physician, especially when there was no medical man resident in that country, will long live in the remembrance of the people.   Since that period the government have endeavoured to supply the want of physicians. A great and beautiful hospital has been built at the Fiord, and a physician who inhabits it is paid for by a small contribution from every inhabitant of the province.   But the province is too extensive for one man; for the jurisdiction of Nordland is equal in extent to the kingdom of Portugal.   If a boat is dispatched for the physician from a place two hundred English miles to the south, he is perhaps then two hundred English miles in the north, and for months cannot visit the patient, whose health cannot suffer such a delay.   It was therefore an excellent plan to convey the patients to the physician; but the plan cannot be carried into execution in a country in which men are so slightly connected with one another.   It is impossible to procure for the patient in the hospital the smallest conveniency, or the strictest necessaries of life.   Nobody disposes of them, and no person has any thing of the sort to dispose of.   How much therefore on this account alone is the successful rise of Hundholm to be wished for.   With the certainty of sale, a spring would soon be given to industry, and the plain of Bodöe, which is so capable of a better cultivation would not in vain appeal to the inhabitants. In the present condition of the country, and with the small advantages which can now be derived from agriculture and grazing, it is not surprising that all the boors deem themselves degraded by cultivating the land, and can only be compelled to it through necessity.   The sea offers them dangers, and frequently a great profit; and they every where deem it nobler and more becoming a man to extort from the waves, amidst storms and tempests, by their courage and skill, what can only be derived from the land through patience, diligence, and constancy.

It is true there is always a small corn-field adjoining each Gaard, but nothing but bear is derived from it, and seldom a return of more than three or four for one. This is too little, say well-informed men, and much more might be obtained in these summers with care. For the support of their cattle in winter they rely more on leaves, fish-heads, and bones, grass wrack, and grounds, than hay; for how can they possibly think of hay-making?

Our two days at Bodöe flew rapidly by. It could not be otherwise, with such a hearty and friendly reception, and hosts so obligingly disposed to anticipate every want as our s were. Every thing breathed gladness around us; even the weather was serene and beautiful. The appearance of the country resembled the mountains of Silesia, if the animated prospects of the Fiord and the high islands of the sea had not recalled to us the neighbourhood of the great ocean. The rooms were no longer heated, and spots of snow lay only here and there along the sea shore.

There can be no want of fire-wood here I thought; for all the hills are covered with birches. But too much is consumed in winter, and these birches are besides too small. They compute that they must procure from inland five hundred loads, and each as heavy as a horse can possibly draw; and from the interior of the Fiord, where high trees grow, another five hundred loads. Such a consumption is truly alarming; it must be felt by the woods, and it must even be strongly inimical to the increase of population in these regions. It occurs to me, however, that this enormous consumption of wood would not be required by the severity of the climate if the houses were more suitable to the climate. But it is singular enough that though they are all well-built, large, and convenient, yet every room is pervious to the smallest breath of air, and I never actually saw a house in the whole way to Finmark in which an endeavour was made to preserve the heat, an object of much more importance than the procuring the greatest abundance of heat from the wood. The winter rooms should be a sort of *Sanctum Sanctorum*, to which we arrive through a number of openings and entries, in which the winter cold requires perhaps months to penetrate into the inmost part. The *preservation of heat* is a problem, the solving of which is highly to be impressed on all the inhabitants of Danish Nordland.

The woods have very much decreased in this country. Formerly timber for building was procured from Beyernfiord. Houses were there built, and sent to Wardöhuus and to Finmark. A number of boats were also built there and exported. But this can no longer be done; for the woods will not admit of it. It also partly happens because Spruce and Scotch firs will not here grow up again where they have been cut down. The birches appear in their place; and the pines *(Nadelholz)* will probably not appear again till the birch-woods have been extirpated. This is not because birches are metamorphosed into pines, and pines into birches, but because it is a botanical law that the same sort of timber, nay, even the same plant, never thrives so well where one of its kind has before taken root.

Almost all the hills which surround Bodöe are mica-slate, with great quartz beds contained in it, but without garnets. The gneiss appears to fail here, and first makes its appearance deeper in the interior of the Fiord. There the mountains become also more striking and elevated. But the south side of the bay does not resemble this northern point. Two chains of mountains rise there with steep and rugged declivities, and sharp ridges above, to a heighth of considerably more than three thousand feet. They inclose the Beyernfiord, and the inmost chain is called Saubnessfield, which is the higher of the two, and the highest view which we have from Bodöe, with the exception perhaps of the Island of Landegode.

GRYDÖE, the 18th of June. In the afternoon of the sixteenth we left Bodöe, and re-embarked at Hundholm, to which place we were accompanied by our hosts of Bodöe. Colours were hoisted on the houses and the brig; and as we entered the boat, and took our departure, we were greeted by the discharge of cannon as long as we remained in sight. Amiable and hospitable practice! Wherever we came the flags were hoisted, and the arrival of strangers was made known by the cannon fired far around. We also hoisted our flags in the prow and poop of our boat whenever we passed a considerable house, or intended to land; and we were always answered by flags and discharges. We reached Kierringö about six o'clock in the evening. Here we found quite a different country. The mountains appeared again with their usual character on the northern coasts. Towards Folden-fiord monstrous rocks with frightful precipices appear on both sides, with a sharp ridge at top which seems scarcely sufficient to afford a resting-place for a bird. It resembles the inaccessible rocks of the *Habcheren* valley towards the Brienzer lake. Similar forms are seen towards the Mistenfiord. The stata have precipitated themselves in all directions, and thereby formed towers and pyramids, which up the Foldenfiord we cannot look on without astonishment. At Kierringö itself there are only hills with small level marshy valleys intervening, in which a stratum of shells lies beneath the peat moss. Gneiss makes its appearance at the shore, in which there are numerous small stripes of felspar which run through the strata.

A violent wind carried us rapidly on the seventeenth across the broad Foldenfiord, one of the greatest Fiords in Nordland. It is less dreaded than the Sör Folden in Nummedalen, although it is one of the most dangerous in the Nordland passage. There is a high mountain on the northern side called Brennöfieldt, which rises abruptly from the Fiord. Hurricanes frequently sweep down from it *(Landkast, Kastevind)* and overturn the unprepared boats. We imagined we should pass at a considerable distance from it, but the wind was too powerful for us. The small boat rocked dreadfully over the deep and broken waves. The sea had quite a black appearance. We were obliged to veer about, and enter the Brennesund at the small Island of Brennö under the Field, and wait for more favourable weather, for the wind had increased to a perfect storm.

The island is very small, and almost a rock, but not altogether without interest. The strata determine the external form of these rocks; they stretch E. and W. (h. 6) and dip seventy degrees towards the south. In this direction we every where see fissures through the whole breadth of the island frequently with deep perpendicular walls; and we every where also see long sharp ridges which rise above one another. Some of these fissures from their breadth may be called vallies, and in the broadest of them the only miserable fishing-hut on the island is situated. The top of the whole is two hundred and ninety-eight English feet above the sea. It is fine slatey gneiss of scaley mica, white fine granular felspar, and quartz, which frequently runs parallel in small beds with the slatey structure of the rock. The most prominent ingredient in the mixture, however, are the garnets, which here appear to be essential. They are to be found in every stratum, and in every part of the island; they are not large, and somewhat like lentiles; and they are mostly distinctly crystallized, or of a tessular form. They frequently appear so finely mixed between the quartz and felspar, that they thereby acquire a blood-red appearance, and the red of the gneiss is in general every where striking through these garnets. In the same manner, only with still larger garnets, and less fine slatey the gneiss again makes its appearance for an extent of many miles in Helsingeland at the Bothnian gulph, and nearly in the same manner also at several places above Crevola, on the southern descent of the Simplon.

We saw here a forcible example of the violence of the winds that blow from the high Brennöfieldt. A hurricane had a few days before Christmas carried off the fisherman's granary, and precipitated it into the sea with all the grain which he had raised on this little spot. Since that the poor man had erected a new granary of logs, which were brought from the interior of the Foldenfiord. This was the first building of the wood of Nordland, and not of that of Nummedal, which we had met with in the whole course of our voyage. The logs of Namsen bear even here a double price compared with the price of Drontheim.

We left Brennöe in the night with an agreeable breeze, and in a few hours afterwards passed Huusö, the Archipelago of small islands, in the middle of which lies Grydöe.

These are also low islands, scarcely two hundred feet in the highest point, like fragments from the high masses of rocks which lie nearer to the land. The strata of these higher islands also dip eastwards into the land; but in Grydöe due south, as if they had fallen off from the former towards the other side. The gneiss of these strata is distinct and beautiful, and in this form by no means common in the north. The large cubical blocks at the foot of the rocks render it evident even at a distance that the slatey ingredients have not here their usual ascendancy. The mica is also rare in it, and not scaley, but in small insulated folia in stripes behind one another. On the other hand, the small granular pale-red felspar shines frequently forth, and often appears entirely of a dark-red. We find but little quartz between, and that not distinct. The mica-slate strata do not appear frequent in the gneiss; but are we to refer it to the older and inferior or to the newer gneiss, by which the mica-slate is covered? As to this its composition is not decisive.

LASKESTAD at STEGEN, the 20th of June. With a favourable wind we arrived in two hours from Grydöe to Stegen, which is reckoned a mile's distance. But such are the miles of Nordland: their length increases as the population of the coast diminishes. Of such miles as these there are only seven to the degree of latitude.

We had already seen the island of Stegen at a great distance; for it is high, and the points which we see from the south are very striking. Three of them rise one above another, and the last and highest is conical at top like a volcano. Westhorn, the first of them, rises in steep precipices over the Fiord; then the Hanenkam, which is higher and completely inaccessible, follows; and lastly Prästekonentind towers on the east side far above the others. They are connected together by a wall of rock several hundred feet in heighth, which appears to be the inclosure of a valley in the high mountain, the opposite side of which is sunk in the sea. We landed beneath Westhorn, not far from the church. From thence we were conducted by an agreeable road beneath the rocks, over green meadows, and past five or six peasant's houses to Laskestad, the mansion of the clergyman, about an English mile distant. Several tumuli by the way brought past times to our recollection, and still more an excellent and very slender *Bautastone (Bautastein)* rising upwards like an obelisk, to the heighth of twelve or fourteen feet. Its Runic inscription is at present obliterated by the moss, but parts of the letters may still be made out. Tradition has not preserved the name of the person to whom this monument was erected: nor has the inscription been preserved by any person while it was still possible to read and preserve it.

In the composition of the rocks mica has decidedly the ascendency. It is mica-slate, not merely as a subordinate formation, but every where over the island. Felspar is so concealed here, that we can with difficulty find any traces of it here and there between the scales of mica. The mica however is not continuous in the manner in which it is usually observed in mica-slate, but appears large in scales lying close above one another, of a silver white and black colour, and a splendent lustre. Pretty large blood-red garnets, often like hazel nuts, are every where to be met with in it in great abundance, and still more at the Westhorn than beneath Hanenkam. There also occur small beds of fine granular white quartz with hornblende, and not unfrequently also beds of small and long granular hornblende, mixed with quartz and felspar.   This is a formation which is characterised by the mica, and not by the felspar, and hence it is the mica-slate formation which separates both the gneiss formations on these coasts.   There is nothing which bears a resemblance to the gneiss to be found here, even on the very highest points.   The highest summit of Prästekonentind displays the very same formation, only with fewer garnets.   Several beds of it also appear on the steep declivities, as black as charcoal. which at first view we might be induced to take for hornblende, but which actually consist of black and extremely fine scaley mica ; and yet beds of hornblende are not unfrequent on these heights.   The way from Laskestad to this high point leads through thickets of birch and aspen to the foot of the range of rocks, and then ascends steeply for more than six hundred feet upwards, by a small and dangerous foot-path.   A valley then opens between the rocks of the Hanenkam and the cone of the Prästekone.   Small streams descend from both sides, and in the bottom there are thickets of birches interspersed among green meadows.   This is an alpine valley in the high mountains.   The cows and horses are sent here in summer, and they remain till they are driven out by the snow.   The present clergyman has erected for them a *Shiel,** (Sennhütte)* called *Sätre* in Norway, and *chalet* on Jura, which is the only thing of the kind in Norway.   The small building is pleasantly concealed among the birches.   There the produce of the cows is now conveniently preserved, whereas formerly they were obliged to bring the milk or the cows down the difficult path among the rocks.   The leaves of the trees below were fully expanded and large, but above, on the cone, they were quite small, and a few hundred feet higher the green points could only be seen.   Where the birches terminate the spring was but just commencing : such is the rapid manner in which the climate changes here as we ascend the heights.   The trees below look like woods ; in the Säter valley somewhat thinner and smaller, but on the cone they shrink down more and more, and spread along the declivity like mountain pines.   The last birch bushes are not two feet in heighth, and long before we reach the top, the smallest trace of these bushes is not to be found. The line of their termination does not vary more than one hundred feet.   It would be a beautiful scale of the mean temperature in the upper regions, if

---

* A Scotch word has been here adopted, because it is believed there is no equivalent English one.—*T.*

we were to find this line of separation constantly the same every where, and on all mountains. On this island, at least, it seems extremely determinate; for even on the steep rocks of the Hanenkam, the birch bushes did not disappear deeper or higher, but exactly at the same level as elsewhere. The declivity was however directed towards an opposite side; but the mean temperature of such open declivities is more determined by winds than by the warmth of the sun. This boundary of vegetation was at one thousand three hundred and sixty English feet: in the latitude of 62° we have to ascend before reaching it about four thousand two hundred and sixty English feet; and in the middle of the Alps, at Graubünden and at St. Gotthardt, nearly six thousand four hundred English feet.

The cone of Prästekone still rises one thousand seven hundred and forty-five English feet higher, quite bare, and like a pyramid: the summit is two thousand one hundred and twenty-six English feet above the sea.* From thence we have an extensive view of the ocean, and over all Lofodden, which is lost in the sea like a high snow-covered chain of mountains. All the mountains and islands from Hindöen to Röst are visible. We every where see points and ridges, and forms of high mountains; and the view bears a resemblance to that of the chain of the Stockhorn in winter from Jura. All the islands of Lofodden certainly belong to the highest mountains of the north. When we turn our view also inland eastwards towards the Fiord, we see mountains rising over mountains, and the snow on them down to their bases seems indestructible and eternal. But there are no glaciers to be seen there; whereas on the high mountains of East and West Vaage in Lofodden these are by no means uncommon appearances.

Prästekonentind is a long extended mountain, with steep declivities on both sides, like a high and ancient Gothic roof. On the summit there is but a small foot-path between the slates, and from both sides the view falls

---

* Laskestad,   h. 8. Bar. 27. 11. 3. Therm. 7. 75. somewhat cloudy, still
  Birch boundary   h. 10. Bar. 26. 7. 7. Therm. 4. 5. height 1276. 8. Par. feet
                  according to Trembley—Formul.
  Prästekonentind h. 12. Bar. 25. 11. 2. Therm. 3. 5. height 1998. 8. Par. feet
  Säter   h. 12. Bar. 27. 2. 1. Therm. 7. 5. height 679. 5. Par. feet
  Laskestad   h. 1. Bar. 27. 11. 4. Therm. 8. 25. Sunshine—faint south-west
                                      wind

directly downwards to the plain, which, is two thousand feet below. Towards the north-west there was a large mound of snow lying, but towards the south there was no longer any, in the same manner as at Luröe.   This does not happen because the sun dissolves all on the south side, and not on the north; but rather because in winter all the snow here, as well as every where else, is mostly driven along by south winds, and consequently being allowed to rest on the north side, it collects in greater abundance.   This is directly proved by the constant overhanging snow mounds on all open ridges and points.   The overhanging of such considerable masses could not originate in melting alone; but through the drifting and heaping up occasioned by the prevailing wind.

The strata of these heights run like the ridge of Prästekonentind itself very determinate N. N. E. and S. S. W. (h. 2.) with an inclination of 40° to the south-east.   It was nearly the same below, and exactly the same at the Westhorn, at the opposite end of the island.   Moreover, at the Mehlberg, opposite the Tind, in the north-east, and exactly of the same heighth with it, this stratification is visible and distinct; and hence it is probably universal over the whole of this mountainous island.   As the mountains on the next adjoining islands of Lundland and Hammaröe are nearly of the same heighth, and run in the same direction, it is probable that the stratification is the same there also.   We may therefore almost venture, after so many observations, and going over such a considerable extent, to lay down as the law of the stratification of these coasts—*that the strata of the outermost larger islands, or of the uppermost summits of the main-land, must always dip inwards towards the land, and towards the higher mountains.*   The strata of the smaller and lower islands are only to be considered as fragments of the larger: their direction and inclination are mostly determined through purely local causes of stratification, like blocks that fall from high rocks.

The Prästegieldt of Stegen, which also includes a part of the main-land, and the annexation of Lennes, contains one thousand six hundred and seventy-five souls, according to the enumeration of 1801.   About twelve or fifteen families of this population, in all forty or fifty individuals, are Laplanders or Fins; the rest are Norwegians.   It is unfair to consider the cultivation of these districts as inferior to that of many of the southern provinces of Norway.   If several of the parishes in Nordland possessed the good fortune of having such distinguished clergymen as Stegen enjoys, we might expect

z

that this province would soon be ranked among the most enlightened of the Danish dominions. M. Simon Kildal, a native of Nordland, is a well-informed, thinking, independent, and prudent man. He knows the people, he has studied the peculiarities of their way of thinking, he knows how to manage them, and what is most for their advantage. This is proved by two beautiful little treatises admirably adapted for their end, *Melk for Börn* (Milk for Children), and a collection with fables, printed at Bergen in 1806, which has been distributed among the parishioners, and has produced the most beneficial effects. The school, which is at the house of the clergyman, is by turns frequented by all the children, who remain there for several weeks before they return again to their parents. They there learn to write and read, arithmetic, and something of geography ; and they learn all this better in the short time of their stay, than they would in the whole year through with the school-master, travelling about from *Gaard* to *Gaard*, as is usual throughout Norway. In the school, the children and schoolmaster are more disposed to occupy themselves with what is the occasion of bringing them together; but this is not the case at the Gaard, where the schoolmaster is seldom a welcome visitor. The school of Stegen possesses even a small library for the pea-santry, the increase of which is provided for by yearly contributions from several private individuals of the country. M. Kildal entertains the bene-volent intention of translating his treatises for children into the language of Lapland, for the use of the Laplanders in Finmark, which is much to be wished for the sake of that unfortunate people ; for they are totally deficient in sources from which they may derive any thing like clear and accurate ideas. There are very few, fewer indeed than we should imagine, who like M. Kildal have studied the language of the Laplanders, and who are capable of conversing with the people in their own tongue. But how is it possible to awaken the confidence of a people, when we begin by holding in contempt their language, their most sacred and unalienable property ?

Lödingen at Hindöe, the 22nd of June. The calm and clear nights of this season of the year lighten the passage over the broad and very difficult Westfiord. The sun stood high in the heavens the whole night through, and gave out a gentle heat, while the rays at mid-day were almost inconveniently hot. We rowed for more than five Nordland miles (about fifty English miles), at first under the steep rocks of Skagstadfieldt at Stegen, and then from Hammaröe, and afterwards quite across the Fiord. About nine o'clock in the evening we left Stegen, and reached Lödingen near ten o'clock on the following morning. Scarcely had the sun begun to ascend, when the calm of the night disappeared. For the last mile we were driven along by a fresh south-west wind, which shortly after our arrival almost changed into a storm, and the water of the Fiord, which was before unruffled, now appeared black, from the deep and wildly beating waves.

It is from these rapid changes and agitations that the Westfiord is so dangerous for the coasting navigation. The Fiord presses like a wedge between the main-land and the high and very extensive islands and mountainous range of Lofodden. The tide urges on at the same time, and the general current from the south to the northern coasts. The narrow sounds between the islands do not afford a sufficiently quick passage for this great mass of water; the ebb returns like a cataract, and the smallest opposition to this motion, such as south winds, occasion immediately broken and irregular waves. A stronger wind, which drives before it the deep waves of the sea, sets the whole Fiord in furious commotion. In all the sounds between the islands of Lofodden, the sea flows in as in the strongest and most rapid rivers, and on that account the outermost bear the name of streams, *Grimström, Napström, Sundström;* and wherever the fall of the ebb cannot extend through such long channels, there arises an actual cataract: for instance, the well-known Malström at Mosken and Väröe. These streams and this fall change their direction therefore four times in the day, as the tide or the ebb drives the water on; but Malström is peculiarly dangerous and terrible to look at when the north-west wind blows in opposition to the ebbing. We then see waves struggling against waves, towering aloft, or wheeling about in whirlpools, which draw down the fish and boats that approach them to the bottom of the abyss. We hear the dashing and roaring of the waves for many miles out at sea. But in summer these violent winds do not prevail; and the stream is then little dreaded, and does not prevent the navigation of the inhabitants of Väröe and Moskenöe. The desire to see here something extraordinary and great is therefore generally disappointed; for travellers for the sake of travelling venture up Norway in summer only, and seldom in winter.*

---

* Moskenström is not in general in that fame in the north which from several descriptions we might be led to expect. This is in a great measure owing to the exaggeration of strangers.

We found less spring at Lödingen than at Stegen. The snow had but left the cultivated lands, and the bear had not been more than six days begun to be sown. It is possible that this may be in consequence of the greater quantity of snow which falls the higher we advance into the north. But the temperature also was not very warm. Fifty-nine degrees of Fahrenheit of extreme heat at mid-day is certainly not much for the solstice; and the thermometer also falls at one or two o'clock in the night to 48° or 50°. Yet M. Schytte, the worthy brother of the professor at Bodöe, says he ha, never seen his very good spirit of wine thermometer above 77° of Fahrenheit in summer, nor below 1° in winter. The winters are milder than we might expect from the high latitude; and it frequently happens that in the middle of winter the thermometer for several days together is at 28°, and even at the freezing point. From this, we cannot, it is true, deduct the medium temperature of these regions; for even several years' observations would not sufficiently enable us to draw accurate conclusions. It is only in tropical climates that the observations of a single year will admit of determining from them the mean temperature. More years are required for that purpose, with almost every degree of increase of latitude, and it is not yet determined whether beyond the polar circle we can rely on an experience of ten years. Nature has furnished us with other means for arriving at this knowledge, shorter, more expeditious, and perhaps also more secure, namely the comparison of the limits of the growth of trees and bushes in the mountains. For although the decrease of warmth on the heights may not exactly follow the same laws as in tropical climates, yet these laws probably differ very little among themselves in high latitudes. For such observations Lödingen furnishes the means. On the rocks, and in the small vallies

---

who would have gladly wished to give rise to a belief of the existence here of some peculiar and unknown natural phenomenon. One of the oldest descriptions of the stream is also the most distinct and accurate, as well as the simplest and clearest. It is to be found in *Jonas Ramus Norriges Bescrivelse*, a book which the author wrote at the end of the seventeenth century. *Saltenström*, at the outlet of the Saltenfiord, and but a few miles from Bodöe, is much more dreaded by the inhabitants. Here also ebbings and flowings are compressed between islands: the water turns round in large and powerful whirlpools, and drags down the boats which approach too near to the bottom. The unfortunate fishermen then attempt to cling fast to the boat; and it has happened more than once that the whirlpool has thrown out the boat and fishermen at a considerable distance from the place where they were drawn down. But they frequently never appear more.

below, there are Scotch firs in abundance, and we see distinctly from below the manner in which they gradually disappear on the height, until they are lost in the birches.   Since leaving Helgeland we had not seen this tree so near to the sea-coast, perhaps because our course was too near to the great ocean. It is astonishing how much the immediate sea air from the great ocean is dreaded by the pine tribe.   All the islands on the Norwegian coast are entirely without trees ; and if a few are still to be seen in vallies where the islands are high, yet the Scotch firs are not to be found among them.   In the interior of the country, they do not venture within fifteen or eighteen English miles of the coast, and in many places they do not approach so near.   But a high sheltering island gives the country a capability of growing Scotch firs; and this is the case at Lödingen.   The long and high sheltering range of islands of Lofodden may be considered as a main-land ; and Hindöe is suf-ficiently large and high to shelter Lödingen fully from the north and west winds.   The only immediate sea winds are from the south-west; but they do not appear alone sufficiently powerful to destroy the Scotch fir.

The birches and alders beneath the high rocks of Lödingsaxel were large and strong, and in full leaf; the firs were also high enough, forty feet and upwards, but their growth was bad, being branchy and crooked. They stand too remote from one another.   That their failure on the heights of the rocks is not owing to accidental causes, is evident from their gradually decreasing heighth and increased sickliness. I found the last of them at a heighth of six hundred and fifty-eight English feet above the Gaard.   Forty feet higher up they could not have easily stood.   The limit of Scotch firs is therefore six hundred and ninety-two English feet above Lödingen.   In the valley of the Driv, below Dovrefieldt, this limit was at least two thousand three hundred and forty English feet above the sea ; and in the south of Norway it rises to more than three thousand two hundred feet.   From several results drawn from the quick progression of the decreasing heat in the higher latitudes, it is probable that the mean temperature of the limit of Scotch firs does not rise higher than 31° of Fahrenheit.   Hence we may estimate the temperature of Lodingen at the sea-coast to be 30°. 2.   More it certainly cannot be. This would give 1°. 44 of thermometrical decrease for every degree of latitude from 63° towards the pole ; or if we reckon the temperature of Helgeland at 36°. 5 of Fahrenheit (which is a great deal for the coast), this decrease would be for every degree of latitude from 65° upwards of 0°. 9 of Fahrenheit.   We can hardly expect it to be less. Even the birches on the mountains visibly lost their free and powerful growth in the vallies. Before they reached the summit they were no longer trees, but dwindled down to bushes ; and at the tops of the mountains they scarcely appeared above the ground : the branches crept along the rocks like claspers. It is true, a tree or bush can never thrive on open mountains : this is prevented by the violence of the winds ; but the progression in the decrease of the birch bushes was a sufficient proof that the limit of birches cannot possibly run much beyond these summits. The nearest point above Lödingen, which is distinguished by vessels navigating the Fiord, from a small observatory on it, is one thousand two hundred and fifty-three English feet above the Gaard ; but a higher summit towards the north-west, which is not visible from below, is some hundred feet higher, and reaches to one thousand five hundred and fifty feet above the Gaard.* The highest birch limit would here consequently be about one thousand five hundred and fifty-five English feet above the sea. We cannot err much when we estimate the mean temperature which birches are capable of bearing at 26°. 94 of Fahrenheit. If this estimate is incorrect, it errs more in being too high than too low: for it is scarcely sufficient when we consider that in these climates one degree of Reaumur, or 2°. 25 of Fahrenheit, is assigned to every five hundred and thirty-two English feet (eighty toises) of elevation ; whereas the decrease in tropical climates is 1°. R. or 2°. 25 F. for every one hundred and twenty-five toises.

The highest summit of this quarter lies still farther towards the north-west, above the commencement of the small Ganstafiord ; it may be about one hundred and thirty-seven English feet higher than Lödingsaxel, or one thousand six hundred and eighty-two English feet above the sea. Farther north, towards the Prästegieldt of Trondene up the Tiellesund, the island rises much higher, and beyond the limit of snow. On this account, a considerable glacier descends from the Jisberg into a small valley, which terminates in the Tiellesund, about an English mile from the Prästegaard of Löding. The

---

* h. 11. a. m.	Lödingen	B. 27. 10. 3. Therm.	11. R. South. Clear.		
h. 12.	Scotch fir limit	B. 27. 2. 4——— 9			
h. 1.	Lödingsaxel	26. 7. 5———6			
h. 2.	North-west summit	B. 26. 4. 1——— 6. 25			
h. 7.	Lödingen	27. 9. 6———10			

mountains themselves are not perhaps much beyond three thousand two hundred English feet. The outgoings of the strata of all these rocks are directed towards the Fiord; the strata themselves therefore dip towards the north-west; and the whole island, as far as we can see from the height, is constantly E. N. E. and W. S. W. (h. 4. 4.) 30° north-west. All is gneiss resembling granite. If there was little gneiss in Stegen between the mica-slate, there was equally little mica-slate here in the gneiss: a singular phenomenon in the north. Moreover, this gneiss is not fine slatey, as in the smaller and lower islands, but every where coarse granular, and more streaked than slatey. The mica appears in small, black, oblong scales, which partly lie collected in groups, and partly in short scales (*Flammen Flatschen*). They surround felspar in long parts; not in such beautiful crystals as at Vevelstadsfieldt; but this felspar is generally of a remarkably beautiful flesh-red, as at Baveno. The quartz in the gneiss is distinguished from its peculiar fine granular concretions, nearly as in the upper granite of St. Gotthardt at the Hospital, which is so much characterised by them. Its colour is generally milk-white; and we therefore easily recognise other smoke-grey quartz crystals, which appear imbedded in this milk-white mass. Very small leek-green hexagonal prisms, foliated in the longitudinal fracture, appear by no means seldom between these ingredients, as well as magnetic iron-stone in small grains, and hornblende in small crystals. Towards the summit of these mountains, a few low beds of hornblende also lie, in a fine granular mixture with a little felspar; but upon the whole this is very rare. The gneiss of Lödingen is therefore easily to be recognised: it bears a peculiar character in the distinctness and purity of the continuation of this mixture. We are not therefore entitled to set it down for the impure gneiss formations of no great thickness which lie on the mica-slate, but must account it one of the oldest formations of these latitudes. The range of islands of Lofodden is a peculiar chain of mountains which connect with the inland chain or Kiölen mountains; an arm of the principal chain of mountains, like the chain of Justedal in Söndfiord and in Sogn, like the primitive chains in the Brescian *Val Camonica*, or like those at the source of the Brenta. It is striking enough that the Kiölen mountains change their northern direction exactly where the chain of Lofodden comes in contact with them, and run in the direction of Lofodden for a number of miles on the north side of the Torneaträsk. Even the change in the stratification at

Hindöe leads us to pronounce this chain older than all the islands which we have yet examined. The strata there generally dipped towards the main-land in the direction of the mountains ; but here they dip outwards into the sea towards the north-west.

Towards the Ganstafiord, the birches soon again made their appearance below ; and in the valley they were excellent, and covered with fresh and young leaves. No use however is made of these trees ; it is too difficult to convey the wood for more than two English miles by land. They find it much easier to bring wood in boats from the Tysfiord, a distance of forty or fifty English miles ; for what they bring by land in a number of days from the smaller distance they receive by water in one day. It is every where the same on their coasts. The inhabitants of the islands, even in Helgeland, bring all the wood which they consume in winter from the interior of the Fiord, and freight yachts with it, or float it down in great quantities, which reach the place of their destination over the sea-straits with the greatest difficulty. The small woods in the interior of the large islands are seldom used, even when quite near ; for who would think of withdrawing so much time from fishing? And who possesses a sufficient number of horses to convey the cut timber home?

Lödingen, the 24th of June. " My house is very happily situated," says Provost Schytter. " Whoever ascends to the north must pass it, and it will not be easy to pass without stopping here." Can hospitality be expressed in nobler terms? His whole establishment, his fame all around, and our reception, bear testimony to the sincerity of this language. We always willingly enter a large and well-contrived house where there are but few to be found ; but how much is our gratitude increased when we are additionally bound by the display of so much friendship and cordiality, and when we derive not only comfort and refreshment, but valuable instruction and information from our hosts.

It was *Hansdagsaften*, the eve of St. John's day. The people flocked together on an adjoining hill, to keep up a St. John's fire till midnight, as is done throughout all Germany and Norway. It burnt very well, but it did not render the night a whit more light. The midnight sun shone bright and clear on the fire, and we scarcely could see it. The St. John's fire has not certainly been invented in these regions, for it loses here all the power and nightly splendour which extends over whole territories in Germany. Notwithstanding this circumstance we surrounded the fire in great good humour, and danced in continual circles the whole night through.

I used frequently to think, when I saw the clear night sun, of the advantage enjoyed by the Nordlands: a perpetual gladsome brightness in summer, and the splendid sparkling *northern lights* in winter.   But I learned with no small astonishment that the northern lights are far from being so frequent as we imagine.   They are among the phenomena which excite astonishment by their appearance, like a storm or thunder in southern summers.   They do not appear here in the least more near than at Bergen or in Scotland.   M. Schytte told me that he never heard either here or in Finmark any hissing, or murmuring, or even the smallest noise.   I have frequently afterwards put the same question to several people in Finmark, all the way to the North Cape, and all concurred in assuring me that they knew of nothing but calm or still northern lights, and never heard of any noise accompanying them.   Northern lights appearing only a few degrees above the horizon, in a quiet position, are forerunners of a great calm in M. Schytte's opinion, and on the other hand, high northern lights, emitting a restless, streaming, and sparkling fire, which spreads even over the zenith, are the precursors of a dreadful storm. But these signs are not quite accurate.

We see here almost incessantly small boats pass by, bound from Tiellesund towards Vaage, to bring from thence the dried fish caught in February, or returning with the fish homewards.   This gives at present an unusual and attractive cheerfulness.   But the unfortunate fishermen do not return with cheerful minds : they have been much disappointed in their expectations.   The dreadful fall of snow reached the heighth of the wooden erection *(Gielde)* on which the fish were hung up.   Many of these erections are yet buried in snow, and in those which have made their appearance, the fish had fallen down and were rotten.   The few who were fortunate enough notwithstanding the storm in February, to have an abundant harvest, through this unforeseen misfortune lost their last hope; for with the few fish which they take along with them they will scarcely be able to pay for the food consumed in their voyage.   The few days during which this fall of snow continued have inflicted a wound on the Nordlands which many years will not be sufficient to heal.   It is not the loss of the fish alone which grieves the inhabitants, but also all which they had acquired on the main-land.   They had never before collected so much provender to supply their cattle till July. All the fish heads, fish bones, grounds, grass wrack, rein-deer moss, or birch branches, with which the cows are generally fed, were destroyed; and the cattle could not live on the little food which remained. There is scarcely a countryman, and the higher north the greater the calamity, who has not lost the whole or the greatest part of his cattle. Such a loss is not easily repaired. Even in good years the price of a cow is much higher than in the much more populous districts of the southern provinces. In Drontheim a good cow was to be purchased for twelve or fifteen dollars; but a bad and lean cow here costs more than twenty dollars.

There is an opinion in very general circulation throughout all the north, in Norway as well as Sweden, and on the sea-coasts as well as the inland vallies, that the climate is perceptibly changing. The summers, they say, are not so warm, and the winters are less cold, though more tedious. Formerly no spring was known in the north, and only winter and summer; but now they have a spring in the time when they formerly used to expect summer, which is very hurtful to the cultivation of the land. Such a generally received and powerfully urged opinion ought to excite great attention, and the more so, as they generally support it by visible signs. At Drontheim, in Helgeland, in Senjen, the seed-time in the youth of the old people was generally over from eight to fourteen days before this time. At Drontheim fruit was formerly had; but they have had none there for a long time. In Hardanger they point out on several mountains of the High Folgefonden, small incipient glaciers, of which a few years before there was not the smallest trace to be seen, and which are at present daily increasing. In Sandtorp, at the outlet of the Tiellesund, we were told by the seventy years old possessor, that formerly the points on the main-land were every year without snow; but for a number of years back they have never lost their snow. When a year like the present comes, they see in it a new confirmation of the continual progression in the change of the climate, and there are many who seriously believe that they will live to see the day when the summers will entirely disappear, and when such a season, nearly as March is in Germany, will be all that will remain. Many a natural philosopher would be easily inclined to consider the whole as a general cosmical phenomenon, the causes of which must be sought for beyond the limits of our globe: perhaps in the variation of the intensity of magnetic powers; in more frequent and larger spots in the sun ; in electrical phenomena ; and in other causes of this nature, which have so often been attempted to be placed in immediate casual connexion with general meteorological phenomena, notwithstanding all such attempts have always hitherto been fruitless. It becomes then a matter of consolation for some, and a sort of disenchantment for others, when we can learn from history that all these changes only apparently follow a regular course; that they never leave the same point, and consequently never return to it. Years like the present have often taken place in Nordland, even many centuries ago; and this has not only happened in the higher latitudes, but even in the greatest part of Norway. Under the reign of Harald Graafeld, in the year 960, the snow lay over the whole country till late in the summer: all thoughts of harvest in Helgeland then were vain, and the fishery also failed. The poet, Evind Skaldaspiller, in the middle of summer, on going out of his house found it snowing very hard; and he gave vent to his feelings in the following complaint :——

Snyr à Svölnis varo
Svà höfom inn, sem Finnar
Birki hind of bundit
Brums at midio sumri.

Nive tegitur Othini uxor (terra)
Ita nos intus, veluti Finni
Betulae pecus (omnes) ligavimus
Fronde pastum, in media aestate.
Heimskringla, I. 185.

In the year 1020, from Drontheim, all the way up to Finmark, nothing whatever was reaped; and this unlucky period lasted for three years. Asbiörn Selsbane, a rich and powerful man, who dwelt on the north side of Hindöe in the Prästegieldt of Trondenäs, had continued from the times of Paganism to Christianity the liberal custom of sharing his superfluity thrice every year with his friends; in harvest, at Christmas, and at Easter. These years did not prevent him from following his usual practice; but in the third year all his stores were exhausted. He proceeded down the country southwards, and at last found grain in the fertile Jedderen, where Stavanger is at present situated; but the joy he felt on being able to assist his distressed province was destroyed by the treacherous conduct of the royal functionaries from Karmöe (Heinskr. II. 183.) Helgeland was very much depopulated by this calamity. We find the Helgelanders shortly afterwards engaged in several extensive war-like expeditions, with Thorer Hund to Biarmaland *(Archangel);* and with him and Hareck of Thiottöe, as distinguished confederates at the siege of Stuklestad against King Oluf the Saint. The good seasons returned. It has no doubt never happened in our times that horses went over from Jutland to Norway, or from Denmark to the Pomeranian coast on ice; but such years have never been common; and centuries elapse before they again make their appearance. Even the present change, which is so universally believed in, and can therefore scarcely be doubted, is confined within such narrow boundaries of time, that we have in reality no ground for attributing it to foreign causes, or to any other than what originates in the most common meteorological phenomena. As Peter Kalm travelled along the coast through Bahuslehn in the year 1742, he was told by an old and experienced man that the winters were equally cold with those of his youth, but the summers were much less warm; but they could not then put a plough in the ground before Sophiaday, the fifteenth of May, whereas now they begin to sow in April, and frequently in March (Kalms Bahuslehns Resa, 167.) Notwithstanding the colder summer, the period of sowing had considerably receded; and now, when the summers are also said to be colder, the sowing time has advanced. In Bahuslehn it will never happen at present (1807) before the end of April; but it will hardly extend to the fifteenth of May, as in the seventeenth century. We may evidently learn from this that no continual cause operates to withdraw its temperature from our globe, and that no northern lights communicate a cold to the winters; but that all these changes are mutable as the element in which they take place. It is very possible that in a few years a belief will again be entertained of an opposite progression in the changes of temperature. It is in fact impossible to adduce with certainty a single fact to show that the mean temperature for several years at the same place has diminished even half a degree. Where is the region, since the earth was inhabited by human beings, where spruce or Scotch firs' could formerly grow and cannot now grow? Or where could oaks and birches grow? *Never beyond the region which the temperature has assigned to these trees.* Should these changes be effected by unusual causes independent, of these which otherwise communicate heat and cold to the earth, it remains to be proved that the latter do not now perform what they did in former years. The north winds would on such a supposition be less cold and the south winds less warm; but this is contrary to experience. When the winter is mild in the north, it never assuredly happens with clear weather and north winds. When the spring is dreary and cold it is the consequence of a daily change of north and south winds. The cold summer on the other hand is occasioned by the west and south winds prevailing over this country more than the others, by which the sun is not allowed to warm the earth and to prepare for the harvest a magazine of heat. All these causes not only lie on the earth's surface, but we have not once occasion to seek for them beyond the usual means which the variation of temperature produces in corresponding days; the atmosphere and its movements. Perhaps it is owing to some slight accident that the south and west winds for years have taken their passage over the Scandinavian peninsula, whereas the returning north winds have gone in the direction of other countries. A grain of sand from the summit of a rock rolls down an avalanche of the heighth of a mountain; a single local thunder-storm in southern latitudes, which does not extend over a surface of twenty square leagues, determines perhaps for years the direction of the winds over a whole hemisphere, and consequently the modification of their temperatures.

———————

Lödingen is scarcely twenty-three English miles from Vaage, the central point and chief place of all the fisheries in the north. The number of boats which passed by gave us a faint idea of the multitude of men who assemble there in winter. Yet the fourth part of them do not pass this place, but only a small number of those who dwell to the north of Lofodden. If we reckon up all the boats which assemble at Vaage, we were told their number approaches nearer to four thousand than to three thousand. Each boat is manned by four or five persons; hence the number of fishers who arrive in boats amounts to more than eighteen thousand. This is the fourth of the whole population, and certainly more than the half of all the grown men who inhabit the Nordlands, or the provinces of Helgeland, Salten, Lofodden, Senjen, and Tromsoe,* a length of nearly a hundred

———————

* The population of these five *Fogderier*, according to the enumeration of 1801, is seventy-one thousand two hundred and thirty-seven.

geographical miles. Besides these, there are more than three hundred
vessels, consisting of yachts, from Bergen, from Sundmör, Christiansand,
and Molde, each filled with seven or eight men : so that the multitude of
men assembled in a small circumference at Vaage, in February and March,
amounts to more than twenty thousand. The banks of Newfoundland, in
the fishing season, scarcely give employment to a greater number of individuals.

The source of the wealth of Bergen lies principally here; for the quantity
of fish which Bergen procures from other places is inconsiderable compared
to the immense supplies from the Fiskevär of Vaage. Every boat at an
average during the few weeks of the fishing season catches three thousand
head of fish; many of them catch fewer, but many also catch as many as
seven thousand, and even ten thousand. If we add to this what is taken in
the yachts and larger vessels, we have the sum of nearly sixteen millions
large torsk or tusk, and cod, yearly caught at these islands. This amounts
to nearly six hundred thousand vog (a vog weighing thirty-six pounds), or
a produce of six hundred thousand dollars; for the vog of tusk and cod in
Bergen is on an average estimated at a dollar. That is certainly one of the
most remarkable points on the face of the earth which is capable of affording such immense results.

The number of fishermen is never limited by the abundance of the fish.
The fishery has hitherto remained an open field for every comer, and no
complaints have even yet been heard that the poor have had their usual
supply diminished by those who were more wealthy than themselves;
neither have the inhabitants of Nordland ever asserted that they were
obliged to share their expected supply with new fishermen from the south.
While almost all the other fisheries of Norway gradually lose their reputation, Lofodden has been in the same high estimation for a thousand years;
and there has never yet been an example of the failure of the fishery. Not
long ago, in the time when Ström published his excellent Topography of
Söndmör (1762), Söndmör and Nordmör were almost in equal repute with
Lofodden; and now we see every year fishermen from the Söndmör proceeding
in several yachts for more than five hundred and fifty English miles to Vaage;
for in their own Fiords and fishing banks the fish for sixteen years at least
have never made their appearance. Vaage was, even in the time of Harald
Haarfager, a well-known and distinguished place of resort of fish, and several

grandees of the country settled in these northern districts to avail themselves of the fish. At a later period, under the government of Oluf the Saint (1020), Vaage was the point of assemblage for the inhabitants of Nordland, where almost all the powerful men of the country annually arrived with their retainers; and hence in the old accounts the returning ships are sometimes called *Vaage-Fleet*—(H. K. II. 202. 242), as we now a-day speak of a Bengal or Jamaica fleet.

The peaceable and benevolent King Eystein, brother of Sigurd the crusader, gave orders, about the year 1120, for building a church here *(nordi i Vogom à Halogalandi)*, and a number of fishing huts around, that poor people might assemble here and procure a living; an erection of which he himself boasts to his brother, and which he extols above all the splendid war-like exploits of Sigurd in Greece, and at the head of the Constantino-politan army; for, adds he, these men will tell, even in distant ages, that a King Eystein once lived in Norway (Heimskringla III. 243.) A town was attempted to be built in later times at Vaage (1384), but the attempt failed, probably on account of the difficulty of the first beginnings. What keeps the fish so constantly among these islands, while they are much less constant in their visits to all the other places on the coast? When we consider the singular situation of Lofodden, the long range of islands, which as it were inclose an inland sea, connected with the great ocean, merely by narrow channels between the islands, it appears evident that the most obvious cause of their arrival is repose, and the protection of the sheltering mountains from the storms of the sea. They make their appearance only at spawning time, when this repose is essentially necessary for them. In summer there are no fish here. In the Söndmör it was also well-known that the fish left the banks, which lay thirty or thirty-five English miles out at sea, for the openings between the islands, when they were too much disturbed by the storms (Ström I. 318). But why do they approach the banks towards the land, where the sea is at least sixty, eighty, and sometimes above one hundred fathoms deep? Why do they not spawn in the bottom of the Northern Ocean itself, as this ocean hardly reaches a greater depth than six or eight hundred fathoms? We can assign no other cause than the greater oppression which they suffer in this depth. Or do they seek in spawning time a warmer water at Vaage, for it is possible that the current which ascends from the south up the coast heats the Westfiord considerably in winter; and it was singular enough that in the night when we crossed the Westfiord, the thermometer, above two English miles from land, stood in the salt water at 42°. 68, when, as in the middle of the Fiord, it stood at 46°. 6 of Fahrenheit. The fish enter from the north by Rastsund, between Hindöe and Oest-Vaage, and less frequently by Grimström between Oest and ,West-Vaage. They then repair to three or four banks in particular, where they collect in millions, and where they wait the coming of the fishermen. Kabelvog, the small island of Skraaven, Henningsvär, are celebrated names on this account in the north; Helli, before Hindöe, is less famous. This arrival of the fish takes place with a sort of regularity: the males frequent the deepest places, and the females select places several fathoms higher. When they reach the ground on which they spawn, the males sink to the bottom, and emit their smelts; the females follow, and let their roes fall into the smelts. They remain there no longer, but immediately return to the ocean, either in quest of the herrings, or to the unknown regions of the great deep. The fishery is therefore confined within the limits of a few weeks. The period at which the fish arrive is not known to a day, but it seldom happens before the middle of January, or later than the end of February. All is over with April. In a diffuse and strict edict for the fisheries of Nordland, dated the 1st of February, 1786, there is an attempt to define the limits of the fishing season more accurately, for it commands that lines *(Liner)* shall not be used before the 4th of March, nor nets before the 26th of February. But M. J. P. Rist*, in a small treatise (Copenh. 1801,) remarks, as it appears to me not without reason, that it is impossible to determine with such accuracy the commencement of the fishery. He asserts that it was issued with the intention of favouring the Helgelanders, and others, who come from more distant places in the south, under a dread that those who live nearer might carry off the whole store, and leave nothing for those who should come after them. This was not probably the cause of the edict; for it is clear, that although the inhabitants of Lofodden might be induced to wait the coming of the Helgelanders, the fish if they made their appearance in January would not do so. Besides, what a foolish policy to endeavour to enrich one at the expence of another, under the apparent pretext of admitting a greater number to a share of the profit! This would be completely in the

---

* He was for some time sorenscriver in Finmark. He died recently.

spirit of those who condemn machinery in manufactures because otherwise a greater number of hands would be employed, or who consider causeways a grievance because by means of them the bread is taken from the mouths of the blacksmiths and wheelwrights who live in the neighbourhood of bad roads.   The intention of the royal edict was no doubt to give the fish time to deposit their smelts and their roes to prevent the whole race of them from being destroyed.   It is not probable, however, that the fish who annually arrive always belong to the same families.   They repair in summer far towards the north, and towards winter may as well take the way to the banks of Newfoundland and Cape Breton as the coast of Norway.   Who could be induced to believe that even any considerable part of the immensity of cod which people the ocean is generated at Vaage? Perhaps they wished to let the fish assemble at the Fiskevär, and to prevent them from being scared back by nets and lines.   The edict prohibiting the drawing of nets before the 26th of February is in fact only directed particularly to Raftesund, the way to the Fiskevärs.   But has it been confirmed by experience that so many millions of fish can be possibly driven back? It it true they feel the obstacles which so frequently oppose them.   As all the fishermen cast their nets a few years ago at a uniform depth of from eighty to one hundred fathoms, they were not a little astonished on drawing them to find that they had caught nothing. An old experienced fisherman of Helgeland, who witnessed this, left them, and placed his net a number of fathoms higher up.   In a short time his boat was filled with males; he placed it still higher, and the following days he caught only females; and he derived a great profit from his undertaking. The fish had perceived the net before them, and proceeded higher up; but still they had not deviated from their course.   Would they not rather attempt every opening to their place of spawning before again hastening to the stormy ocean from which they have but just escaped? Does ever the spawning not take place in the rivers even when they are wholly covered with nets and baskets?

Three methods are principally used to catch these fish;—nets, lines (liner), and hand lines.   The most important undoubtedly is the net fishing, and it is at present in general use.   The net is about twenty fathoms in length, with mashes of a few inches in size; the depth is not above seven or eight feet.   From the under border of this grated wall a number of cords hang with stones fastened to them, which sink the net in the sea; the stones

fasten the net to the ground, and the length of the dependent cords determines the depth in the water at which the net is to be kept; for on lengthen-
ing the cords the net rises higher, and on shortening them it sinks nearer to
the bottom.   Similar cords are fastened to the upper part of the net, which
are preserved on the surface, where they come in contact with it by fastening
pieces of light wood to them (*Kavler*), by which means the net is also kept
in a perpendicular direction in the water.   The net is thus placed like a wall
against the course of the fish.   The inland inhabitants of a country would
hardly at first suppose it possible to catch a great number of fish in this
manner without enclosing them as in a sack, and extracting them environed
in such a manner.   The reason of the capture lies in the violence of the tusk
in its course: he runs with all the force of his course among the mashes;
but his bulky body cannot follow the head; the fish then wishes to get
back; but the long and small pectoral fins oppose this like two steel
springs, by which means he remains caught in the mash.   The fisher
generally sets his nets in the evening at the twilight, and draws them in the
morning at day-break; the whole net is then frequently covered with fish,
and the boat is speedily filled.   If the net were higher it would frequently
be unable to sustain the weight of the fish.   Even at present they find every
precaution necessary in the drawing of their nets.   In the water the fish lose
the greatest part of their weight: it appears then an easy thing to draw up
the net; but as soon as the cod comes above the water, he employs his whole
strength against the mashes, and would tear them and effect his escape, if a
second fisherman did not as soon as he touches the surface of the water
drive an iron hook into his belly, and drag him with it into the boat.   This
net fishing is impossible by day; for even when the nets are sixty or eighty
fathoms beneath the surface, they are seen by the fish, who avoid them.   It
has therefore always been the custom, and it is expressly ordered in the
royal edict, that nets shall only be set in the evening, and drawn in the
morning; but this has no great influence at Vaage; for the days in February
under the 68th degree of latitude are not very long.   It is owing to these nets
that the number of fishers in Lofodden is limited.   Since their introduction
(not yet a full half century) they have begun to want room.   Every one
endeavours to set his nets in the most advantageous places; some from
south to north, others away from these from east to west.   Hence arise
numerous disorders and quarrels; and the government was under the neces-

sity, as it appears for the first time since Lofodden was frequented by fishermen, of regulating by a law of police the order of the fishery. Till 1786 it
had never been found necessary to reconcile such considerable numbers of
people assembling there for the same end. Overseers are now appointed, who
point out to every man where he is to set his nets, and what direction he is
to take, that there may be no confusion. But this has not put an end to all
the quarrels: it appears that of late years, from the increasing numbers arriving from the south, that they have rather multiplied. We heard a general
wish expressed at Vaage that the government should dispatch in the fishing
season an officer to Vaage as a general overseer; and the fishers themselves
have requested such a presiding magistrate.

In Ost-Vaage, all the way to the Malstr m and Röst, the compla   are
not so general; and this is easily accounted for, because the fishers do
not frequent that place in such abundance, and are consequently not so
crowded.

The introduction of nets in fishing has very considerably changed the
condition of the inhabitants of the Norwegian coast; for by that means the
quantity of fish taken has been at least doubled. This custom is not old.
Claus Niels Sliningen, a merchant in Borgund in the Söndmör, was the first
who introduced nets, in the year 1685. It immediately excited a general
outcry against him. The advantage was undoubted and clear, and could
never be equalled by the hook. But to procure nets was attended with an
expence which could only be borne by the rich. All the fishermen therefore
stood out against them; and all their acuteness was displayed in pointing
out the prejudicial effects of the net fishery. But the richer people and the
merchants did not regard the outcry, and the use of the nets became daily
more general. The practice became at length almost universal; and the war
for and against the nets was long carried on with an unusual degree of fury. It
came to a law-suit: and after an accurate investigation into all the circumstances, it was decided that nets were not only not prejudicial but also
more useful.* The poorer sort were at length compelled by necessity to
resort to means for the procuring of nets; and in a short time in the Söndmör and the neighbouring provinces they became the usual fishing implements. In later times experience has amply confirmed the wisdom of the

---

* Ström Norske Videnskab. Gelsk. Nye Scrifter, I. 403

sentence of the law.   The Söndmör for the space of full twenty years, from 1740 to 1760, long after the introduction of this mode of fishing, enjoyed an abundance and superfluity of fish which the province neither before nor afterwards ever attained.   One would suppose that the sentence respecting net-fishing would then be generally acquiesced in as just; but we see with astonishment that the same dispute extended itself northwards, from one province to another: the same reasons were always again adduced, and evidence had always again to be taken on the spot, as the provinces did not lie near one another, but scattered in distant parts of the world.   In the year 1762, when Störm published his topography, it was still prohibited on the coast of Drontheim to set nets either in the common fishing places or in the open sea, because by such means the poor fishermen, who were unable to purchase such dear tackle, would suffer very much from it; for it was also asserted that the fish will no longer bite a hook where cod nets are placed.   This is a very particular and incessant sympathy for the poor.   For them, every spur to industry was completely removed, and men of active dispositions were degraded, like machines or cattle, to follow the same track, without improving, and without advancing.

When the poor are unable to procure the means from which the rich derive advantage, they must unite with others.   They ought then to keep up their net in common, and share their capture.   By this means they will be stimulated to profit, to frugality, and will at last be enabled to procure the sole possession of a net; and it will then become evident to them that they also may raise themselves by their endeavours to prosperity.   When they cannot find means to attain by uniting together what they cannot alone do, they must then serve the rich, till they can not only stand on their own bottoms, but get forwards.   They do not by this means lose their freedom; for he is only free who has the means of procuring more than what his strict necessities require, and not he who vegetates from morning to evening, and from evening to morning.   It is better to enjoy limited freedom in service, than none in independence.

Nets made their way very lately up to Nordland.   Bishop Gunnerus, in 1768, relates* that they had been but a very short time in use there, and especially in Rafftesund.   In 1788, when Ström wrote his last treatise on

---

* Leem om Laperne.

net-fishing, the contest was still carried on here with all the violence against its prejudicial influence that took place in the beginning of the century in the Söndmör.   They are now pacified respecting it in Nordland; but at this present time (1807), the question which has been so often decided is now under investigation in Finmark.   The presiding magistrate *(Amtman)* there, who is since dead, deemed it necessary in the year 1806 to prohibit nets from being set in the rich fishing station at Loppen.   They had been only introduced there a few years before: they have not probably yet made their way beyond Wardöehuus; and even the Russians, in other respects the best and most assiduous fishermen of the north, make but little use of them.

The fishing with lines (Liner) is much inferior to the net-fishing.   The net overtakes the fish in his unsuspecting progress, and admits of no choice of avoiding or escaping the evil, but the line is for the purpose of enticing the fish, and leaves it in some degree to his option whether he will give into the snare or not.   Lines can only be used at the bottom of the sea, whereas nets can be set at any depth.   A line consists of three pieces, of which the middle one lies extended along the bottom of the sea, and kept down with stones: the two other pieces ascend from the ends of the other, and indicate at the surface the place where the line is lying in the sea.   This middle piece is generally some hundred fathoms, and often whole miles in length, and at the distance of every second fathom a strong hook is fixed, so that in general several hundred hooks are fastened to one line.   The line is left to remain for a day or a night at the bottom of the sea, and then drawn up with the fish, which have swallowed the hook, and are caught by it in the belly. In this manner several hundred fish may be caught by one line; and hence this mode of fishing is also very general.   It appears however to be less productive in the beginning of the fishery of the Spring Tusk *(Vaar Torsk,* the proper cod)* than towards the end; probably because the fish does not go to such great depths before spawning; but more so after returning from the fishing stations.

---

* The name Torsk or Tusk in Scotland is applied exclusively to the Gadus Brosme, a species completely different from the common cod, the Gadus Morhua.   The common cod has three dorsal fins; the Tusk or Scotch Torsk but one.   In Norway different species of fish appear to be included under the appellation Torsk: thus the Gadus Callarius, the dorse or dursch, is named *Torsk*; the Gadus Barbatus or Whiting Pout of Pennant, *Sma Torsk*; and it would appear from Von Buch, that the Gadus Morhua, the common cod, is denominated *Vaar or Spring Torsk.—J.*

But in general the cod which is penetrated by the hook is in much less estimation than the one entangled in the mashes of the net. Throughout the whole of the north there is but one opinion on this subject; the former, it is said, neither equals that of the net in size or goodness. It is even asserted, that, supposing an equal number of fish those caught in the net are always heavier than the others by at least a half. Why? M. Rist says, with great probability, because the fish which follow the bait at the end of the hook from that very circumstance prove that they are lean and in want of nourishment, whereas those which are fat keep themselves higher, though they cannot escape the mashes of the net. This is certainly no small advantage which the net possesses. There are also times in which it is altogether impossible to catch with lines. The fishing with nets does not however then also fail. This happens in particular when the herring appears in the south of Nordland, or the Lodde in the north;* for the hook then offers nothing to the fish which it cannot better find in the open sea. Both the herring and lodde are the most esteemed baits *(madding)* to fasten upon the hooks; so much so that the fishing edict of 1786 expressly prohibits the use of lodde as a bait in Helgeland, for the very singular reason that but few persons can procure lodde (as the fish is peculiar to more northern latitudes than Helgeland) and that the cod, by getting accustomed to it, will no longer be easily caught with the usual baits. This is another restraint on industry, to please those who would fain have things with the least possible trouble. The lodde however does not drive away the cod.

The sea harbours, however, to the great annoyance of the fisher, a number of other animals, which, as well as the fish, snap at the bait on the hook, and which are sometimes not caught, but at other times remain on the hook to the great increase of the fisher's distress, on account of the failure of his endeavours. The most troublesome of them appears to be a small species of sea-crab called *Aat†*, the general food of the fish, which completely devour the bait from the hook. To prevent this, the line is not suffered to remain

---

* The Lodde of the Norwegians is the Salmo Esperlanus of naturalists, the Smelt, Spirling, or Sparling.—*J.*

† It was the late celebrated Professor Fabricius who first examined scientifically the aat of the Norwegians, and discovered that under that title different species of crustaceous animals were included. They are as follows.

long at a time on the ground, where the aat is supposed to be in great
numbers. It is impossible, however, to prevent even in that case a number of
star-fish *(Korstrold,* Crossdevil*)* from fastening themselves on the hooks and
keeping back the fish, instead of which the fisher draws them up along with
the line. This is another disagreeable consequence not experienced in the
net fishing. It may be a matter of wonder, therefore, that nets have not
every where superseded the use of lines; but the latter are still used on
account of their being less expensive. The nets, too, are exposed to many
risks by which the fishers not unfrequently lose them altogether: the storms
also very often find their way to the bottom where the net lies, upset the
stones, and carry all along with them into the sea; or the *Kavler,* which keep
the upper ends of the net at the surface, get penetrated by the salt water,
when they sink, and draw along with them the net to the bottom. It is very
frequently too torn from the multitude and size of the fish. It often
happens, also, that the capture with lines is more rich and considerable than
with nets, and particularly in summer, when a perpetual day prevents the
setting of the nets, and also towards harvest. The fish then find no longer
any herring or lodde in the upper depths, and are obliged to keep themselves
at the bottom, where they probably swim along the ground, and straight
against the hooks on the lines.

The form of the hook is by no means indifferent, but is, according to the
account of some experienced fishermen, of no inconsiderable influence. It
appears to have undergone no change in Norway for many centuries, and on
that account, the northern hooks are very different from those of the rest of
Europe. They are procured from Bergen, where their manufacture gives
employment to four or five master tradesmen. The size of the hook is pro-
portioned to the nature of the fishery in which it is to be employed. The
largest are perhaps a hand in length, of the same breadth nearly in the
opening of the hook, and nearly of the strength of a rope. These are destined
for the catching of the great plaise,* *(queite, helleflynder)* a fish which is

1. Astacus Homari—the *Hummer Aat* of the Norwegians.
2. Astacus Harengum—the *Silaat* of the Norwegians.
3. Gammarus Esca—the *Aat* of the Norwegians.
These animals are particularly described in Fabricius's travels through Norway.—*J.*

* The Queite of the Danes is the Pleuronectes Hippoglossus of naturalists, the true Holibut.
The Plaise is the Pleuronectes Platessa Lin.—*J.*

frequently as broad as the boat itself.   The form of the hook, however, does not vary notwithstanding of the different sizes.

The shorter arm is always inclined towards the longer, and perpendicular, in an angle of nearly forty-five decades; and the barb at the end of it goes with a considerable point towards the inside of the hook.   In the Russian hook, on the contrary, both arms run down parallel, and are connected with another by a horizontal arm : the shorter does not reach more than the half of the larger, and the barb at the point is less frightful, and less distant from the main arm than in the hooks of Bergen.   The English again are at first sight of a singular construction.   Both arms run along in two perpendicular planes, and are connected together by a small horizontal piece, like the Russian.   But when we look at a section of the hook, so that the longest arm is the farthest from the eye, and the shorter the nearest, the point of the shorter deviates something to the right, so that the angle of this deviation with the perpendicular line may scarcely amount to more than ten degrees. The Norwegian hooks are also all tinned over.   This is not deemed necessary in England and Russia.   Experience has shown that the Russian hooks catch much more than the Norwegian, and the English much more than the Russian.   It may be difficult to assign the true reason of this.   Is it that the fish glide more easily off the Northern hooks, and hold faster on the other ? Or is it that the bait in the latter is more secure from being gnawed off by marine insects ?   Fishermen themselves are not unanimous respecting the causes ; but they do not entertain a doubt of the fact.   On this account the English hooks are in great request in the north, notwithstanding their higher price, although those of Bergen are by far the most generally used.   It is objected to the English that they are more easily broken.   This prevents poor fishermen from purchasing them ; for they generally calculate on what they have to give in addition for the hooks, and not on the superior profit which they are thereby enabled to derive, and which they do not themselves call in question.   Does the tin covering contribute any thing to the greater firmness of the hook, in the same way as a coating of blue oxidised steel increases the elasticity of common steel ?

The fishing with hand lines *(Haand Snöre)* is the simplest, and, compared with the other modes, never very considerable : a hook attached to a single line in the sea.   Many fishermen throw out hand lines while they are rowing, and others where the fish collect together.   When they are in heaps together, they grasp at the first bait which comes in their way, though it should only be a dazzling tin fly.   It would appear that this was the only mode of fishing known in old times ; but that it was varied very much according to circumstances.   But lines and nets have almost wholly exploded the fishing with a single line ; and it is scarcely now attempted to make use of any but the simplest hand-line with a single hook.   In such a case the English is also better than the Bergen hooks.

Were the fishermen allowed to proceed immediately with their capture homewards, or to barter it on the spot, as in Finmark, the fishery of Lofodden would derive an infinite advantage.   They are, however, at present obliged to land for the purpose of drying their fish on wooden erections for that purpose.   Here they are left exposed to the air and the winds for two or three months, and they must then determine on a new voyage to bring away the dried fish, and convey them to the merchant, or freight the yacht of the district with them.   This is a great loss of most valuable time ; for the land is not free like the sea: each inch-breadth is enclosed property, and the fishermen must treat with the proprietors for every place which they occupy for their operations.   The ground-rent of a place for this erection is of no small importance with respect to the profit which the fisher expects from his labours; and notwithstanding the edict of 1786 accurately prescribes the amount of the ground-rent which is to be taken, the fishers are perpetually complaining of the difficulties which the peasants contrive to throw in their way.   It is strictly forbidden to take down the fish from the erections *(Gielden)* before the first of June; for the fish cannot be expected to be fully dried before this period; and a half-dried fish not only becomes itself soon putrified, but gradually sets the whole mass in fermentation in which it happens to be, and thus destroys whole cargoes and magazines. For this reason the preparation of round or cod-fish is permitted till the 14th of April, whereas, after that period, it is only permitted to prepare *Rothskiär*, or *Fläkefish*.   In March the air is dry ; and if little of the moisture of the fish is withdrawn from it, on account of the very low temperature, still, however, there is less danger of rain than on the coast of Bergen, by which the fish regains all the moisture that it lost.   The cod has time to get thoroughly dried ; but not after the rainy season in May commences.   It then becomes necessary to cut up the fish and make a *Rothskiär* of it, for the purpose of increasing the points of contact.   *Klippfish* (salted fish) is seldom prepared in the Nordlands, and perhaps never at Lofodden. Salt is too dear, and too difficult to be procured; and very probably the fishers who come to Lo fodden, would not have time to carry on all the operations requisite in the preparation of salt fish.

It were to be wished, that some King Eystein would again turn his bene-volent thoughts towards Lofodden, and not confine his views to the mere preservation of harmony among these active and courageous men, which the edict of 1786 attempted to do, but also contrive something for their comfort and advantage. The life of such men is an object of importance. Not only are they every moment exposed, in the fishing season, to all the dangers of an ocean perpetually tempestuous, but when they come to land, they can scarcely find a roof to shelter them from the cold and inclemency of a polar winter. The country people admit but few fishermen to live in their houses, and the scanty room in them is their justification. There are slight buildings called *Boder*, erected for the fishermen, which are calculated to afford but small shelter. The men live closely crowded together in these places; and neither find rest, dry clothes, nor heat, after their dangerous voyage. Even the nature of a Nordlander is not always capable of enduring such incessant hardships. The preceding winter gave rise to a destructive distemper, with which the fishermen returned home, and which they spread over the whole coast. These slow fevers have frequently depopulated Nordland, and after an interval of repose for some years, they generally break out again from the fishing stations. These fevers are quite independent of the venereal dis-tempers, which the crews from Bergen bring with them from the capital, and which rage very much at Vaage. It is nearly the general opinion that they originate in the excessive hardships suffered during continual storms: probably from the fishers being constantly soaked with salt water, without having it ever in their power to have their cloathes dried; for it is well known that those who suffer shipwreck, and cling to the wreck, which the salt water incessantly washes and again leaves, generally end their lives in a few hours in fever and delirium.

The Arab and the Persian build caravansaries for those who travel through the deserts; the inhabitant of the Alps founds hospitals on the heights of the mountain-passes, and the Norwegian Fieldstuer on the *Dovre* and *Fille-fieldt*. Why then should not houses also be erected for the roofless multitude of Lofodden? At Bodöe there is a beautiful and large hospital for the patients of Nordland. Would it not be proper and praise-worthy to erect similar houses of assembly for the fishers at Lofodden, which would reduce the numbers who people the hospitals and the church-yards?

LÖDINGEN, the 24th of June. The Prästegieldt of Lodingen runs for a great way upwards towards the Swedish boundaries, and contains the whole of the extensive Tysfiord, which is well-known in Sweden by the name of Titisfiord, and appears under that name in the Swedish maps. In 1801, the inhabitants of this Prästegieldt were found to amount to two thousand two hundred and fifty-seven: in 1769 they amounted only to one thousand nine hundred. Since the last enumeration, however, the population has considerably diminished, and this in a great measure from the distempers originating at the fishing stations. To this parish several hundred Laplanders settled in the interior of the Tysfiord also belong; but the Laplanders who descend from Sweden in summer are looked upon as strangers, and not included in the enumeration of the inhabitants of Lödingen. They belong to the Swedish Pastorates of Gellivara and Jockmock, both in Luleo-Lappmarck. They generally cross the mountains about the 14th of April, and draw downwards towards the water. When they approach the sea-shore, the rein deer run in crowds to the Fiord, and immediately drink the salt water with great eagerness. This is believed by the Laplanders to be necessary for the health of their rein deer; but notwithstanding the pleasure which they find in it, these animals never drink this water more than once. The Lap landers then drive them upwards towards the mountains and upland vallies not inhabited by the Norwegians; and as the summer advances, and the snow melts, they ascend higher and higher up the mountains. By St. Oluf's day, in the middle of August, they again leave these regions, hover for a few weeks on the borders, and bury themselves at last, in harvest, in the woods which surround the church and clergyman's house of the pastorate. They preserve the principal part of their property, however, at their winter station, as winter is in general more convenient to transport themselves and baggage on *Skyer* and *Pulker**, when vallies and hills are levelled, and lakes and marshes are firm. Hence every father of a family generally possesses a small building

---

* Skates or snow shoes, and Lapland sledges.

in the neighbourhood of the church, in which he deposits, during summer his valuable goods and his winter implements. It may be easily conceived, therefore, that they consider their journey over the mountains as a real absence from their home, and consider themselves as only resident where they pass their winter, in the same manner as the citizen who passes four or five months of the year at his country-house near town looks upon himself as then absent, and does not believe himself at home again till he gets to his house in town. Those men who thus cross the mountains are called in Norway Laplanders *(Lappen)*, probably because they are so called in Sweden ; for it appears highly singular to a stranger, that in all Norway no Laplanders are known. The people who receive this appellation from the rest of the world are called Finns *(Finner)* by the Norwegians ; not in one small district alone, but from Röraas (the southernmost point inhabited by Laplanders) up to the North Cape. Moreover, as far back as the most ancient accounts go, this custom has always prevailed ; and the inhabitants of the north side of the Kiölen mountains, from the White Sea down to Drontheim, have never been named Laplanders by any writer of the country, or any foreign writers who have followed them. Should we suppose therefore that the Swedish name is new, and was not used in ancient times ? This is not possible, however ; for Fundinn Noregur, an old Saga, in whom *Schiönning* and *Suhm* place great reliance, relates that Norr on his passage from Finland to Drontheim was obliged to combat the Laplanders to the north of the Bothnian gulph.* If this name was not known in the old poems from which the Saga is believed to be composed, still it was known in the twelfth century, the age of the supposed author. It has not therefore been invented in Sweden ; for in that time the Swedes did not ascend as high as Lapland. Both appellations are unknown to the people themselves. It is certain, however, that if we cannot trace the origin of the custom, it creates a great deal of embarrassment when we speak of the same people, differing among themselves in no respect, under two different names. Two nations thus receive the common name of Finns, which at present, at least, have nothing in common with each other. It is an error, although affirmed by Schiönning,† that in Norway those of the nation who dwell by the sea are only called Finns, and that all those among

---

* Schiönning Forsöck til Norges Gamle Geographie Seite 13.   † Gamle Geographie, Seite 122

the mountains are called Laplanders; for those who live on the mountains of Drontheim, at Röraas, and in Nummedalen, are not called Laplanders, but Finns, although they never come near the sea; and the inhabitants of Kautokeino are metamorphosed, from Swedish Laplanders, which they were formerly, to Norwegian Finns.   All the Finns are Norwegian subjects, and all the Laplanders belong to Sweden.   But where the people are spoken of in general, it is no longer now allowed to call them Finns.   The active and industrious inhabitants of the great principality of Finland, who have an equal right and prescription to use this name, would with reason feel a repugnance to be thrown into the same class with Laplanders.

CASSNESS, the 27th of June.   We were soon driven by the violence of the tide from the hospitable house in Lödingen, through the Tiellesund.   For the first league or two we were opposite great collossuses of snow, which rise from Hindöe towards the north, in form of pyramids hanging together. Hills then advance before these greater masses, as before on the south side. They were covered with alder and birch to the top, and the Gaards at the foot follow close upon one another.   The country assumes a lively appearance, and may well, in comparison of the foregoing, be called beautiful; an appellation which it receives from all the Nordlanders who sail through the Tiellesund.   The mountains of Hindöe run towards Trondenäs, and dip there deeply and perpendicularly into the sea.   In the nearest to Lödingen, notwithstanding the snow, the stratification is easily recognizable: at the *Korring-Tind* the nearest it is N. N. E. and S. S. W. (h. 2.) with an inclination of thirty degrees to the north-west; and the same at the *Fisketind* from which the glacier descends, and the *Lia* and *Säter-Tind* which follow it, and are all named from the Gaards which lie at their feet.   The stratification is thus always the same as at Lödingen itself.   In the evening we were driven by wind, current, and rain, towards Sandtorv, a sort of peninsula, on which there are only low hills, with some plains, which may be reckoned considerable for that country, and which were altogether covered with trees. The mica-slate again makes its appearance there, not only at the sea-side, but on all the hills up the country.   Middle sized garnets lie in great multitudes between the mica-folia of the formation; and preponderating layers of fine granular quartz alternate with the slaty mica.   There is no trace of felspar visible. These strata stretch N. N. W. and S. S. E. (h. 10.) and dip thirty degrees towards the south-west, contrary to the dip of the strata of gneiss. Has the mica-slate here thrust itself between the gneiss rocks of the main-land and Hindöe, and in this manner raised a part of Tiellesund in hills above the level of the sea?

Sandtorv lies in the province *(Amt)* of Finmark. Salten in the south, and Senjen in the north, separate not far from the Gaard; and at this boundary the province of Nordland terminates. We have not therefore yet reached Finmark. Both districts, Senjen and Tromsöe, were in the year 1787 separated from Nordland and annexed to Finmark. As their political condition was only altered in so far that their presiding magistrate was to reside in Finmark instead of Salten as formerly, and as they received no share of the immunities enjoyed by Finmark, it has always been customary in common language to consider Senjen and Tromsöe as part of Nordland. There is also something of vanity in this. Nordland is believed to be a better and more advantageous country to live in than Finmark, and distant from the Finns, for whom they entertain so much contempt. Very few Finns actually inhabit these districts; they keep to the interior of the Fiord, and are nowhere to be seen on the islands.

From Sandtorv we still continued to avail ourselves of the tide in our way upwards till we came opposite to the flat and small woody island of Rogla, where the tide enters the great Astafiord, and spreads itself over a great extent. We we were now driven by a gentle south wind towards the high rocks of Rollenö, and for about three leagues we proceeded along the west side of the islands under the precipices. The rocks on this side sink so abruptly into the sea, that for their whole length there are but few points where a boat can land, and there are hardly any habitations here on that account. Water-falls descend from the heights as in the vallies of the Alps. The outgoings of the strata are on this side, and their inclination enters the land towards the south-east. On that side also there are larger and level spots, Gaards and cultivated places, and the church, and clergyman's house of Ibbestad cr Astafiord, also lie there. The mountains of the small island may be at least three thousand two hundred English feet in heighth. And-orgöe is perhaps still higher. After we had proceeded round the island, and had anchored about two English miles from its northernmost point at Cassness, the summit of the mountains on the east side of Andorgöe appeared not only covered with snow but with actual masses of ice, small incipient glaciers.

A single island of small extent in this intersected country reaches a greater heighth than most of the considerable mountains in Sweden.

Cassness has in reality a very pleasant situation. The mountains every where covered with birches present the most picturesque forms; and the ascent from the Gaard up the mountains is green and cheerful. A considerable stream rushes with violence out of the rocky cliffs, and runs close by the houses into the sea; and the high and lofty form of the Faxefield rises above the place like one of the *Aiguilles* of Chamouny. It is almost perpendicular from the base to the summit, and the trees in the fissures terminate at about the third of the height. It is an immense mountain, at least upwards of four thousand two hundred and sixty English feet in heighth. It is seen far out at sea above the other islands, and serves as a landmark (Landemärke) for many miles around. It is not connected with other chains, but stands insulated between the Fiords, and on that account there are no glaciers perhaps on its perpendicular declivities. The snow is perpetually collecting on its summit and sides; for this mountain belongs to the highest beyond the Polar Circle, and its equal is scarcely to be found farther north, even in the interior of the country.

Fishing and agriculture do not at all coalesce. The weather was fine, the snow had left the fields; but there were no hands to cultivate them. At the very time when all hands are wanted for agriculture the sea calls them to the holibut *(queite)*, and the *Sey*-fishery, and the coasts are as desolate as in February. The *Queite* or holibut *(Schollen, Helleflynder)*, and *Ling (Gradus Molva)* collect together several thousand men in the neighbourhood of the small island of Hovden at Langö to the West of Hundöe; and the *Sey** (Gadus-virens)* carries out the fishermen for more than five leagues into the open sea. Hence, what is to be done on land is left to those who have neither

---

* It would appear that the name sey is applied to two different species of gadus. In Norway, according to Von Buch, the gadus virens is named sey, but in Shetland the same name is applied to the young of the gadus carbonarius, the coal fish. In Norway the gadus virens when but one year old is named *mort*: it is employed as an article of food for cattle, and the liver affords a considerable quantity of oil, three hundred weight of livers affording two hundred weight of oil. When two years old this species is named *Drotte-mort;* when three years old *Middel-sey;* when four or five years old *Holufs-sey;* when six years old *Half-sey-ufs;* and when seven years old and upwards *Sey-ufs.* This fish when two years old and upwards is dried, and exorted in great quantities to the different provinces of Norway, and also into Sweden.—*J.*

spirit nor strength to try their fortune at sea. Nature however has not con-
demned this country to remain for ever uncultivated like the moss-fields
of Spitzbergen and Greenland. Where birches and aspens thrive so well,
and attain so beautiful a growth, the diligence of man can never be thrown
away. We have been assured that grain generally ripens well here, and does
not perish by the frost, but that it does not yield at most more than four-fold;
but this is a great deal here in ground which may be almost said to be left
to itself. Potatoes do not succeed: they are too small; but that is ascribed
more to the want of soil than the climate, and this even might be got the
better of. They pay somewhat more attention to grazing, as it requires
fewer hands; but it still requires too many if they wish to get in as much
provender as is sufficient for the cattle during the eight months which they
must pass under cover. This Gaard keeps thirty cows, and several hundred
goats and sheep. The bears are a great obstacle to their operations. They
commit numerous devastations, and easily fall upon the sheep in the woods.
They have even swam more than two English miles across the Sound to the
beautiful and green island of Dyröe, which appeared in sight. Here they
have lived, and multiplied for more than six years, and commit great havoc
in the neighbourhood of the Gaard. They know well enough that their
number does not exceed four or five; and the island is not five English
miles in circumference; but they cannot prevail on themselves to go in
pursuit of these bears on land. If they had fins they would have all long
ago been destroyed.

The fine slatey gneiss is the prevailing rock in these hills; not the gneiss
of Lödingen, but probably the formation of Saltensfiord and the lower
islands. The mica of the gneiss is not continuous, but always scaley; and
yet the folia lie so close on one another, that they form a plane without any
interruption in the stratum, and not detached particles (Flatschen or Flammen).
Small granular grey felspar lies between the mica, and but very little quartz.
Every where, however, there is a great multitude of red garnets, like pease
and cherries, and even frequently as large as wallnuts. These strata stretch
N. N. W. and S. S. E. (h. 11.) and dip from 30° to 40° towards the East.
Small granite veins traverse them very frequently; and the granite is composed of yellowish, white, coarse granular felspar, a little silver-white mica
and grey quartz. It is worthy of remark, that whenever we see granite form

itself, the felspar immediately increases: the mica disappears when any repose is allowed to the gneiss formation, as in vein fissures. By this means the great truth, to which all geological phenomena lead us, is ever more and more supported, that all the diversities of formations arise only through *external movements*, which modify the internal powers of attraction, and which, at last, when they have reached the highest conflict against each other, give thereby rise to the *vivifying power*. This diversity of formation is not however a consequence of polarities, of separations from opposition among powers; for in the remotest members of the series of formations, in conglomerates, and in sand-stones, these materials, and consequently the internal powers also, are entirely *passive*, and almost in no respects longer active: they are driven together merely through external movements, the cause of which, *necessarily wise*, has nothing at all in common with that which gave rise to granite and gneiss mountains.

At the shore of Cassness we are completely reminded of the rock of St. Gotthardt. Large blocks of gneiss lie every where also here with frequent and large garnets; but there are many blocks, however, of a greenish grey *continuous* shining mica, on which there are beautiful, large scopiform hornblende crystals, as at *Airolo;* and not unfrequently single prisms of hornblende are also found in the mass. Farther on, blocks of small and fine granular marble make their appearance, white, and almost semi-transparent. The beds of these stones are soon found to the south of the Gaard, and not more than half an English mile distant from it. There the white lime-stone rises immediately out of the water; the waves sprinkle it, wash deep cavities in the bed, and carry the soft stone along with them. Several blue stripes in it are firmer, and rise prominant above the remainder. The bed may be four or five feet in thickness, and is to be found all the way along the shore. Perhaps there may be more of them. The strata dip exactly like the gneiss strata towards the *south-east;* and form a connected whole with them. Hence it follows that all the gneiss of the hills at Cassness rests on this mica-slate and lime-stone beds. The mica-slate above the lime-stone contains a number of cavities, which are lined all round with beautiful druses: principally long *epidote* crystals, and next garnet, hornblende crystals, and felspar. If we could follow the marble bed farther we should certainly not miss the tremolite above it.

KLÖWEN at SENJEN, the 29th of June. We proceeded in the night
through the Sound between green bushy hills, with Senjen on the one
hand, and the main-land on the other, and about six o'clock in the morning
we anchored at the beautiful trading station of *Klöwen.* Senjen is here
rocky enough, but not high. I ascended the nearest hills over great masses
of snow, in which they lay almost entirely enveloped. These hills were only six
hundred and ninety-two English feet in heighth*; and yet they were almost
the highest in a wide circumference. The northern part of this large island
differs very much in this respect from the southern. In the south there are
only heights, and no distinguished points; but towards the norththere are
true Alpine *Horns.* The small *Fieldts* round Klöwen are sufficiently steep,
and are covered along the declivities with woods of birch and Scotch fir.
The firs become all visibly feeble towards the height and at the top they
do not exceed the heighth of from ten to fifteen feet; the branches are
drooping, and the tops are bare. They seem very near to their limit, and
we can scarcely place it at much above six hundred and thirty feet; somewhat less in truth than at Lödingen, but still not so much as to render the
influence on vegetation in the vallies very remarkable.

The formations of these heights, and their stratification, coincide
almost entirely with those at Cassness; above there is fine slatey gneiss,
interspersed with garnets, with isolated scales of mica, fine granular felspar,
and fine granular quartz; but below, at the sea, and in the neighbourhood of
the Gaard, the *mica* is continuous, the felspar almost entirely fails, and the
quartz no longer appears common, but garnets more so. This is again *mica-
slate,* and the gneiss rests above it. A fine granular marble, with blue stripes,
and several feet in thickness, also presses through between the strata as at
Cassness. The strata stretch N. N. W. and S. S. E. (h. 11.) and dip very
little towards the west. Southwards, at the shore, the lime-stone bed
becomes thicker; and in one precipice, which advances into the sea, the
white bed attains a thickness of ten feet. Above the bed lies a bed of ex-

---

* h. 6. a. m. Klöwen, Bar. 27. 11. 4. North wind, clouds at 2500 feet heighth.
  h. 10. a. m. Klöwensfieldt, Bar. 27. 4. 3. 6. 25. North wind, 560 feet.
  h. 12.   Klöwen - - - - - - - 8. 25.   gentle N. clouds
  h. 4.   Klöwen - - - 27. 11. 4. 8. Clouds at a heighth of 3000 feet

cellent divergent fibrous *tremolite* of an inch in thickness. It is seen upon
the surface of the numerous blocks which have tumbled down the precipice
into the sea, and it is beautiful to see the manner in which we can here
follow the large stars, and fasciculi, and the manner in which they are
linked together, and follow one another. Immediately above the tremolite
there is a very firm and dark stratum, which for the most part is nothing but
massive garnet, with but little mica, and no felspar or quartz. This stone has
a strong effect on the magnetic needle; not only attracting, but also possessing
polarity. The north-pole of the needle sometimes stands towards the east,
and then veers round to the south, or it remains closely attached to the
bottom of the compass; the south pole follows it with similar movements.
It would be here an endless and very useless labour to determine the position of all the poles; for they appear to be changed by every fissure which
traverses the garnet bed. Strata of mica-slate repose above, as at the Gaard,
with small garnet crystals in them.

The Gysund between Senjen and the main-land is the only passage which
connects Tromsöe and Finmark with the southern regions; for the sailing
round Senjen on the sea side would be very tedious, useless, and attended
with danger. Hence this sound is always very cheerful as well as Klöwen,
near to which all ships and boats must pass. In winter, it is said, that about
three hundred boats pass this place on their way to Lofodden. These may
contain nearly fourteen or fifteen hundred men, and this affords a scale for
ascertaining the numbers who frequent the fishing station of Lofodden from
the north.

LENVIG, the 30th of June. The Sound on both sides is only surrounded
by hills; and Gräsholm, a round prominent cape, is even full of plains, and
thickly covered with birches and alders. At the foot of these hills, where
the Sound is narrowest, lies Gebostad, a fishing station, and also an inn.
Here about five hundred rein deer annually swim over from the main-land
to feed in summer on the Alps of Senjen. Senjen cannot support them the
whole winter through, and on that account the Finns transport them towards
winter to Sweden. They are but poor and miserable beings; the rein deer
scarcely suffice to support them, and defend them from the cravings of hunger.
And yet, when their whole stock consists merely of a few skins or rein deer
horns, and cheese, they go down immediately to the public-house, and consume all their property in brandy on the spot. If we wish to be informed re-

specting the manners, customs, and dispositions of these people, we must not, therefore, take those we find in the public-houses, and among the merchants, as a specimen; for in such a case we might be easily led to believe the sentence pronounced by the Norwegians respecting the Finns not altogether unfounded. " *They are the scum of the human race,*" said a grave Norwegian one day to us, as three Finns came reeling towards us at Gebostad. No, indeed, that they are not; but they are children whose ideas never extend far beyond their rein deer, and whose pleasures are limited to the merest enjoyments of the moment. The contempt with which they are treated by the Norwegians is inconceivable; they will hardly be prevailed on to allow them to set a foot in their houses; and they even endeavour to shun the remotest connexion with them. " *I care no more for him than a Finn,*" was even in Helgeland an expression of the most sovereign contempt; and we have frequently heard it said that " *a Finn is not worth more than a dog.*" Adjoining nations are always jealous of, and inimical to one another. The Norwegian boasts of his advantage over the Swede, and the latter again believes himself superior to the Norwegian, and greatly superior to the Russian. The Poles, Russians, and Germans, entertain similar opinions respecting each other. Every nation reckons itself highly favoured in comparison with others; but their contempt seldom goes so far as to induce them to believe that humanity is only to be found amongst them, and that all the rest of the world are only to be considered as a sort of inferior creatures; for these nations have frequently been at war with one another, and frequently been conquered and conquerors. The *Laplanders,* however, have never attempted to oppose the attacks of the Norwegians. They have never been successful in the smallest attack; and seldom even is the least trace of opposition to be found among that peaceful people. Hence the great reluctance which the Norwegians have to reckon them as men: and in fact, if any one would take the trouble to prove to them that the Laplanders never were men, they would willingly believe him. Unfortunate the people subject to such masters.*

---

* Linnæus agrees with our author, in stating the Laplanders to be fond of spirits. " Lappones et rustici nonnulli septentrionales, sœpius nimiam spiritus frumenti copiam ingerunt," (Flor. Lap. p. 165. London, 1792); but it is curious to contrast the different estimate formed by the young Swedish philosopher of their condition. " O felix Lappo, qui in ultimo angulo mundi

In four hours we crossed the Sound from Gebostad to Lenvig, where we were received in a friendly manner by the clergyman M. Heyberg. The view of the mountains of Senjen became always more and more grand and sublime ; and when we reached Lenvig, the view of the horns of Medfiord and Oyfiord, opposite the Sound, exhibited the appearance of high Alpine vallies on projecting horns, like the Alpine vallies of *Lugnetzer* on the points of the *Hinter rhein.* These are not the mountains which surround Lenvig in its vicinity. The mountains of Senjen in question rise far beyond the limits of snow, whereas the hills of Lenvig scarcely leave the region of trees beneath them ; but the former are not many hundred feet higher, and scarcely sixteen or eighteen hundred feet in heighth. The mica-slate predominates here also along the shore : the mica did not appear altogether continuous, but fine scaley foliated ; but there was no felspar between, and no garnets. Several beds of very white fine granular, almost friable dolomite, lie in it ; the very rock of Campo longo : and above these there was scopiform fibrous tremolite, with green mica resembling talc. In the limestone there are great streaks *(Flammen)* of thirty feet in length, and one in thickness, of nothing but irregular promiscuous crystals of tremolite. In the drusey cavities there appeared prisms of epidote, and not unfrequently red metallic octahedrons.* The strata stretch (N. N. W. and

sic bene lates contentus et innocens. Tu nec times annonæ caritatem, nec Martis prœlia, quæ ad tuas oras pervenire nequeunt, sed florentissimas Europæ provincias et urbes unico momento sæpe dejiciunt delent. Tu dormis hic sub tua pelle ab omnibus curis content ionibus rixis, liber, ignorans quid sit invidia. Tu nulla nosti nisi tonantis Jovis fulmina. Tu ducis innocentissimos tuos annos ultra centenarium numerum cum facili senectute et summa sanitate. Te latent myrriades morborum nobis Europæis communes. Tu vivis in sylvis avis instar, nec sementem facis, nec metis, tamen alit te Deus optimus optime. Tua ornamenta sunt tremula arborum folia, graminosi que luci. Tua potus aqua crystallinæ pelluciditatis, quæ nec cerebrum insania adficit nec strumas in Alpibus tuis producit....Te non obruit scorbutus, nec febris intermittens, nec obesitas, nec podagra, fibroso gaudes corpore et alacri, animo que libero. O sancta innocentia, estne hic tuus thronus inter Faunos in summo Septentrione, in que vilissima habita terra? (p. 277.)" If this view is not philosophical, it is at least highly poetical. But have not the improvements of society served to multiply our enjoyments, and consequently advance us in the scale of happiness? And yet how difficult it is to prevent the most civilized European nations in the neighbourhood of the American savages from *squatting* down among them !—*T.*

* What mineral is to be understood here ?—*J.*

S. S. E. (h. 10.) and dip seventy degrees towards the east. It is the same all the way up the sound which surrounds Senjen, and even before we reach it the gneiss on the coasts appears only on the heights, and the mica-slate seems to be almost predominant in extent. In the stratification of both formations there is in fact something definite ; and their order is not dependent on accidents which do not fall within any general rule. The mica-slate of Senjen is particularly characterised by the frequent beds of lime-stone, and the layers of tremolite in them. In the mica-slate, on the coast of Bergen, we find also beds of lime-stone ; but they are small, and upon the whole not frequent, and tremolite is there altogether unknown.

The Prästegieldt of Lenvig is very extensive. On one side it stretches over a great part of Senjen, and runs out as far as the sea between Senjen and Hvalöe ; for Hellesöe, one of the outermost islands, is an annexation to Lenvig. On the other side, its boundaries reach inland along the Fiord the whole way to Sweden. Yet in 1801 it was not inhabited by more than one thousand five hundred and fifty souls ; of whom five families were Laplanders.

This population has not been increased since on the coast, and M. Heyberg even fears that during the present year it will be lowered. The people have brought from Lofodden, a slow, infectious disease, occasioned by the unfavourable weather, which has brought a number of men to the grave. Add to this, that from the length of the winter, most of the cattle at the Gaards have perished from hunger, and the men follow them through poverty and want. The snow has not even yet left the arable land, and the meadows, and in the midst of summer they are obliged to look out for provender to feed their cattle under the roof. It is true, most of the snow fell in the Fogderies of Senjen and Tromsöe, and it lasted incessantly from Christmas to April. It was affirmed in Gebostad, that the snow lay to the depth of ten ells (twenty feet) ; and in Lenvig, at most, only twelve feet. But even that is an extreme ; for more does not fall in the highest Norwegian vallies ; and on the coast of Bergen there has never been seen more than four feet of snow, even in the interior of the Fiord.*

---

* As the well-informed minister, Neils Herzberg, informed me, who was born on that coast, and who has long made meteorological phenomena his study.

If the population and the prosperity of the coast be diminished by such unforeseen, and, fortunately, passing calamities, on the other hand, better, and we may almost venture to say, splendid prospects begin to open in the interior of the country; for here, and only here, in these distant regions, under the sixty-ninth degree of latitude, have the foundation of new colonies been attended with success.   This has been effected by the zeal and the perseverance of one man.   The Foged Holmboe in Tromsoe had long been in the possession of a well-merited fame in the Nordlands, for his economical knowledge, and its successful application in different parts of that province. It did not escape him, that it might be very possible to turn the immense woods in the interior of the extensive Malangerfiord to account; and that not only the trees for many centuries have come in vain there to maturity, but also the other productions of the vegetable kingdom.   His plan for cutting deals and logs from the woods for exportation, as in the southern part of the country, did not succeed, notwithstanding he was supported by the powerful Chamberlain Berndt Ancker in Christiania; for it was reasonably dreaded that Nordland could not support such an exportation, which was also prohibited on that account by old edicts.   This active man was, however, much more successful in bringing land under the plough.   In the year 1796 several families from the south actually made their appearance, for the most part, from Guldbrandsdalen.   Even the Nordlanders, fortunately for the new undertaking, refused to leave the coast.   The strangers were only acquainted with agriculture.   The sea life could not so easily seduce them ; for they would, like children, have been obliged to learn the first elements, and even the courage, to brave the Nordland waves.   They were conveyed into the great valley, through which the great Monsenelv tumbles along into the Fiord, not above fifteen English miles from Lenvig, and deep in the interior of the broad Malangerfiord.   The colonists soon found here that their hopes had not been excited in vain.   They built Gaards along the river, they cleared the woods, and the culture of grain succeeded admirably.   They were then joined by more of their countrymen.   They ascended farther up the country, twenty English miles from the former settlement, to a broad and level valley watered by the Bardonelv; and there also their pains were rewarded with success.   In the year 1800, only five years after the first settlement, thirty families, consisting of one hundred and ninety-six individuals, lived in these wastes, where formerly scarcely a Laplander set foot.   They maintained three hundred and thirteen head of oxen and cows, five hundred and eleven sheep and goats, and thirty-nine horses.*   In the year 1807 there were thirty families on the banks of the Monsenelv, and sixteen families in the valley of Bardou. The grain had never yet failed, nor been destroyed by frost since the period of their settlement.   Hitherto they have stood in need of no foreign assistance for their support.   Their diligence has not confined itself to the dwelling-places at first taken in by them: new situations are annually brought under the plough, and new Gaards erected, even up the mountains, to the borders of the kingdom. At present, they are occupied in changing a *Rydningsplads* (a cleared spot) into a Gaard at the Rosto Jaure. Rosto Jaure sends its waters westwards, to the Northern Ocean, and eastwards towards Sweden and to the Baltic, and consequently the new habitations lie in fact on the greatest height between both kingdoms and both seas. The manners of these men have improved in this situation, and they are very justly esteemed the best in the whole parish. They are not corrupted by the brandy which is so inimical to every thing good on the coast, and opposes so strongly the improvement and prosperity of the inhabitants. At the beginning of winter, when their field and forest labours are ended, they find sufficient employment in their houses, in converting the wool of their sheep and the hides of their cows into articles of cloathing : they are not stimulated by the uncertain profits of sea-fishing, and they thence escape those prejudicial consequences which are but too much felt by the inhabitants of the coast of Nordland.   Their enthusiasm also retains them in this confined and insulated life ; for like all men who live in retired situations, they are particularly disposed to the reception of high flown religious ideas ; and the enthusiast Hans Niels Houg of Tunöe at Frederickstadt never found more numerous and fervent adherents than here, when he visited the colony in the year 1800. He unites a pietistical Moravian doctrine of the immediate operation of God in human affairs with exortations to domestic industry, and a secluded life in a family circle ; a doctrine which here, at least, has improved the condition of men.   An intelligent preacher knows both how to make the proper impression and to turn it to utility. It is a pity, however, that this people has not received a proper and well-qualified preacher of their own ; for they stand very much in want of such a source of comfort, exortation, and instruction among them.   Their present preachers live at too great a distance from them; they are too little known to one another, and therefore cannot altogether feel mutual confidence and trust.   Such individuals ought to be kept altogether separate from the inhabitants of the coast. If there was a man among them like Simon Kildal, or M. Normann in Transöe, the importance and influence of this colony would perhaps be felt throughout all Norway.   Even at present, experience shews how little the usual regulations respecting Prästegieldts can be here applied, for the inhabitants of Monsenelv only frequent the church of Lenvig.   Those of Bardonjord, on the other hand, notwithstanding they are also within the bounds of Lenvig, go the more easy and commodious road towards Salangfiord, and still farther towards Astafiord, to the church of Ibbestad.

Foged Holmboe is dead, and even before he died, the government saw themselves under the necessity of renouncing his services; but Bardonjord will remain an enviable monument to his memory, which neither the treasury assets nor time will easily destroy.

TROMSÖE, the 2nd of July.   Senjen, almost the greatest island of all those which lie on the northern coast, terminates not far from opposite Lenvig. We had scarcely proceeded round the Gaard of Vang on Senjen, when we saw the open sea before us, through the sound which separates Hvalöe from Senjen.   This sound receives, singularly enough, from the fishermen here, the appellation of *Vangs Hafsöie* (Vangs *view into the sea)*, and we could not learn here that it had any other name.   A light wind from the north-west soon drove us through the straits, and along the coast of Hvalöe to beyond Malangerfiord.   We might expect that the high and rocky Alps of Senjen should be continued along Hvalöe; but this is not the case, at least, on the southern side of the island.   The mountains are all round in section, like large cupolas, without rocks, and of long extent, and we should, perhaps, be induced to account them low, if we did not see the trees disappear before reaching their tops, or even the half of their heighth.   This would give these mountains a heighth of about two thousand feet.   But what rocks and masses on the east side of the Malangerfiord !   We imagined we were again in view of the gigantic shapes of Kunnen.   Andenäss, the outermost cape, rises up like a pyramid, on which the dark rocky masses form a wonderful contrast with the deep snow which every where surrounds the base of this collossus; and these rocks are connected, with a range of others of almost

E E

equal heigth, which are lost deep in the interior of the Malangerfiord. We could only pass here with the utmost exertions; for the current from the Tromsund came with such force against us under the rocks as the Rhine does at Basle. Four hours were scarcely sufficient to enable us to clear about two English miles, from Andenäss to Bensjord. The ebb flows back into the great ocean. We may easily conceive that in all the Fiords, the stream of the tide carries us into the Fiords, and that the ebb carries us out of them. In the same definite manner in the straits *(Sunde)* which run nearly from south to north, the tide enters from the south, and fills the Fiord in the interior, and the ebb returns again from the north; for the general motion of the tide in the great ocean is not in the higher latitudes from east to west, but rather from south to north, probably because the larger tides of lower latitudes flow where the tides, on account of the smaller elevation to which the moon rises, must also be smaller.

With these rocks along the Malangerfiord, the *Fogderie* of Senjen on the main-land also terminates, and borders with Tromsöe. These bounds, since the settlement of Monsenelv and Bardonjord, have become of more importance; for Senjen only is the property of the state. All Tromsöe, which was not already settled by Norwegians, all the woods, mountains, and fields, every thing which could be turned to advantage, were sold by King Frederick the Fourth to an individual, who is here called Baron *Petersen*, with almost the same jurisdictions as those exercised by the counts in the counties *(Grafschaften)* of Jarlsberg and Laurvig. Some say Baron Petersen was a Dutchman: he possessed also a great deal of property in Helgeland; but he appears to have completely settled himself in Tromsöe after getting possession of it, for most of his relations still live here. He probably was better acquainted with these provinces than those who had the valuing of them formerly at Copenhagen. It is supposed that the first saw-mills in these latitudes were introduced by him. The boundaries of these possessions were hitherto always pretty indefinite. As the spreading of the colonists in Bardonjord made it necessary, however, to know with certainty where the royal property terminate, and where the property of the private individual began, a commission was appointed in 1806, by whom the boundaries through the woods were accurately laid down with what are called *Rösen*. They run now on the east side of the Malangerfiord, and from thence in a straight line through the wood to Mittel Rosto Jaure, on the Swedish borders. This considerable property did not long remain in the hands of one man ; it was soon brought by daughters into other families ; and it is now divided among three masters, who, by way of distinction, are here called the *Proprietors*. They all three live in the province and in the midst of their property, and one of them, M. Maursund, at Bensjord, where we landed.

It was not without pleasure we saw the beautiful saw-mill which was here in motion close to the house; for saw-mills are of importance in a country where nothing but wood is used for building, and still more so in a province that otherwise would be obliged to procure its logs from a distance of near five hundred English miles to the south. The saw very naturally only cuts what is wanted for the use of the surrounding country; but in this respect it is not limited. It receives the trees from the interior of the Malangerfiord, from whence they are floated down in rafts, and conveyed to Bensjord, not without danger. It has not, however, exclusively the liberty of supplying Senjen and Tromsöe with deals ; for in the Prästegieldt of Astafiord, in the interior of the Salangfiord, there is a saw at work on the government account; and each of the remaining proprietors also possesses a saw, the one in Lyngen, the other in Reisenfiord, which do not, however, supply the inhabitants with any considerable quantity of deals.

The mass of the Storhorn above Bensjord rises frightfully steep and rocky, like the *Aiguille de Midi* above *Aigle* and *Bex*. Every thing else disappears before this prodigious heighth, and even at a great distance this point far surpasses in grandeur every thing which surrounds it. The whole chain is narrowed between Balsand Malangerfiord to a breadth of little more than two English miles ; and yet it is still so high and steep ! But it is every where the same in this wonderful country. The mountains of the interior are not the highest, but those which lie near to their base, and which on both sides are followed by two Fiords. The Fiord rises on the main-land to the valley in the heighth, and immediately the chain decreases in heighth and steepness, and spreads itself out, and for several miles inland valley and mountain are connected together in a mountainous heighth, intersected in no great degree, and full of undulations. We are hence enabled to perceive with clearness the manner in which chain, and Fiord, and valley below, depend on one another, as cause and effect. The chain sinks into the one Fiord and rises out of the other, by which both are opened.

E E 2

Here also at Bensjord, mica-slate every where appears with continuous mica, containing garnets. Is the whole chain only mica-slate ? In the lower strata we not unfrequently find the white, fine granular dolomite ; and single blocks of it lie frequently along the shore, or appear in walls. Hornblende seldom appears here ; it does not even lie in the mica-slate in single crystals, and as little does grenatite (staurolite).

The violent current in the Sound was so hostile to all navigation from this place to Tromsöe that we could scarcely proceed up the Sound with the most favourable wind. We were obliged to wait the returning tide ; but we were then driven quickly by the waves farther, past the small island of Strömen, and beyond Balsfiord ; and at the end of four hours we cast anchor at Storstennäss, opposite the town of Tromsöe.

It was about eight o'clock in the evening, with admirable weather and sun. Our boat had already hoisted its flags before we saw Storstenness.   As soon as they perceived us, the flags there were also immediately hoisted, and guns fired to welcome our arrival.   We saw the town beyond, and almost every where flags were hoisted from the houses.   They seemed eagerly to contend who should receive their new magistrate *(Amtman)* with the greatest demonstrations of joy.   It was in reality a magnificent and delightful moment. The numerous flags from the houses and vessels were briskly agitated by the wind, and shone at a distance in the sun. Scarcely had the guns of Storstenness ceased firing, when they were again discharged in the town, and filled with their echo the Sound and the mountains. New vessels in the bottom of the sound of Hansjordnes succeeded; then the ships in the harbour, and then again the town : an incessant tumult and jubilee. We were all of us powerfully beset: we scarcely knew where to turn our eyes and our ears, and proceeded between flags and discharges, half deafened, and lost in admiration, to the shore at Storstenness, when the elegant and polished *Sorenscriver* Aaars introduced us to the circle of his numerous and worthy family.

Tromsöe, the 3rd of July. The view of Tromsöe is certainly not unpleasant across the narrow sound, and bears a resemblance to a small town ; for there is no want of buildings ; and several very considerable houses are situated, partly on a small elevation, and partly along the shore, and are taken in by the eye at one glance.   A number of packet-boats and *Bryggen* (wooden quays) appear in the water, and ships in the harbour.   No place in the course of our voyage had excited in us so strong an idea of a brisk trade ; and this impression is not altogether lost when we approach nearer, and enter the town itself.   It may bear a comparison with Egersund and Jedderen.   The houses of the merchants, Figensko, Giäver, and Lorck, although constructed of wood, would be an ornament even to Drontheim : the custom-house in the midst of a peninsula appears like a castle, and the habitation of Hans Jordnäss the *Foged* closes the row in an impressive manner.

The place is situated on a small island, of between four and five English miles in length, and from six to seven hundred feet in heighth at its greatest elevation in the middle of the sound between the main-land and Hvalöe. As the communication from Finmark towards the south runs through the sound, it is not to be wondered at that the island was peopled in very early times by Norwegians.   We find even in old poems that Thrumu is mentioned along with Senjen as one of the most distinguished islands of Norway; and yet it was almost the most northern point inhabited by the Norwegians, not far from which was the commencement of Finmark.   Hence the more southern inhabitants, as always happens in the case of places very little known, placed the seat of the mountain spirits and wizards beyond Tromsöe, as the Greeks did in Thule.*   When the Moguls inundated Russia in the middle of the thirteenth century, they also penetrated as far as the coasts of the Frozen Ocean, plundered the celebrated and terrified Biarmaland at the mouth of the Dwina, and compelled the inhabitants to have recourse to flight.   They sought protection from the Norwegian king, Hakon Hakonson, and received permission from him to settle in Malangerfiord.   The king, however, required from them that they should all avow themselves as christians ; and other Finns, dwelling on the coast, were also, probably, prevailed on to become christians.   The king built at that time, about the year 1260, two churches for the new community ; one to the southwards in Ofoten, and the other in Tromsöe.   The former long remained merely an annexation to Lödingen, but Tromsöe at once became a head church, and the seat of the clergyman, and it has since not only continued to be so, but at present the parish is one of the most considerable of the north.   In 1801 it

---

* Schönning Game Geographie, 46.

consisted of three thousand and twenty-four souls. Yet the whole Präste-gieldt has only this single church; and the parish is spread wide out along the interior of the Fiord, over Hvalöe, and over the small islands in the sea. Many of the parishioners when they frequent the church have a journey to perform of five Nordland miles and upwards, equal to above fifty English miles. This cannot be performed in one day, particularly as the passage by sea renders them dependent on the wind and weather. They therefore come a day or two before hand, and remain at Tromsöe perhaps a whole day after the Sunday. On this account every proprietor has built a hut in the neigh-bourhood of the church, consisting of a single room of logs, which receives himself and family during that period, and in which they find shelter against the cold. These *Kirkestuer*, to the number of about a hundred, are situated irregularly around the new town, and give it a very singular appearance. On a Sunday, when hands are not demanded for the fishery, all these little habitations are full of life. Young and old pass through the numerous streets ; here old acquaintances meet together ; there new ones are made; in one place may be seen buying and selling, in another a bargain concluded. The young folks swarm about, and trip along the meadows, others collect round a lottery table, or with a glass in their hand recount the success of their fishing. The church is the middle point, and almost the only point of union among these widely scattered men. This conflux has evidently been the means of inducing several merchants to settle here, for the more conveniently purchasing the produce of the inhabitants, and disposing of their own ; and the *Sorenscriver* and *Foged*, the royal functionaries, have from similar causes always lived in the neighbourhood.

In this manner a sort of town rose of itself, at least for several days in the week, and this appears to have induced the government to select Tromsöe, when they resolved in 1787 by the foundation of new towns to infuse new life into Nordland and Finmark. Merchants were invited to carry on trade immediately from this place to foreign countries, and they were promised an immunity from customs for a period of twenty years. Mechanics, both natives and foreigners, were to receive the freedom of the town and corpo-ration, and an assurance for their apprentices that they should be considered as corporation apprentices in all the Danish towns. They were also to receive a personal freedom from the taxes that were not requisite for the preservation of the town, such as stamp duties and the like, for a period of twenty years. There was a premium to be given of two rix dollars annually for each commercial last (twelve tons) of every ship between fifteen and twenty lasts, the property of a townsman, which should pass the winter in one of the harbours of Tromsöe or Finmark ; and premiums for ships fitted out for the whale fishing below the island of Beren, Jan Main, or Hoppens Eyland. There actually came several burgesses who commenced building; and at the end of thirteen years, about twenty persons had received grants of freedom of the town, of whom there were still in the place thirteen merchants, three shoemakers, one taylor, and one smith. The whole inhabitants might amount, perhaps, to one hundred and fifty. This was, however, far below the expectations formed. The twenty years of freedom from custom were almost elapsed, and the foreign trade from the place was still by no means established, nay not even attempted; for the merchants Figensko and Giäver had but just purchased a sloop of twenty-five lasts in Fensbury, with the intention of proceeding directly with it to the Spanish harbours. Hitherto all the merchants, as well as the country dealers, had dispatched their yachts from hence to Bergen. There is in fact an almost indispensible difficulty in the way of liberating the merchants of Nordland from the voyage to Bergen, which does not press on the people of Hundholm. The owners of the Bergen craft being little skilled in speculation, seldom succeed in coming to a complete settlement with the merchant at Bergen. The Nordlander is always in his debt, and every voyage, instead of diminishing it, generally adds to it very considerably. Hence Bergen exercises an almost unlimited despotism over Nordland. If the inhabitants of Nordland wish to liberate themselves from this subjection, the merchants of Bergen demand payment of the old debt, and this completely ruins the merchants of Nordland. This debt, however, which is never paid, has been long more than compensated for to the people of Bergen by the dearness of their commodities. In Bergen, an anker of common brandy, an object of such importance for Nordland, was sold in the current year for thirty-two rix dollars ; whereas the Flensburgers sold double brandy on the spot, at Loppen in Finmark, for not more than thirty-two rix dollars. The price of many other commodities of Bergen is in the same proportion. This is therefore one of the chief reasons why Tromsöe has not increased faster, and which perhaps will never admit the town to attain that degree of prosperity which was expected at its first foundation. The Nordlanders themselves will scarcely be ever able fully to extricate themselves from the shackles of Bergen, and foreign and independent capitalists will not easily be prevailed on to settle as far up as Trómsöe, since Hundholm has shewn them that their speculations may be carried on from other places. Tromsöe is also in point of situation in no respect 'to be compared with Hundholm. In the Tromsund the current is uniformly so violent and strong, that a vessel can only work against it with great difficulty, and cannot even be so secure in the harbour as one might wish. The harbour itself is also, when considered in the light of a secure winter station, very small, and scarcely sufficient to hold ten vessels; whereas Hundholm suffers nothing from currents, and has a harbour like a bay. Tromsöe is deeply buried among the islands; and towards the sea, from Hvalöe outwards, there are a great number of cliffs and *Skiärs* before the entrance into the Sound. On this account no vessel ventures to expose itself to the cliffs and the current between Senjen and Hvalöe, and to enter Tromsöe from the south. Almost all of them sail northwards, as far as the bounds of Finmark, a whole degree of latitude beyond Tromsöe: there they pass the high Fuglöe on their way to Lyngenfiord, from whence they descend to the Tromsund. They take the same course on parting. It has been attempted several times recently to go through Hvalsund along the northern end of Hvalöe. This course is, however, never taken by the usual craft to Bergen, who always rather go within the islands, but by the vessels freighted by the house of Fridish in Copenhagen, to Tromsöe and Finmark, to return laden with fish, not for Spain and Italy, but Bergen.

The Sound between Tromsöe and Storstennäss is but one thousand five hundred and seventy-six English feet in breadth, and not more than twelve fathoms in depth. This may contribute a great deal to the violence with which the current passes the island. To one accustomed to it, it appears truly astonishing to see that a boat in crossing must be piloted and guided as if it was over the Limmat at Zurich, or over the Rhone; and that in the course of a few hours the opposite course of the stream requires a complete change in the direction of the boat; for the ebb flows back with as much violence as that with which the tide enters the Sound. The harbour is deeper. Vessels anchor in from sixteen to eighteen fathoms water, with a good bottom. The town itself is not situated in a rocky bottom, but this is no advantage. All the buildings are erected on white bivalve shells, like those of Luröe and Bodöe, and on Senjen and Gebostad. The cellars are all excavated in the loose, broken shells, and in none of these cellars have they reached the bottom of the bed.   The humidity every where penetrates through the shells, carries along with it the calcareous particles, and covers every thing with stalactites and calcareous incrustations.   The vessels and wood are all soon covered with and destroyed by green *fungi*.   This is a great subject of complaint among the inhabitants of Tromsöe, and it is one which it is not easy to remedy; for we never saw the shell bed more extensive or thicker: it occupies a space of several hundred paces in extent all the way to where the island begins to rise, and its thickness undoubtedly amounts to from ten to twelve feet.   If the cellars were therefore to be dug out where this bed terminates, they would be situated at too great a distance from the harbour and the houses.   When we see it dug out for the erection of houses and cellars, it appears covered with streaks caused by inundations, such as we see formed by the waves along the shore, and as we observe almost always in loam beds on the banks of great streams.   Large and distinct shells are very seldom interspersed, but all appear as if purposely diminished in size, and broken.   Along the present sea-shore at the *Fiäre*, such collections of shell are no where to be found; nor will they be longer found when we search for them, at only twenty feet above the present highest level of the sea.   This is a singular phenomenon!   The bed runs through the whole Tromsund, and it is uninterrupted so far as we go along the Sound, and even beyond the Sound.   There the peninsula of Storstennäss lies at the mouth of a considerable stream from the mountains, and we might be easily led to suppose that they were brought down by the water from the higher grounds.   This may have been the case at first; but there is now also found here every where an uppermost stratum of pure shells, lying above one another.   This land therefore owes also its last elevation to the sea, and not to the stream.

The whole phenomenon appears of the greatest importance to us, when we consider that it by no means belongs to what we generally observe in geology; but that it seems entirely limited to this country, and is local, and has reference to a change in the land after the termination of all the considerable geological processes: to a depression of the surface of the sea, or perhaps more correctly to an elevation of the land.   Does it belong to the phenomenon of the apparent decrease of water in Sweden? If we were to apply the computation of Celsius, according to which the land rises four feet and a half in a century, we would not require more than four centuries to arrive at the beginning of the phenomenon; for this shell-bed in Tromsöe no where lies higher than twenty feet; and even at Luröe, it was scarcely forty feet above the surface of the sea. Hence if the rule of Celsius were accurate, the level part of Luröe, and probably also Tiotöe and Sör-Herröe must have been for the most part covered by the sea in the tenth century, that is in the times of Harald Haartager and his sons; but we know with certainty that this was not the case. It follows from hence that the shell-bed must owe its origin to other causes than the one which is gradually raising the whole of Sweden above the sea. The belief in the decrease of the water is not general among the people of Nordland, as it is in Sweden and Finland, where facts have given rise to it, and where in Bahuslähn and Gottenburg, they are as firmly convinced of it as at Carlskrona and at Gefle, Torneo, and Wasa. Are such pure, undisturbed, insulated beds of shells to be found on the other northern coasts? Are they in Scotland? or are they in Sweden?*

The interior of Tromsöe consists of mica-slate without felspar, and with frequent beds of dolomite, in which the white fascicles and stars of tremolite not unfrequently appear.

TROMSÖE, the 4th of July. The perpetual clearness and brightness of the sun gives at present an indescribable charm to the days. When it takes its course along the heavens about midnight towards the north, the country enjoys the evening repose, as in southern latitudes; and when it again rises higher, the inhabitants in the same manner experience the charms of the

---

* Similar beds of shells occur in different parts of Scotland, as mentioned in a preceding note. In the coal field of Clackmananshire, according to Mr. Bald, in the first volume of the Memoirs of the Wernerian Natural History Society, there occurs a bed of sand mixed with sea-shells, immediately below a thick cover of alluvial soil. He noticed the following shells.

1. Ostrea edulis. 2. Mytilus edulis. 3. Cardium edule. 4. Turbo littoreus. 5. Donax trunculus. 6. Patella vulgata. Some time ago the shell of a common crab was found in the clay near Alloa, when digging an arc for a mill wheel. The Rev. Mr. F eming, in a short communication to the Wernerian Society, describes a bed of fossil shells he observed on the banks of the Forth, to the west of Borrowstoness. It is about three feet thick, and has been traced upwards of three miles from east to west. It rests upon gravel, is covered with clay, and is about thirty feet above the present level of the Forth. All the shells it contains are of marine origin, and most of the species are found at present in abundance on the neighbouring shores. The most abundant species is the *common oyster.*—J.

morning. The feeling of gladness is undisturbed ; for it is not embittered by the melancholy and dreary sensation which takes place when the sun sinks beneath the horizon. When the sun re-ascends, new warmth is diffused over the country, and we scarcely perceive the approach of evening when we find from the thermometer that midnight is already past. Every thing begins now slowly to move ; the clouds ascend from the earth, and flit about in the air, and on the tops of the mountains. Small waves on the water of the Sound shew that the wind from the north is beginning to pass more and more downwards ; the sun ascends higher, its rays operate powerfully on the earth, and streams gush from the snow, with which the ground is every where covered around. When the north wind has fully risen, it no longer blows in squalls, but regularly down the Sound. About eight o'clock in the evening, every thing is again still ; there are no clouds in the sky ; no north wind in the Sound ; and we only feel the gentle warmth of the sun throughout the night.

Few places on this coast beyond the Polar Circle enjoy like Tromsöe the advantage of always seeing the sun in its course in the heavens. Almost every where to the east or the west some rock interposes itself, and hides the sun for several hours ; and when it again appears, it seems as if it were rising above the horizon. The temperature falls, and does not rise again till an hour after the new sun-rise, whether at five or seven o'clock in the morning. This was the case at Lödingen ; but there are also places which twice see the sun ascend, and which twice experience this oscillation of temperature at sun-rise. But at Tromsöe, the warmth of the air till two o'clock rose to 61°, and sometimes even to 63° of Fahrenheit. It then fell slowly till eight o'clock, more rapidly till ten o'clock, and very slowly again still lower to 50° or 52° ; but between midnight and one o'clock it began again to rise, after which it never fell.

The sun remains here for two whole months above the horizon ; from the middle of May, till towards the end of July. Tromsöe lies in 69° 38' of north latitude, equal to that in which Cook and Clarke were driven back by the ice fields in their northern voyage, and as high as the most northern colonies of Greenland, or the entrance to Baffin's Bay. Hence Tromsöe has a prodigious advantage in point of climate. It is true, it was no very pleasant spectacle to us to see spots of snow in the streets of the town, before the houses, and every where over the gardens and meadows ; but who would form an estimate of the climate from such unusual seasons ? The island is here covered to the highest point with trees. In the valley of Storstennäss there are excellent birches, and they rise to a great heighth on the declivities of the high and steep mountains of the main-land. I did not succeed, however, in measuring the immediate heighth of the limit of trees. The snow on the Dramfieldt, the nearest mountain above Storstennäss, was in general soft in the woods, and innumerable streams run in hollows underneath. Higher up it was covered with an ice border by the frosts of the preceding nights, on which the foot could find no resting-place to prevent precipitation down the whole of the declivity. The birches remained vigorous and beautiful more than six hundred feet upwards, and they probably would have endured another six hundred feet before disappearing ; but more than this they could hardly have borne, for Dramfieldt is little more than one thousand four hundred and fifty English feet in heighth, and the trees do not reach the summit. Hence the climate of Tromsöe is somewhat inferior to that of Lödingen, though it may not be much. The Scotch fir could grow in the valley of Storstennäss between the birches for several hundred feet perhaps up the declivities ; and hence the mean temperature at the foot of the mountain may be stated at the freezing point : probably an average of several years would be sometimes a little above, and sometimes a little under this point.

Agriculture is no longer compatible with such a temperature, and the inhabitants must confine themselves to grazing ; but even the meadows here are of no great importance, for the flat shore of the island is of too small extent, and the upper part, on the other hand, is a morass. On the mainland the mountains rise with too steep an ascent, and leave also there too little room between their foot and the Sound. The land in the interior of the Fiord is better cultivated ; for what is singular enough, every mile we trave into such Fiords the climate improves ; and in many of them we might imagine ourselves transported several degrees farther south, were we to judge from the culture and vegetation, though the northern latitude is in nothing changed. In the Lyngenfiord there had been no longer any snow for more than fourteen days ; and at Lyngenseid, the small tongue of land which connects together the Ulsfiord and Lyngenfiord, and which lies scarcely farther south than Tromsöe, the grain was above a hand in heighth, as if this Fiord did not lie three degrees beyond the Polar Circle, but in the latitude of Helgeland. But then it lies deep in the land, and remote from the sea. Tromsöe, on the other hand, is alone protected by Hvalöe, which is but of small breadth, from the passage of the cold clouds from the sea. If, therefore, cultivation and population have increased in any part of this country, it is deep in the bottom of the Fiord and in the mountains, and not on the sea coast; for the land is only there grateful, and capable of repaying the toil and the diligence of the cultivator. Hence the industrious Finns or *Quäns*, as they are here called, from Swedish Finland, have principally settled there. About thirty years ago only five families inhabited Balsfiord, and now the inhabitants exceed seven hundred. Lyngensfiord was so uninhabited in the beginning of the former century, that in 1720 Lyngen and Ulsfiord were provided for from Carlsöe, and Carlsöe itself was but an annexation of Tromsöe.* The praiseworthy zeal of the missionaries first caused Carlsöe to be formed into a separate Prästegieldt, independent of Tromsöe; and then the arrival of the Quäns was the cause of separating Lyngen from Carlsöe. In 1801, the parish of Lyngen contained one thousand seven hundred and twenty-eight inhabitants, and hence it is none of the smallest in the north: of them, only one hundred and four were Norwegians : the rest were mostly Finns, who dwell in houses like the Norwegians, and cultivate the ground. The Laplanders, who in summer cover the mountains of Lyngen, belong almost all to the Prästegieldt of Kautokeino, and not to Lyngen. I was told at Tromsöe that *Lyngen is a blessed corn country.* This must be relatively understood. But who would have expected in the south to hear the agriculture of a country praised under the seventieth degree of latitude. The potatoe cultivation has become quite general for several years in Lyngen, through the endeavours of M. Junghans, the present clergyman of Tromsöe, according to the account of the people of Lyngen themselves. The merits of Dr. Monrad, the physician of Lyngen, in diffusing the cultivation of this vegetable, are well known. Potatoes have not been long generally known in Norway. They came every year, it is true, along with the other garden produce, from Holland to Bergen ; but they were used merely as a rare foreign production, as a dish which could only be set before the guests on particular festivals, and at weddings. The first who strenuously inculcated the cultivation of the potatoe was Provost Peter Herzberg in Findaas, in Sondhordlehn, a man

* Hammond Missions Historie.

zealous for every thing useful. He did not succeed in his first attempts. As in the year 1762, on the occasion of the Danish preparations against Russia, German troops were sent to Bergen ; these soldiers wished for pota- toes, to which they had been accustomed, and the few which they could obtain from the country people were paid for by them at a high price. This induced Provost Herzberg to publish his experiments in a small treatise ; and this, and the pains taken by him to send the root wherever it was wanted for cultivation, have now extended it over the whole of the southern part of the province *(Stift)* of Bergen.* The book was re-published in 1773, and 1774, and it contributed in no small degree to excite also the public attention in the provinces of Drontheim and Christiania to the culti- vation of this vegetable. Are we to wonder that the cultivation of the potatoe was twenty years in ascending from Bergen to the Nordlands, and that it did not penetrate into the Lyngenfiord before the year 1790 ?

TROMSÖE, the 5th of July. Not only the island of Tromsoe consists of mica-slate, but in the high mountains of the main-land we see nothing else. The mica is continuous, and contains a great multitude of garnets. Here also, and particularly at the Dramfieldt, there is a thick bed of white fine granular dolomite in it, covered above with tremolite of several inches in thickness. On the uppermost surface, the tremolite is fascicular and stellar- fibrous, and deeper down appears in promiscuously aggregated crystals. It seems as if gneiss would not again make its appearance here ; for since leaving Tiellesund we no longer saw gneiss on the main-land.

But the main-land opposite Tromsoe is itself very little different from an island. The Balsfiord is only separated from the interior part of Ulsfiord by a narrow *Eid* (a tongue of land), and the small Ravenfiord is scarcely two English miles distant from Ulsfiord. The mountains above Storstennäss may therefore þe looked on as an individual and insulated mountain range ; and even in this point of view they make no great impression. The mountains rise higher and higher, and Ravensfiordsfieldt, the highest cupola, passed the limits of perpetual snow. These heights are not however striking, owing to needles and ridges ; probably their precipices are in the direction of the Strö- menfiord, the southern half of Ulsfiord. The strata in that case dip towards

---

* Provost Peter Herzberg's Biographie af Niels Herzberg, 1803. 30.

the north-west, and down towards Tromsoe, and may there only exhibit more rockless declivities. The northern part of Hvalöe is more distinguished, almost resembling the horns of Senjen, only not so high; but it rises upwards in an equally insulated, pointed, and sharp manner. The manner in which these points from the main-land seem to rise above the thickets of Tromsöe is frequently very singular: we imagine we perceive before us the *Wetterhorns* and the *Schreckhorns,* with their icy points, rising above the green declivities of the *Haslithal.*

MAURSUND, the 8th of July. The high mountains of Ringsvadsöe were perpetually before our eyes, as we rowed along on the afternoon of the sixth, in bright and clear weather. We crossed the Qualsund, passed the eighty feet high island of Hoegholm, on which numerous birds deposit their eggs, and about ten o'clock in the evening we landed at Finnkrog. Here they were still in the expectation of spring: the snow was only beginning to leave the lowest parts of the coast, and still lay on the hills, and in the mountain vallies. The cows were still obliged to be satisfied, as in winter, with wrack leaves and fish heads; at most they could only receive young birchen boughs, the outer bark of which was greedily and skilfully peeled off by them.

Ringsvadsöe is large, but not much inhabited. The island is too high. Formerly the Laplanders used to cross over from the main-land to the mountains; but this is no longer permitted to them. These Laplanders are very inconvenient and dangerous guests in the neighbourhood of a cultivated property; and at Tromsöe, and in the Balsfiord, their coming is always carefully prevented; for they pay very little respect to the property of the ground, break down the fences of meadows and arable fields, and allow the rein-deer to feed in places where grass was expected for the cattle. Although the rein-deer do not eat the grass, a single passage of a flock of them over a meadow is sufficient to render it useless for the whole year; for the cows will in general touch neither grass nor hay over which for months before the foot of a rein-deer has passed. This appears very singular and exaggerated, but it is the unanimous account given by all the Norwegians who inhabit the coast. We may easily conceive that this involves the inhabitants of the main-land in numerous quarrels every year, and that these quarrels are very prejudicial to the improved cultivation of the land. In solitary and remote Gaards, they are obliged to bear patiently with the incursions of the Laplanders; for they have too great a dread of their vindictiveness to venture to oppose them: a dread by no means unfounded. A Laplander was threatened by a proprietor in Lyngen with punishment and reparation of the damage he had done. The Laplander struck him dead on the spot;—and yet this Laplander comes yearly down towards Lyngen. It is not without reason, therefore, that their entrance into the smaller islands is interdicted. On Ringsvadsöe about twenty rein-deer run about at present wild, the remains of the former herds: the Laplanders still account them their property, and cross over in summer for the purpose of shooting some of them. They belong, with all the other Laplanders at Tromsöe, and on the mountains of Balsfiord, to the pastorate of Enontekis in Sweden.

All the hills at Finnkrog consist also of mica-slate, with continuous mica without gneiss.

The beautiful, clear, warm, and calm weather allowed us on the following morning to double very quietly two points, very generally dreaded, and to cross two of the greatest Fiords, the Ulfs and the Lyngenfiord. Both Fiords open widely into the sea. The sea winds easily penetrate into them, and contend with the land winds, which descend the Fiords, or blow through the Sounds. It frequently happens that a vessel leaves Tromsöe with a good south-west wind, which continues to Ulfstind, the extreme western point of the Ulfsfiord, when the north wind immediately blows from the sea in their teeth, and they are imprisoned between the two winds, and must wait at the steep cape for a more favourable occasion. The cape is uninhabited; and the winds may continue unchanged for days. Lyngen's Klubb, between Ulfs and Lyngenfiord, is still much more steep, high, and striking, but not so dangerous; for this range of rocks is out in the open sea, and far from the Sounds and the main-land, and consequently the land winds do not reach so far. The steep and rocky Fuglö stands entirely insulated on the outside; a rock known and avoided by all the vessels to and from Archangel, and on whose cliffs many a ship has been dashed to pieces. It is called in the English charts *Rock Huygens;* and it is perhaps two thousand feet in heighth. Vanöe rises nearer to us with two distinguished peaks, and Arende, of an equal heighth, opposite; both about three thousand feet in heighth.

The water which is surrounded by these high islands resembles an inland sea. The whales sported every where about in it as we passed along, and threw their shining columns of water into the air with a hissing noise: a noble sight, which gives every where life and variety to the otherwise monotonous immensity of water. Those great giants rush along with amazing rapidity under the water, raise themselves again above the calm surface, and shoot to a great heighth in the air their aqueous columns. They are seen on all sides, and the jets of water may be distinguished at a distance of more than two English miles. Should the waves be thrown into the slightest motion, they all immediately disappear, and do not again make their appearance. The whales do not enter the great opening of Fuglöe to no purpose. They follow the fish, which in spring collect here in great multitudes; not for the same purpose as at Lofodden, to spawn. It is a general chace of the inhabitants of the ocean against one another. The *Lodde* (the smelt) enters first in numerous swarms from the sea in pursuit of the aat (sea crabs); and they again are followed by the cod and sey, which are in their turns followed by the whales. The fishermen are therefore in universal motion when the lodde (smelt) appears; for the cod and sey may be hourly expected. The impatient pont *(Smaae Torsk, Gadus Barbatus)* darts upon the lodde, and drives the whole swarm into the Fiord, where they are easily caught by the fishers. The larger *Sey*, on the other hand *(Gadus virens)*, endeavours to surround the lodde; and before they can enter the Fiord, their passage is closed up by the sey, who again drive them out into the sea, and follows them. Hence the fishermen frequently lose, through the artifices of the sey, in one night the entire hopes of a rich harvest, notwithstanding on the evening before the boats could hardly make their way through the droves of fish. For this reason, the fishers all assemble with the utmost rapidity from the innermost parts of the Fiord, whenever the lodde appears, which is generally in May or June. What fish is the *Lodde?* Opinions are not unanimous on that point. With the exception of the single time when it is driven by the large fish into the Fiord, it never makes its appearance; and in summer it is not even caught in Finmark, nor many miles out at sea. In the spring also, it is not seen deeper than in the Fiords beyond the Polar Circle; never at Helgeland; still less consequently on the coast of Bergen; and even at Lofodden it comes very seldom. Bishop Gunnerus, in his northern visitation, did not appear early enough in these countries to examine the lodde when it was caught; he received only an imperfect specimen in brandy, which appeared to him to be the *Salmo Eperlanus (Leem om Lupperne)*. This was also Ström's opinion (Söndmör's Topogr. 1. 294.) It is true, he had never seen the fish; but the fishermen told him that they sometimes caught it several miles out at sea. From its characteristic penetrating smell, both these excellent naturalists believed it could only be this fish, so well-known for its odour.* We were actually told that when the lodde enters from the sea, the fishermen smell them at a distance of nearly ten English miles, and immediately set off in their boats in quest of them. If it is the *Salmo Eperlanus*, why should this lodde be exclusively peculiar to the remotest north? It is true, there are several sorts of lodde in these regions, which have nothing in common with this *Queitte* or *Järnlodde*, and do not appear in such swarms. The *Sild* or *Vatslodde*, which ascends the rivers, is the *Scolopendra plana* according to Gunnerus. On the eastern coast of the island Vanöe, which lies opposite to Lyngensklubb, there is a high and white cape, which from its colour is called Huitnäss or Queitenäss, according to the hard Nordland pronunciation. There the swarms of lodde most generally appear, and it is also the general rendezvous for the fishers in June. But as all these fish, the lodde, pont, and sey, are only accidentally driven into these situations, and do not in preference select them as the cod does the fishing station at Lofodden, we may easily conceive that the arrival and the numbers of the fish must be very uncertain and changeable. In the three years, from 1799 to 1801, Lyngen, and Carlsöe, and Skiervoe were choaked full of fish; but almost none have been caught there since until 1807. How is it conceivable that the lodde pursued from the North Pole should always light on the same point of the Norwegian coast, for the sake of escaping into the Fiords? We too often forget the immensity of sea which these fish have to go over, and the small proportion which single Fiords and coasts bear to such an extent, a negligence which often leads us to draw the most erroneous conclusions. For instance, in the year 1806, the burning of kelp *(Tangashe)* was prohibited at Christiansund, a branch of industry which produced more than five thousand dollars to the town, and that merely because a poet had heard the fishermen say that the smell of the burning kelp drove the cod from the coast. Did the fish only stay away when kelp

---

* M. Brünnich, superior overseer of the mines in Kongsberg, demonstrated to me that in all probability Mohr, in his voyage to Iceland, described the Lodde under the name of the *Salmo Arcticus.* Pallas gives also the same name to a Salmon of a span long, which is not found in the sea, but in the streams running into the Frozen Ocean. *Reise.* iii. 706. M. Brünnich is of opinion that the Lodde receives its name from the lateral line, which is Lodden.

was burned? or did it not also leave those places where this preparation was not made? Surely the smoke of the kelp could not reach all the way to the Söndmör. Poets and fishers are assuredly not the most competent judges in an investigation of the general causes of the course of the fish: this belongs to the naturalist, who follows the fish every where throughout his extensive element, and does not limit his knowledge to the mere coast.

The gigantic chain of jagged rocks of Lyngen struck us immediately on our doubling the high Cape of Lyngen's Klub, and seeing the long range stretch on the east side up from the Fiord, even more than it had done before; for the rocks are almost perpendicular on this side, and their forms are truly wonderful. They enter the Fiord to an amazing depth, and they rise high above the region of snow. We saw from Ulfstind a glacier descend from the summits towards the small Fiord of Sör Lenangen; and scarcely had we passed the Klub when other glaciers appeared before us on the very tops of the rocks, and descended about a fourth part of the heighth like the glacier of *Grua* in *Chamouny.* Others still shone in the depth of the Fiord. Opposite to the small Island of Strubben there is actually a valley in which a glacier descends from the rocks, and runs along the valley almost all the way to the sea shore. About five miles onwards the tongue of land called *Lyngenseid* follows, which, like a channel, cuts through this high range of mountains; and the mountains do not again rise to the glacier heighth. They are too near to the main-land.

The highest points of this range certainly rise to upwards of four thousand feet; for the limit of perpetual snow lies here above three thousand feet; and to form such considerable glaciers, above which the naked rocks rise to a considerable heighth, the rocks must ascend many hundred feet above this limit of snow. These are the highest mountains between 69° and 72° of latitude; for we no where afterwards find such masses in this high latitude, even if we search the whole hemisphere through Siberia to Bering's Straits.

How narrowly this chain is enclosed between two great Fiords! Their heighth and steepness is only so excessive when they are accompanied by the two Fiords. The strata of the chain probably dip towards the west; this at least may be conjectured from the precipices, notwithstanding the strata dip determinately towards the north and north-east at the Lyngen's Klub itself.

At the Klubben we were overtaken by a fresh north-west wind from the sea, that drove us quickly over the Fiord, which is between four and five

English miles in breadth, against the low green and bushy Vorteröe, on whose sides several waterfalls descend in a most agreeable manner from the rocks, and on whose hills we saw green pastures and Gaards. We passed between Fälles and Vorteröe, and made for the southern cape of the high and rocky Kaagsöe, which even at a great distance rises above all the islands before it. Maursund separates this island from the main-land; and the trading station in it, a beautiful convenient and elegant house, is scarcely an English mile distant from the western outlet of the Sound.

We arrived at Maursund from the west at the very time when the yacht of the proprietor, M. Giäver, entered from the north: a new proof of the severity with which the Bergen voyages press on Nordland. M. Giäver had laden the yacht in Skiervöe, and wished to proceed with it to Bergen. But the crew was deficient, and men were no where to be had. Still less could he procure a pilot who knew accurately this cliffy coast for about three hundred leagues to Bergen. The yacht consequently could not proceed to Bergen without being exposed to the most manifest danger; so that he was obliged to unload the fish, by which the whole country was disappointed of the provisions and grain which they expected would be returned with this yacht. Pilots are every where wanting in the Nordlands, and many vessels have been lost from no other cause, but that for want of better they were obliged to put up with inexperienced pilots.

M. Giäver's yacht, the *Northern Star*, is the most northern and at the same time the largest of all those which frequent Bergen. It carries about eight thousand *Vog Fisk* (of thirty-six pounds each); so that it is larger than many a brig. Still it is nothing but a yacht with square yards *(Raaseil)*, which requires a large crew, but in the opinion of the Nordlanders is lighter and more secure than a yacht with *Bom-seil*.

MAURSUND, the 10th. The tide rose to a heighth which it had not reached for four years: the water almost entered the storehouses. This is a dreadful western storm at sea, said the fishermen. It appeared in the Sound to come from the north; but its direction is changed by the islands. The rain fell in powerful torrents, with flakes of snow between; and but a few hundred feet above the Gaard and on the mountains it snowed very hard. Is the winter never to disappear here? At Michaelmas-day, in the former year, it had begun to snow; and fourteen days afterwards the cattle could not leave the house. It is now scarcely two weeks since they were first able to take out the cattle. To feed a cow for nine months in the house makes the keeping of cows a luxury; and it would come to this if all years were like the present. But all years are not of that description. Reissford is only separated from Maursund by a narrow tongue of land; and the agriculture of Reissford is in a flourishing state. This could not be the case if a winter of nine months were not an extraordinary pheno- menon. The Quäns in Reissfiord not only successfully cultivate bear, but also raise it in such abundance, that frequently they stand in need of no foreign assistance. The woods in the interior of the Fiord not only furnish them with a sufficiency of fire-wood, but also with logs and deals for building; and in the considerable rivers they might carry on perhaps a pro- fitable salmon fishery if the incessant floats of wood did not prevent the erection of the necessary constructions for that purpose, and in some degree drive back the salmon. Potatoes have been also of late successfully culti- vated. Provost Junghans has excited a very general desire to cultivate them. Reissford also receives annually new inhabitants as well as Lyngen; Quäns who arrive from Sweden. The land and woods, however, belong to the proprietor M. Lyng, in Rotsund, one of the heirs of Baron Petersen.

ALT-EID, the 12th of July. Our departure from Maursund was favoured by wind and tide, so much so, that we were obliged to lie for some time at Taskebye at the mouth of the Sound, and also at Kaugsöe, till the wind should somewhat fall. The appearance of the island from this place has something dreadful in it. The strata stretch from south to north, and dip very strongly towards the west. Hence the precipices towards the east appear like huge entirely smooth and perpendicular walls, which rise in order one above another. The ridge at top appears as sharp as a knife, and the snow can only remain on a few insulated spots. These strata appear to be a continuation of the tongue of land between Reissfiord and Rotsund, which runs in the same direction; and the high rocks of Skiervöe, beneath which the church and the mansion-house of the clergyman are situated, belong also to this range of rocks. The eastern part of Kaagsöe is lower, as well as the tongue of land towards Reissfiord, which lies opposite. Horn- blende slate only appears there; hordblende and felspar in almost fine granu- lar straight slatey mixture; a rock which belongs to gneiss. But the higher rocks of Kaagsöe are probably mica-slate, like the rocks on Vorteröe, and at Maursund.

The rocks which separate Quänanger and Reissfiord have a wonderful appearance. They are not high, but look like a row of needle points crowded together, and as if cut with knives; or if we compare small things with great, like the hacked quartz in the veins of Freyberg. It looks as if nature wished to exhaust all her forms in Nordland; such is the endless diversity of singular shapes.

We passed along the small Högöe, and soon crossed the Quänanger Fiord, which from its being open to the sea is very frequently agitated with storms from the north-west and with high waves; and in an hour's time we passed Spilderen at the entrance of Alt-Eidsfiord. There we appeared to be suddenly transported to a new climate. Vassness, the extreme point of the Fiord, was every where covered with high and fresh birches, and a shining green was visible under the trees; and as we rowed into the high and narrowly enclosed Fiord, the snow every where disappeared even on the heights. Thickets of aspens, birches, and alders, in agreeable alternation, run along the shore with small meadows between, and streams which tumble down from the hills; and the rein-deer of the Laplanders in hundreds were feeding along the declivities. The Fiord was as smooth as glass. It appeared as if we had landed in a bay of the lake of the *Vierwaldstäter*. In such a situation, how was it possible to think of the snow and ice mountains which we had only left in the morning? In the bottom of the bay lies Alt-eids-Gaard, on the gentle and green declivity, and embosomed among the high mountains around, which completely shelter it from the winds and the waves of the sea. Every where here, grass, flowers, and trees, were shooting out vigorously: and the canopy of clouds which the north-west wind drove over us up the waters of Skiervöe, and to the entrance of Altensfiord, had now given place to the purest brightness. The sun the whole night through pursued its course behind the mountains, and gilded by turns the summits of all the surrounding heights.

The birch wood was so thick and so beautiful in the valley which run up the tongue of land, and the birches ascended so high up the mountains, that it in fact appeared as if we had here lighted on a climate which far surpassed all that we had hitherto met with in Nordland. But what affords a better decision than the barometer? I ascended therefore through the wood, the mountains on the south side of the Fiord, and immediately above the new house of the merchant Morten-Gams. The trees remained close and large for more than eight hundred feet up; the mountain then rose in steeper declivities, and the rocks supplanted the trees. But on insulated spots they still continued to grow with freshness and vigour, and at a heighth of one thousand two hundred and twenty-six English feet we found a bush which we could scarcely have expected to find in better condition in the valley below. The birches then however began to contract, the alders disappeared, and the aspen-tree had long left us beneath the mountain. Its sides were now covered with large patches of snow, and on the unsheltered rocks the small birches could no longer keep their footing. On this account the line of their cessation withdrew itself from immediate observation; but we could recognize it to be higher than one thousand four hundred and ninety English feet, and it may well reach as high as one thousand seven hundred English feet, where birches of a foot in heighth may creep along the ground. This heighth consequently proves that the Fiord of Alt-eid actually enjoys a better temperature than either Senjen, Tromsöe, or Salten; a temperature under which excellent Scotch firs would succeed in the valley and up the Eid, if they had but more room to spread. The mountain, Morten-Gams-Tind, rises to the heighth of one thousand nine hundred and eleven English feet; the top, however, is naked rock, strata rising above one another. This summit is not insulated, but merely the extreme point of a ridge which runs southwards, always increasing in heighth till at a distance of about nine English miles it joins with the range of mountains of the main-land. The next distinguished summit above the Fiord is about an English mile from the former, and two thousand one hundred and eighty-three English feet in heighth.*

The view from these heights is remarkable, and highly conducive to a knowledge of the country. All the mountains, all the points of the *Fieldts*, appear only islands; deep vallies separate different parts of the mountains, and supply the place of Sounds between the islands and the sea. The

---

* 12th of July, h. 4. p. m. Morten-Gams, Alt-Eid, Baromet. 28. 1. 5.

       Therm. 9. R. clear

       h. 5. high birch thickets on the declivity 26. 10. 7.

       h. 6. Morten-Gams-Tind 26. 3 1. 4. 5. strong east wind.

       h. 6½. highest eastern summit 26. C. 1. 4.

       h. 8. Morten-Gams, Alt-Eid 28. 1. 5. 7. 5. clear sun behind the mountains.

whole *Eid*, or the valley between Alt-eid and Langfiord, surrounded by high mountains, gives in this manner the appearance of an island to a great part of the main-land to the north of the Eid. A deep valley runs from Langfiord towards the south, the Beina valley, which is finally connected with vallies, which ascend from Talvig out of the Altenfiord, by which the high mountains between Talvig and Langfiord are again formed into islands. Higher mountains, deeply covered with snow, appear opposite on the north of Alt-Eidsfiord. The nearest cupola is perhaps two thousand seven hundred and seventy English feet in heighth; then the deep hollow of Jöckuls-fiord, above which rises the extensive snow mass of the Jöckulsfield. The manner in which the glaciers separate from these regions of snow, and descend into the narrow and deeply surrounded Jöckulsfiord, was very distinctly visible. They remain pendent in the middle above the steep and almost perpendicular rocks, and in summer the great masses of ice are incessantly precipitated from above into the Fiord, and frequently in such abundance and with such force that the motion thereby communicated to the water of the Fiord causes it to rise for miles several feet above its banks, and not un-frequently to carry along with it the huts of the Finns. This was the case two years ago. These Jöckulfielde like Folgefonden, and the mountains of Justedal, have no peculiarly prominent points and summits. The whole range of mountains rises gently and almost imperceptibly from the perpendi-cular rocks which surround the Fiord, and the snow roof covers it every where like a carpet. These are the highest mountains of this region; but they are much inferior in heighth to the *Fieldt-points* of Lyngen, and hardly exceed three thousand seven hundred English feet in heighth. This height is not reached by the mountains to the south of the Fiord, and hence the glaciers of Jöckulsfiord on the main-land are the most northern glaciers of the world, with the exception perhaps of those in Greenland, supposing the temperature of Greenland in this latitude to be such as to form glaciers. The glaciers of Jöckulsfiord are exactly in the latitude of 70° north.

The heights which surround Alt-Eid are rich in different formations. As we first entered the Fiord we saw on the point of Vassness beautiful and distinct gneiss, straight slatey, and rich in felspar. At the Gaard also the blocks of gneiss lay in abundance along the shore; but in equal abundance hornblende and felspar in fine granular mixture. Farther upwards, towards the valley of the Isthmus, mica-slate appeared above the surface with continuous mica and small garnets; and here also for the first time it contained small *Staurolites* (grenatites), and black scopiform divergent hornblende on the folia of the mica. Small beds of white small-granular lime-stone lay abundantly between the strata. The formation of the mica-slate appears here also distinguished even in its accidental individualities. All the rocks on the north side consist of it in the same manner; and probably to the greatest heighth beneath the snow; for we see no other formation among the blocks at the foot of the rocks. But at the Morten-Gams-Tind, along the whole of the south side of the Fiord, we no longer find this mica-slate. Even the very first strata at the foot of the mountain are pure quartz of various colours, with a little white mica, which is contained in the quartz in the manner of a bed. There are rocks of it in the wood, and a great number of blocks lie every where along the declivity which have been detached from these rocks. In some places the abundance of these blocks is so great that the trees cannot penetrate through them. Nearer to the summit, and where the mountain begins to rise more rapidly above the birch-wood, the quartz disappears; the strata now consist of a fine granular mixture of hornblende and felspar, which from its dark colour forms a strong contrast with the white quartz on which this formation reposes. On the summit must be added to this red garnet, which, from the general smallness of the crystals, appears compact: the whole mixture itself above becomes so fine granular that we can only discern the ingredients with difficulty. These strata are so polished by the weather, the rain, the melting snow, and sand, that they become shining, and bear a deceitful resemblance to the basalt of the numerou svases of the Vatican or the Egyptian monuments in Velletri. Insulated larger and white crystals of felspar lie in the black mass as in a porphyry; and the fundamental rock itself is in this polished condition no where to be distinguished.* All the strata stretch N. N. W. and S. S. E. (h 11.) and dip towards the east, but never more than about 30 degrees. It is clear and

---

* The polished and shining appearance which rocks acquire by the influence of the weather is often very striking, and has been confounded with volcanic appearances by some observers. Thus serpentine by the action of the weather, acquires at the same time a vesicular form and shining surface, and has been taken for lava: basalt, by the agency of running water, or by the long continued blowing of sand over its surface, obtains a groved and shining surface not unlike that exhibited by certain varieties of Lava.—*J.*

evident that both this formation and the quartz beneath lie on the mica-slate; for this appears in the valley in the streams which run from the Isthmus to the Fiord: but are these formations subordinate to the mica-slate? Or are they the commencement of a new and independent formation?* This is the first time since our leaving Drontheim that we have met with any other formations than gneiss and mica-slate on our way.

They are not perhaps the only ones in this Fiord; for to the east of Alt-eids-gaard, on the way towards Langfiord, the foot of the mountain is thickly covered with large dark blocks, the fissures of which are covered with brown rust, and the corners of which are rounded in a manner which we never observe in blocks of gneiss and mica-slate. They are a mixture of dark brown *Diallage* (Smaragdite) and of white felspar. The former preponderates: it is thick foliated, the cleavage of the folia single, by which the distinction from hornblende is so easy and striking; in the cross-fracture almost conchoidal, and semihard. The mixture seldom goes beyond the coarse granular; generally it is small granular. We must go a great way up this mountain, which is more than a thousand feet in heighth, before we discover this rock above the surface; it first appears in the last third part of the height, and there lies distinctly above the mica-slate, and sharply separated from it. We see that they are two formations, quite different from one another. But the diallage formation does not extend beyond the top, and it is not to be found in the high valley of Alt-Eid, towards Jockulsfiord. It probably runs along Alt-Eid itself, farther on towards Langfiord, and there forms other points and summits.

This remarkable *Diallage* rock is not very unfrequent in Norway. M. Esmark first discovered it in the Thron mountains, between Tönset and Foldalen, and not far from the road from Christiania to Röraas; and it was found there to the very tops of the mountains, four thousand seven hundred and ninety-two English feet above the sea.† The *Diallage* is then very coarse granular, generally greenish grey, and of a single cleavage of the folia it is true, but with a disposition to the two-fold cleavage, which was evident

---

* The trap rock described in the text occurs in different parts of Scotland, and appears in beds of considerable magnitude in mica-slate. Fine examples of it appear in the country around the noted mountain of Schihallion in Perthshire.—*J*.

† Pfaff und Scheel nordisches Archiv für Naturkunde, III. Bd. 3. S.-I-. 199.

in the form of the fragments. The rhombs of the fragments are much less oblique-angled than in hornblende. The grey felspar with which the diallage is mixed appears to have the smaller preponderancy in this mixture. Moreover, green scales of talc mica, and probably also a great deal of magnetic ironstone, are intermixed, for the whole rock, according to Esmark, possesses great polarity. This formation is equally extensive in the neighbourhood of Bergen. About fifteen or sixteen English miles to the south-west of the town a small chain of mountains runs along the Samnangerfiord, from Samnanger to Ous. The road from Hougdal to Vaage along the Fiord goes over these mountains, and rises to a heighth of about one thousand and twenty English feet: and to this heighth mica-slate continues to appear above the surface. But as soon as we begin to descend on the other side, great blocks of diallage make their appearance, which continue half the way down the mountains. The most of them have fallen from a very steep range of rocks which rise above the greatest height of the pass, and which are distinguished a great way off from the darkness of their colour. The diallage is here for the most part coarse granular, greenish grey, and extremely fine foliated, by which it bears a stronger resemblance to the green smaragdite of the Saaser valley in the Valais. A two-fold cleavage of the folia is here not to be mistaken. The felspar of the mixture appears yellowish, and greenish-white, beautiful, and distinctly foliated. When the mixture however becomes fine granular, which it not unfrequently does, the felspar in it generally bears a resemblance to large needles which shoot through the dark diallage. Here also this formation is immediately above the mica-slate; but it does not however appear to stand in such close relation to it. At the end of the chain between Klyve and Ous, where the Samnangerfiord runs into the Biörnefiord, these geological relations are disclosed somewhat more accurately. Distinct clay-slate first appears in the rocks at Hatvig, and then fine granular diallage-rock, which forms the uppermost points of the mountains. The gneiss again appears under both, towards Ous, at the stream which runs down there. But on the way from Ous to Bergen, towards Bierkeland, and before we come to Kallandseid, the clay-slate rises again into considerable hills, and shortly afterwards the diallage of new is seen with distinguished beauty. In many large blocks it is large granular, with separate pieces of half a hand in size, clear greenish-grey and shining. Here it no longer puts us in mind of hornblende: the magnitude of the surfaces, the colour, the splendour, are completely different. This formation lies here also on the clay-slate, and appears to be actually connected with clay-slate. At least it is certainly newer than the latter, and one of the last rocks in the series of the primitive formation. Hence we may without any great error assign also the formations of Alteid, which contain diallage, a place in the neighbourhood of clay-slate, and remote from mica-slate. It is possible, perhaps, that the quartz of Mortengamstind here takes the place of the clay-slate, and that the diallage formation lies immediately upon it; for the district of Bergen affords us some analogical instances. In the huge and steep mountains between Hardanger and Hallingdal, beyond Ullenswang and the Söefiord, mica-slate appears in the undermost strata, with numerous hornblende beds as usual; then till we ascend two-third parts of the heighth, there is fine granular hornblende-slate, with quartz alternating in thin streaks; a formation which in the whole of Upper Walders supplies the place of the mica-slate, and is very extended; then follows lastly, to the highest points extending over the whole mountains, pure and very coarse splintery quartz in large rocks and blocks. At Revildseggen, four thousand four hundred and ninety-four English feet above the Fiord, for many hundred feet no other blocks but these are visible. This quartz is also then a formation which follows the mica-slate; and it is extended and deep enough to be considered as self-existent, and not merely subordinate to the mica-slate. It is also very near to the clay-slate. In the middle parts of Guldbrandsdalen, also, the mica slate was only separated from the clay-slate by a formation of pure quartz.

ALTENGAARD, the 14th of July. It is a great advantage for those who travel from the south to Altensfiord to leave the boats at Alteids, cross the isthmus, and embark again in Langfiord; for by this means they not only save a considerable part of a long sea-passage, but this passage towards Loppen, and through the Stiernesund, is not without considerable danger. Unfavourable winds and strong waves from the sea continue here for weeks, whereas there is nothing at Alteid and in the narrow Fiords to prevent the continuation of the passage even into the great Altensfiord. Hence Alteid is very generally frequented in summer. The isthmus between the two Fiords is but two English miles in breadth, and is soon passed. The trees in one continued wood are large and beautiful, and at present in full leaf; but in other respects it is a perfect wilderness. Even the way which we ought to follow is not visible till we come to the last third part of the isthmus, where the active magistrate M. Sommerfeldt ordered a road to be dug, which he was unfortunately, however, prevented from completing. In the middle of the Eid, between the trees, a solitary Gaard, Brodskyflet, is situated exactly on the highest point, which, however, does not reach two hundred English feet. Such is the small elevation of the Eid. The separation of the waters in the valley is not perhaps half that heighth. If the Swedish laws respecting the diminution of water were to be admitted as applicableto Finmark, this Eid must have been a Sound in the times of Saint Oluf, and the Jöckulsfieldt an island: but this is incredible; for such a change would never certainly have been omitted in the northern annals. In the last English mile we finally reach the boundaries between Nordland and Finmark, when the waters descend towards Langfiord, and shortly afterwards we arrive at the Fiord. It resembles a fresh-water lake, without motion. Towards the south, and almost in a right angle with its former direction, we see it continued through a deep valley, the Beina valley, which for more than fourteen English miles does not rise in any considerable degree. In this valley it is said there are excellent plains, with good trees growing on them; and perhaps it is not impossible for a new Bardonjord to make its appearance in it. At present, however, it is only an uncultivated morass.

So long as we continued on the north side of the valley, we saw also the strata composed of hornblende-slate, or very fine slatey gneiss; formations which here probably belong to the mica-slate. But on our crossing over to the south side, and reaching Siockhammerbacken, over which the boundaries run, we saw nothing but quartz, partly greyish-white, and partly red, with a little mica, and every where a great abundance of irregular drusy cavities. This formation did not leave us all the way to the Langfiord; and the hills on the north side of the Fiord were similar. Not so, however, the higher mountains which surround the Fiord on this side; for at Subsnäss, a trading house on a cape which advances far into the sea-bay, the strata again become hornblende and felspar in small granular, and then afterwards in slatey mixture. This cape is but an arm of the higher mountains. The strata stretch N. N. W. and S. S. E. (h. 11.) and dip 70° towards the east.

The passage down the Langfiord is agreeable; for the mountains and rocks along both sides, and the waterfalls and vallies up these mountains, not only exhibit the utmost variety of form, but the Fiord is also wonderfully animated by a number of habitations along its banks. Wherever a stream pours down the ravines, there we may see several *Gammes*, the clay-huts of the Finns; with their small wooden storehouses beside them, and the house by the side of the water which secures the boats from wind and weather. The stores and boats are better residents than the proprietors. The nearer we approach to Altensfiord, the houses appear in greater numbers. There Norwegians inhabit the country. We had never hardly seen a Fiord or a Sound so thickly inhabited.

The south side is still more so than the north side; for it is less steep, and the mountains are green for a considerable way up; and towards Altensfiord these mountains gradually decline into hills. But on the north side most of the rocks descend perpendicularly into the sea, and the mountains above rise beyond the limits of perpetual snow. They are connected with the *Jockulsfielden*. Towards the point where Langfiord and Stiernesund flow into Altensfiord, the whole mass of the mountains is nothing but bare and naked rocks, from the summit downwards to the Fiord, with only dreary heaps and avalanches of stone scattered here and there.

We entered Altensfiord in a beautiful night. The sun never left us. The thermometer, even at midnight, and about one o'clock, had never sunk below 54° of Fahrenheit. About three o'clock, the rays of the sun became so powerful, that we were obliged to recur to the shade for protection. The water of the Fiord was calm, and the distances four or five miles beyond the banks, and the view of the Gaards, rocks, and waterfalls towards Talvig, were uncommonly beautiful. We could not have expected a finer morning in the south of Norway. The circular bay, and the amphitheatre of Talvig, as they suddenly burst upon us through the narrow channel by which we passed, were highly attractive. The church stood in the middle of the green and animated slope, with the large clergyman's house above, and on the sides there were two considerable Gaards, with Quäns and peasants along the banks, and picturesque rocks, with a majestic foaming waterfall above. Add to all this, the animation of summer; ships in the harbour, a Copenhagen and a Flensburg brig, with a Russian vessel from the coast of Archangel, and Finns and Norwegians in continual motion backwards and forwards in the bay, with fresh fish to the Russians, and dry to the merchant, and returning with meal and grain. Who would represent Finmark as dreary and miserable, if he saw the beautiful situation of the bay of Talvig?

About mid-day we crossed the nine short English miles from Talvig to Altengaard, the seat of the head magistrate *(Amtman)*, in the inmost part of the Fiord. This Gaard is also an object of surprise. It is situated in the middle of a wood of high Scotch firs, in a green meadow, with noble views through the trees of the Fiord of the points which project beyond one another into the water, and finally of the *Fieldts* of Seyland and Langfiord. The trees around are so beautiful and so diversified! We see through the boughs on the opposite side of the water the foaming stream which descends from the rocks, and communicates perpetual motion to the saw-mills; and in the Fiord and in Refsbotn, every hour of the sun in its progress lights up some new Gaard to us. This habitation is a villa, not a country-house built for the dust of law papers, or for the management of law suits. It appears when we enter the wood from the beach as if we were transported to the *Thier Garten* at Berlin; and when the perspectives down the Fiord open on us, it then seems as if we were viewing Italian distances, or one of the lakes of Switzerland.

CHAPTER VII.

FINMARK.

View of Alten.—Climate.—Quäns in Alten.—Origin of the Quäns.—Finns in Ag-gershuusstift.—The Norwegians inhabit the Islands, the Quäns and Laplanders the Interior.—Clay-Slate and Quartz at Alten.—Voyage to Hammerfest.—Object of the Institution.—Climate.—Tyvefieldt, Gneiss.—Trade and Fisheries of the Russians in the North.—Influence and Consequences.—Voyage to Mansöe.—Vivacity of the Lap-landers.—Danger of a Residence in these Islands.—Norwegian Dwellings on Ma-geröe.—Rein-Deer on Mageröe.—Winds.—Passage to Repvog.—Hooks of the Lap-landers. — Olderfiord. — Sea-Laplanders. — Passage to Reppefiord. — Merchants in Finmark. — Their Appearance and Influence. — Rage for Brandy among the Lap-landers.—Causes of it.—Return to Alten.—Thunder Storm.—Talvig.—Mountains.—Boundaries of Vegetation.—Clay-Slate.—Mica-Slate.—Akka Solki.

ALTENGAARD, the 21st of July, 1807. What could lead us to suppose in the summer months of July that we were in a latitude of 70°? At this season it is not warmer at either Christiania or Upsal. On the 13th the thermometer rose to 80°. 3. of Fahrenheit: it generally stood at mid-day at 70° or 72°, and the mean temperature of the month rose to near 63°, as high as in the best districts of Sweden or Norway. The poverty of the vegetation no doubt ought to indicate the northern latitudes; but even that little is so charmingly disposed by nature, that it almost appears superfluity. How beautifully rural Elvebacken appears at the mouth of the Altens Elv! It looks like a Danish village. The houses, to the number of about twenty, lie up the banks of the great stream in the midst of green fields and meadows, and surrounded with high Scotch firs in every direction. How majestic is the appearance of the steep hills on which the fir wood ascends ! The stream is seen at a great distance, bursting through between the mountains, winds through the plains between islands, and thickets, and Gaards, and then, as large as the Main, directs its proud course along the bottom of the hills towards Elvebacken. The Gaards are most romantically situated in small retired vallies, which run from the wood towards the stream, and in the midst of charming meadows, by small rivulets or lakes, surrounded by alders and aspens. What appear in the distance only wood and wilderness, on a nearer approach immediately opens and displays to us meadows with houses scattered among them. These are not inanimate views, we everywhere see softness united with grandeur.

Transeuntibus, say the inhabitants. These charms of July are almost obliterated by nine uninterrupted winter months. Possibly enough; but how few places can compare their summers with that of Alten! And what a contrast if we follow it in the same degree of latitude over the globe! On the southern point of Nova Zembla, at the mouths of the Jenisey and the Kolyma, no trees will grow, not even birches; and the pine tribe disappear at 67°. Even in the interior of America, according to Mackenzie, the last Scotch firs appear in 69 latitude, before reaching the shores of the sea. But in Alten we not unfrequently find in the valley Scotch firs of more than 60 feet in heighth.

But to ascertain with more accuracy the climate of Alten, we are assisted with the different limits of the vegetation on the heights: these limits not only instruct us sooner, but actually much more accurately than even the observations of the thermometer for several years can possibly do; for the growth of the trees is determined by the true medium of the temperature, and not by the extraordinary average of a few months or years. A steep and high cape, at the distance of an English mile westwards from Alten Gaard, seems destined for such observations. It projects far into the Fiord, and divides the bottom of the Fiord *(Fiords-Botn)* into two halves, from one of which the Alten's Elv in the east takes its course; a circumstance, perhaps, that renders it so shallow. But in the western half larger ships find a sufficient depth of water to enable them to approach the land, and even to anchor along the shore. This part (Whalebay) is therefore considered as the proper harbour of Alten; the other is only navigable for boats, but is called, from the royal mansion on the shore, *Kongshavn*, and the steep Fieldt which separates the two bays is called Kongshavnfieldt. This singular mountain resembles a fortification: on three of its sides it falls quite perpendicularly into the water, and it is only accessible from the wood. When we are, however, among the firs, as soon as we get over several masses of rock, we see on this side also large

fissures between the rocks over the whole breadth of the isthmus, with perpendicular walls on both sides; and others, especially towards the foot of the
mountain, like abysses, from 80 to 100 feet in depth. The summit of the
mountain appears quite near: we see no obstacle; but the channel between
the rocks immediately appears under our feet, and we search in vain for some
means of descending this rocky wall, or of ascending again on the opposite
side. Several of these ravines resemble glens, being 20 or 30 feet in breadth;
and there we find a few pieces of rock tumbled down, which allow of a
descent. Others of them are only six or eight feet in breadth, and on that
account almost impassable. Others still, and those the deepest, may be conveniently leaped over, the edges being only a foot or two separate. These
singular fissures run parallel with one another from both sides, down to the
shore of the two bays. On the entrance of the mountain we see nothing
from which we should be led to suppose their existence; for even at the
largest of them the declivity ascends with the same inclination as before.
They are not strata which may have been carried along, for the direction of
the fissures traverses the direction of the strata almost at right angles. They
are evidently *veins*, real open veins, that have not been filled up. This we
may also perceive below, on the sides of the mountain; for we can there
often enter into fissures of two feet in breadth, and follow them for a considerable length and heighth. We seldom, perhaps, find a mountain so instructive as this for the theory of veins; for even the cause why these veins were here
opened, and in this direction, is evident and clear. They were probably occasioned by the fall of the whole mass towards the Fiord. If the rocks sink to
one side, they must be separated on the opposite from the solid mountain,
which can take place in no other way but by fissures, which become Fiords
and vallies, when the falling masses extend over square leagues; or ravines
and veins, and open fissures, when only single mountains or rocks sink down.*
The declivity of the mountain between the firs is remarkably dry and hard;
the bottom of the fissures, on the other hand, is a morass, over which a con-

---

* The interesting facts mentioned in the text are particularly valuable in a geognostic point
of view. Similar appearances are mentioned by Werner, in his celebrated Treatise on Veins;
and I have observed many very striking and extensive rents and fissures in the trap and sienite
rocks of the islands of Skye and Mull, and the sand-stone districts of the main-land of Scotland,
and of the Orkney and Shetland islands.—*J.*

tinuous covering of mountain bramble is floating. Alders *(Betula, Alnus incana)* and birches also grow there, quite a different vegetation from what we see forty feet higher above the edge of the fissure; and this gives a still more singular appearance to these openings.

The summit of the mountain rose only 560 English feet above the Fiord,[*] and therefore it was not to be wondered at that the Scotch firs ascended to its greatest heighth, and that their sickliness above seemed rather the consequence of the soil of the naked and dry cliffs, than the influence of the climate; for when the declivity is gentle, the trees immediately look fresh and vigorous, and rise to a considerable heighth. Hence this mountain does not denote with accuracy the mean temperature of Alten, but merely demonstrates that it must at least be placed a full degree Reaum. or more than 2° of Fahr. above the extreme of the Scotch fir temperature, and that it nearly approaches the temperature of the Tannen-fir.

The mountains which surround the Fiord, and the valley of Alten on both sides, ascend to a greater heighth, and visibly a great way above the Scotch firs. The nearest, and most accessible, are immediately above the small and narrow *Kaafiord*, which enters Altensfiord. There a number of small vallies, abounding in streams, run up the mountains, the bottoms of which are covered with alders, and beneath the alders with an innumerable quantity of mountain brambles *(Rubus Chamaemorus)*. But on the declivities, and above these birches, the Scotch firs rise like a thick wood, and the firs are succeeded by birches. When the firs are contracted by the cold of the elevation, the birches look still strong and large, and seem to be particularly fond of this temperature. Close even to the summit of Skaane-Vara, the nearest and highest of the mountains which are situated to the south of the Kaafiord, the birches appear in a thriving condition, and they would probably have also covered the summit, if the violence of the wind would have permitted them. Skaane-Vara is, however, 1406 English feet above the Fiord.[†]

In Nordland, at a heighth of 1491 English feet, the birches were sickly bushes, but here they would have risen to a greater heighth in the form

---

[*] Altengaard, h. 4. 27. 11. 1. 18, cloudy, south-east. Kongshavnsfieldt, h. 6. 27. 3. 4. 17.
[†] Skaane-vara, h. 5. 26. 2. 8. Therm. 7. Strong west wind.
 At the sea, h. 7. 27. 7. 4. Therm. 9 25. Calm. Barometer rising.

of trees. At a heighth of 640 English feet the Scotch firs appeared large and beautiful, and did not disappear till we came to between seven and eight hundred feet above the sea. Hence the temperature of Alten must nearly reach -¦- 1° of R. or 34°. 25 of F. a higher temperature than is necessary to the thriving of spruce firs; for the spruce does not disappear till the mean temperature falls below ─ 0. 75 R. or 33°. 68 F. Still, however, we may clearly see that the hope of seeing woods of spruce fir in the Valley of Alten would have small chance of being realized. The quarter of a degree R. which Altën possesses above the spruce climate, would no doubt be sufficient to allow a few trees to attain a large size, where they were sheltered from wind and weather, but not whole woods; for woods are exposed to all the changes of the climate; and these changes are immediately destructive to the trees, when they are reduced to live within a quarter of a degree R. of their mean temperature. The most destructive accidents are every year to be dreaded in a climate where the mean temperature of the year frequently differs from one another in a number of degrees. It would be nearly the same as if we were to plant birch woods at Christiania, or nut-woods in Denmark.

Skaane-Vara, though of no very considerable heighth, domineers over almost all the mountains which surround Alten. Bonasfieldt alone, on the east side of Alten, rises to a greater heighth in its farther course; the summit, visible from below, is not so high as Skaane-Vara. All the heights in the neighbourhood round Refs-Botn, or down the east side of Altenfiord, are still less high. The lofty mountains at Talvig, and above Langfiord, are in fact the last remains of the great range of mountains which has hitherto continually divided the northern peninsula. From this place, and onwards, between Finmark and Sweden, and towards Russia, the mountains lose altogether the distinguished rocky and divided form which was hitherto peculiar to them, especially along the Western Ocean; and the mountains in the interior of the country appear all along only hills, when we compare them with the masses in Helgeland, and perhaps in Nordland. Single branches, however, run in the direction of East Finmark, and divide themselves between the long Fiords; but Alten's Elv penetrates through all these arms, and at last issues out of the straits in the great valley of Alten, about nine English miles before its entering the Fiord.

From thence the fir-wood begins, and the better climate, and almost at the same time the hamlets of the inhabitants, with their corn fields; for almost every hamlet is surrounded by a small corn field. They are situated on hills along both sides of the stream, till they are at last collected into a village at its outlet at Elvebacken.

Alten is not only the most agreeable, the most populous, and the most fertile district in Finmark, but also the only one in which agriculture is carried on—the most nortnern agriculture of the world. This merit is due to the Quäns in Alten. Before they appeared the cultivation of grain had never been tried. They may now have inhabited these vallies for nearly a century; and they brought along with them diligence and industry into the country. They were very probably driven out by the wars of Charles XII. and especially by the cruel havoc made by the Russians in Finland of their flocks and herds. They went higher and higher north, till at last they passed beyond Torneo, and first descended into Alten about the year 1708. The first emigrations were followed by others; and since that period they have to the great advantage of Lapland perpetually continued, to such a degree that the Laplanders themselves, not without reason, are in fear that the Quäns will at last take possession of the whole of their country, and drive them completely out. This they might easily prevent if they were to follow the example of the Quäns, and select constant habitations, and cultivate the ground. The Quäns still resemble their ancestors; they live in the very same manner, and observe the same customs. They speak exactly the same Finnish language which is spoken throughout all Finland, and which bears less resemblance to the Laplandic (or the *Finnish*, as it is called in Finmark) than the Swedish bears to the German (Leem. S. S. 10, 11.) Their houses are wholly constructed, for the most part, like those in Finland, and in quite a different manner from those of the Norwegians. The greatest part of the house consists in a large room of logs, the *Perte*, which reaches up to the roof. On one side there is a large furnace, without a chimney, which takes in the greatest part of the wall. The smoke from the furnace rises up towards the roof, descends along the walls, and issues out through several quadrangular openings in the remaining walls, about three feet from the ground. When the fire is burnt out, they shut up the furnace and collect a Syrian warmth in the *Perte*. The upper part of the furnace serves for the sweating baths, everywhere used in Finland and Russia. In their dress alone the Quäns do not differ from the Laplanders; in their manners they completely differ. The Quäns are the most civilized inhabitants of Finmark, not even excepting the Norwegians.* They are distinguished for their understanding: their comprehension is easy and rapid, and they do not dislike to work.   Hence they easily learn all the trades which are necessary for ordinary establishments; and the progress they are capable of yet making in agriculture, and, consequently, in the arts of life, is proved by the peasants of Torneo, Uleoborg, and Cajaneborg.   Even the pernicious influence of a sea life, the expectation of profit, without laying by any thing for times of want, has never manifested itself among the Quäns to the extent which it has among the Norwegians and Finns; and hence it is possible enough that they will in time not only drive the Finns from their districts, but also the Norwegians themselves. The prosperity of the country will lose nothing by it.   Why this people is called *Quäns* here is as little known as the origin of *Lappe* and *Finner*; but they are all equally ancient.   The oldest Icelandic *Sages* speak of *Quäns* and *Quänland:* even *Eigla* (Torfäus I. 160) lays down the situation of the country pretty accurately. She says: *Eastwards from Nummedalen (at Drontheim,) lies Jämteland, then, farther eastwards, Helsingeland, then Quänland, then Finland, and lastly Carelen.* Under this was probably understood the greatest part of the present Finland; and it has been so laid down by Schöning and Bayer in their maps.   The name disappeared after King Erich the Saint took possession of the country in the middle of the twelfth century, and subjected it to his authority; and now the general name of Finland and Fin is all that remains in Sweden.   The oldest geographer of the North, *Adam Von Bremen,* had heard something of this country, but being unacquainted with the correct Icelandic writers, he was deceived by the name: he transformed *Quäner* into *Quiner* (Women,) and *Quänland* into *Quindeland* (the Land of Women;) and he was hence induced to lay down here an Amazonian country, which the native writers never dreamt of. This was eagerly laid hold of by Rudbeck and his scholars, who imputed to this Amazon land all that the Greek writers had related of the Scythian Amazons. Schöning has hardly been able to extirpate these romantic notions by his excellent treatises (*Gamle Geographie, p.* 64.); for even in recent times a Magister *Eneroth* wished to prove that the Amazons did not inhabit *Osterbottn,* but the Swedish

---

Quäner ere uden modsigelse de duelligste af Vestfinmarckens Indbyggere, says M. Dahl, the late clergyman of Talvig, in his manuscript *Chorographie* of *Vestfinmarcken.*

province of *Norrland*; and we cannot help being grieved at seeing similar things repeated in the last edition of Tuneld's Swedish Geography, notwithstanding the learned Giörwell is given out as its editor.

The Quäns were a quarrelsome people; they frequently came over from the Bothnian Gulph to Finmark and Nordland, and committed depredations on the Norwegians and Finns, which they in turn endeavoured to repay by their prædatory incursions as far as *Carelen*. Is it to this momentary appearance that they owe their name in this country? Or must the Finlanders have another name, as the Laplanders had already taken possession of that of Finn? The name actually disappears when both people are no longer in immediate contact. Even in Helgeland nothing is known of Quäns, and still less is known of them in the South of Norway, or of the country inhabited by the Laplanders. The Swedish practice is there followed, and the *Finlanders* are called *Finns*; and this sometimes gives rise to misconceptions and errors. Several thousands of Finlanders, perhaps, live in the western part of Dalecarlia, and among the mountains of Orsa Socker, above the Eastern *Dal Elv*, who were invited there, it is said, by Charles IX. and who still retain their language and customs, notwithstanding they are surrounded by Sweden, and far distant from their original country. The country which they inhabit is in Sweden called Finmark. This appears to have misled Tuneld to connect this Finmark with the Norwegian Finmark, and the Norwegian Finns. Finmark, says he (*Geographie* 1 111.) is a name given to a tract of country which runs from Bahuslehn along the Norwegian Frontiers, all the way to Lapmark. There the remains of the first inhabitants of the country still live, who, driven out by *Othin's* conductor, advanced farther and farther northwards, and now alone inhabit Lapmark. What errors! Neither in Bahuslehn, nor in Dalsland, nor in Elvedal, nor Herjeadalen, are there any Finlanders, and consequently there is no Finmark there. And how is it possible to join the Finns of Orsa with the Norwegian Finns, or the Laplanders? Nomades with diligent Agriculturists? A people who yet speak the language of *Abo*, with Laplanders who do not understand a word of Finnish? The Finlanders of Dalecarlia have also advanced into Norway, and have peopled and brought under the plough several districts in the Prastegieldts of Tryssild, Grue, Elverum, and Vinger. They rooted out, and set fire to the woods, cultivated rye among the ashes, and procured in this way rich harvests. But they remained in the Gaards, which they first constructed in the valley where they still dwell. The indefatigable missionary, Thomas Von Westen, heard of these Finns in Drontheim in 1719, on his return from his third journey to Finmark : and his zeal would not allow him to rest till he had also converted them to Christians. But the college of missionaries in Copenhagen dissuaded him from his purpose, as the place inhabited by the Finns belonged to the bishoprick of Christiania, and they were afraid of the powerful and highly-dreaded Bishop Deichmann of Christiania, who, secure of the king's favour, acted in every thing in a very arbitrary and disrespectful manner. This, however, with M. Von Westen, was but throwing oil into the fire. With a true zeal for proselytism, he boldly threw the souls of all these *Rugfinns* on the bishop's conscience, and proved that he would be answerable for them at the last day. The Missionary College could oppose nothing to such weighty reasons. With the utmost caution, almost trembling, and after several years consideration of the matter, they ventured, in 1727, to represent to the mighty bishop that it would be useful and necessary to dispatch missionaries also among these Finns, which were independent of the bishop, and belonged to the Missionary College. They received the unlooked for answer, that their zeal was here perfectly unnecessary, as the Finns of Christiania lived like Norwegians, and as far back as any thing was known of them, had always been christians, like the Norwegians.* Had they been called Quäns, or merely Finlanders, they would have spared M. Von Westen many a sad hour towards the end of his life, and the Missionary College many a consultation how they should represent so ticklish an affair to the bishop, for none of them would have thought of doubting that the Finlanders had been christians for centuries. These Finns then were of opinion, in 1727, that they had come over from Sweden more than a hundred years before, and previously from Tawastehuus in Finland : both very probable circumstances.

At present the Quäns actually constitute by far the greatest part of the population of Alten, and in the valley of Altens-Elv they are almost the only inhabitants. In the year 1801, of one thousand seven hundred and ninety-three souls who inhabited the Prästegieldt of Alten, only four hundred and seventy-five were Norwegians. Of the remaining one thousand four hundred and ninety-three, a few hundreds only were sea Laplanders *(See-Finner)* who lived in Langfiord and in Stiernesund; the remaining one thousand two

---

* Hammond's Missions Historie. S. 504.

two hundred, and consequently nearly two-thirds of the whole population were Quäns. A cruel, slow, and infectious disorder, it is true, considerably diminished their number in 1806. Whole hamlets died : in remote places, men were found dead before their houses and on the roads, who in the midst of winter had gone out in quest of assistance for their suffering families, but being themselves attacked by the malady, for want of strength had fallen down by the way. The disease spread over all Finmark, and gave a very considerable shock to the population of that province. Alten will more easily, however, recover from it ; for the meadows and fish of Alten are still sufficient to protect the industrious Quäns from the gripe of hunger: and of the rapid increase of the population afterwards, such striking examples are furnished by Cajaneborg, Kusamo, Sodankylä, that they deserve the utmost attention of every politician. The Norwegians in Finmark have in general never inhabited the interior of the country. They remained always on the extreme islands towards the sea, for the sake of more easily carrying on their fisheries. Hence there are yet a much greater number of Norwegian habitations at the North Cape than in the interior of the Fiords; but they did not settle there either till the flourishing trade of Bergen began to turn the fisheries of Finmark to account, that is, till after the year 1305. We find no account before this of any Norwegians settling themselves beyond Tromsöe. Finmark was looked upon and treated as a tributary province ; and the Laplanders as a people who were bound to deliver, not only to the lord of the country and his vassal, the skins, feathers, and furs, which were the produce of the country, but the same also to many a powerful individual of the country, who expected a similar tribute. *Other* tells King *Alfred of England* that in his *Periplus* (in the year 850) the richer Laplanders were compelled to deliver yearly to him fifteen martin skins, five rein-deer, a bear skin, ten bundles of feathers, a bear skin coat, another of otter skin, and lastly, two ship cables of sixty ells long each, the one prepared from whale skin, and the other from the skin of the sea-dog. The poor Laplanders were to contribute in proportion.* One is astonished at the patience of a whole people, who could

* *Schöning Forsöck til Norges gamle Geographie Soröe,* 1763, and *J. R. Forster's Geschichte der Entdeckungen und Schiffahrten in Norden,* p. 85. It is singular that neither Forster nor Sprengel seem to have held Schöning's excellent treatise in any degree of consideration. Both are silent respecting it, though both of them quote from other works of Schöning. Hence

good-naturedly allow such heavy burthens to be imposed on them ; but we may conceive that the Norwegians would not have derived much advantage in driving out the Laplanders.　Other says that no Norwegian lived beyond Senjen, and that it was the most norther settlement of his nation.

The Norwegians who at present live in the interior of Altensfiord are for the most part descendants of persons who were banished there: they have occupied the whole coast from Langfiord above Talvig to Kaafiord and Bosecop.　Some of them also live in Refsbottn, and not far from Altens Elv. They have now forgotten that they are not natives ; they live tolerably well on their Gaards, and would attain a sort of prosperity, if they carried on their operations with the industry of the Quäns.

If we reckon the Prästegieldt ten miles (forty-six En lish) in length, and nine in breadth (forty-one and a half English), its contents will be ninety square miles, (one thousand nine hundred and nine square English miles), and the number of inhabitants to the square mile about twenty-two, (nearly one to the English square mile).　We no where find the same again, neither in Danish, nor in Swedish Lapland, a few districts of Sodankylä perhaps exeepted.　The whole area of Finmark, according to Pontoppidan, contains one thousand two hundred and forty-four square miles (twenty-six thousand three hundred and twenty-three square English miles) ; but the population of the province in 1801 was only seven thousand eight hundred and two. The whole province consequently contained only six inhabitants to the square German mile.　The area of Swedish Lapland, according to Baron Hermelin's Statistical tables, contains one thousand six hundred and sixty square miles (thirty-five thousand one hundred and twenty-five English) ; and the population of the country amounted in 1799 to eleven thousand one hundred and sixty-two ; so that the number of inhabitants there to the German square mile amounts to about seven.

Alten is not only different in its civil and political relations, and has something peculiar in the external appearance of the country and the mountains, to dis-

Forster places *Other's* risidence in what is now called Helgeland, notwithstanding it is pretty distinctly proved by Schöning that he must have l.ved beyond Senjen.　But when Schöning, Forster, and others, suppose that Other did not catch whales in Finmark, but morses, they are evidently under a mistake ; for as far as I know, morses have no where been yet seen on the coasts of Finmark.

tinguish it, especially from Nordland, but the internal composition of the mountains is also very different in the two districts. We may take a wide circuit in Alten before we light upon gneiss, which we do not find till we proceed out to the Fiord. In the hills, of at least two hundred feet in heighth, formed of materials rolled from the mountains which surround Altengaard like a mound, a great variety of stones are mixed together, and almost all of them stones which we did not see on our way: but pieces of gneiss are very rare among them.

The hills run for a like heighth for the space of an English mile from Elvebacken, and from Altens Elv onwards to Kongshavnsfieldt. We may consider these rolled heaps as a collection of all that appears in the mountains up to the source of the Altens Elv, that is, more than ninety English miles up the mountains; and if this is the case, we learn that in the whole way to the borders, and into the interior of the mountains, the older primitive formations are but rare. This is also a confirmation of the splitting of the Kiölen mountains before reaching Finmark, and the probability of their running between Quänanger and Altensfiord, and above Stiernöe and Seylandt to the North Cape; for the Kiölen mountains in Nordland are gneiss and mica-slate mountains; but in the rolled hills of Altengaard we see scarcely any thing but coarse splintery quartz of various colours, black and fine granular lime-stones, pieces of clay-slate, quartzy sand-stones, fine granular hornblende, and not unfrequently grey diallage, and white felspar, in small granular mixture, or still more frequently grey diallage, fine granular and almost imperceptible, and then scarcely felspar between. All these stones point to a formation in which every thing is unknown, and in which all the marks which characterise the separate fossils are concealed in the extreme minuteness of the individual parts: they lead us to the *transition formation*, but not, however, into it. For a true and distinct transition stone is not to be found among these rolled blocks; and even the nearest of the prevailing mountain-rocks around Altengaard do not belong to it.

Kongshavnsfieldt to the very top seems a quartz rock; the most of the strata are in fact nothing but pure, smoke-grey, very coarse splintery quartz, but little transparent, red in some of the strata, and in others of thin reddish brown. Its being always found here coarse splintery, so little transparent, and so much coloured, and never white-greyish, or reddish-white conchoidal, though ever so imperfectly or fully transparent, is sufficient to distinguish these quartz rocks from bed quartz, which sometimes rises in mica-slate to high and extended rocks, and which is often considered as a peculiar formation. In this last there is also sometimes found some felspar, or at least mica in large and distinct folia. In the quartz of Kongshavnsfieldt, however, there is no where distinct mica, still less druses, notwithstanding small white quartz veins not unfrequently traverse the strata. This quartz at the foot of the mountain, especially towards Bosecop, lies on a dark blackish-grey slate, which shines little, and which is neither mica-slate, nor clay-slate. It is true, the rock towards Urnäss, a charming little island to the eastwards beneath the Fieldt, appears to bear some resemblance to mica-slate, but the mica does not free itself from the quartz: it appears here only accidental, and is also not extensively distributed as pure mica-slate. The quartz may consequently be here like the quartz of Alt-Eid; it lies there where clay-slate might also be, and is nearer related to it than to the mica-slate. This is more fully disclosed when we proceed farther up the valley. In the first rocks on the Altens-Elv, and not far above Kongshofmarck, the residence of the *Foged*, nothing but quartz seems to make its appearance. When we consider it more narrowly, however, its mass is soon discovered to be actually a *quartzy sandstone* : dark smoke-grey grains, which a pure quartz mass binds together; in fact, such a stone we should not be surprised to find in greywacke.* Similar stones lie along the declivities of Skaane-Vara. Although this stone does not belong to greywacke, yet it is very remote from mica-slate : it is probably above the clay-slate. And although the situation of these masses in the valley here does not seem to lead convincingly to these conclusions, the constitution of the opposite bank will not allow us to entertain a doubt on the subject. On the heights there which surround the Pors-Elv, the noisy stream of the saw-mills, and at the tops of them, lie considerable strata of greenish-black, and very fine granular *diallage rock* without felspar. The strata are separated by innumerable fissures, and there is not unfrequently a covering of grass-green epidote over the fissures, scarcely

---

* Many of the rocks described by mineralogists and travellers as quartzy sand-stone appear to me to be but varieties of quartz, having a porphyritic structure. Indeed, it would not be surprising if it should be proved that many great tracts of rock at present described as sand-stone are but varieties of arenaceous quartz. Probably the quartzy sand-stone described in the text may be a porphyritic quartz.—*J.*

thicker than paper. In the midst of this rock, however, lime-stone beds make their appearance, and this lime-stone immediately betrays how short a way it, and all the substances connected with it, ascend in the primitive formation; for this limestone appears broken as in the alum-slate at Christiania, or as at Storsjö in Jämteland. It is *dark smoke-grey, fine splintery*, but faintly translucent, and scarcely more than fine granular, even in the light of the sun. How different from the white dolomite in Senjen, or even from the limestone beds in the mica-slate at Alt-Eid! It is certainly singular that it should lie in diallage stone, or in fine granular greenstone; but it proves from this how much we ought to consider these stones as belonging to the clay-slate formation. Even the small and very steep island of Bratholm, in the middle of the Fiord between Pors-Elv and Bosecop, consists also of this fine granular diallage, as do probably also three or four other smaller islands (Holme) towards Talvig. Hence the district of Alten presents us with a series of stones, which connect the primitive with the newer formations; but these newer formations themselves remain behind in middle latitudes, and have not penetrated so far northwards.

HAMMERFEST, the 23rd of July. Early in the morning of the twenty-second I left Altengaard in a small boat, and proceeded down the Fiord. A gentle and warm south wind filled the sails, and carried the boat in a few hours to Altennäss, a cape which projects a great way into the Fiord. We then began to have the *Havkirlje*, the north wind, from the sea in our teeth, which the cape had hitherto sheltered us from. The Quäns flew to their oars, and by the shelter of the islands, and the points beneath the rocks, they succeeded in making a pretty rapid progress. The east side of the Fiord here is every where steeper, and consequently less cultivated, than the opposite coast of Langfiord above Talvig, all the way to Altengaard. From Refsbotn to Näss there is but a single habitation in a bay, Storvig, which is occupied by a boat-builder; and yet the space is about five English miles. In this bay we see the last Scotch firs: they are the most northern firs of Europe. They are by no means dwarfish, but still far from the beauty of the firs in the valley of Alten. The climate has visibly changed in the small number of miles we have proceeded in getting out of the Fiord. The three large sounds of Stiernesund, Rognsund, and Vargsund, which meet together as in a centre, between the two capes of Altennäss and Korsnäss, and which drive back the winds, the clouds, and the fogs of the sea, may perhaps contribute in no small degree to produce this. Korsnäss and Korsfiord *(Cross Cape, Cross Fiord)* have probably derived their name from this circumstance ; and in fact, we get out of all the three sounds from thence at the very same time : and the view is the more striking, as all these straits are every where surrounded by high mountains. Between Langfiord and Stiernesund, the mountains approach very near to the limits of snow : they are higher still on Stiernöe, and on Seylandt they rise a great way beyond perpetual snow. We proceeded through the Vargsund, between Seylandt and the main-land : the water on the Seylandt side appeared completely black from the frightful rocks which fall perpendicularly into the Sound. Where the Beckerfiord opened to us a view into the interior of the island, the crown of eternal snow shone above the black rocks; a splendid canopy, which was spread over the whole island. From this place the island appears altogether uninhabitable and desert ; for there is not even the smallest spot of green visible between the mountains. Towards the north, however, they decline more gently, on which side there are even rows of hills and plains between the mountains and the sea. It is probable that the strata ascend towards the south-east, and dip towards the north-west; and it is also probable the Vargsund has been opened by the elevation of these strata.* In this point of view, the Sound is a continuation of the Langfiord, which actually enters the land in an equal direction; for Langfiord also probably arose from the elevation of the Jockuls-fieldt, and this Fieldt and the mountains of Seylandt would form but one and the same chain, if they were not interrupted by Rognsund, Stiernesund, and

---

* Von Buch, in different parts of this work, speaks of the strata of mountains being *elevated* and changed from their original position. For some time I was in the belief that he attributed this elevation to the action of a force acting from below : in short, that he had adopted the ancient opinion of a central fire, which in our times has been supported with so much address by Dr Hutton, Sir James Hall, and Professor Playfair ; but very lately, on reading a paper of our author's in the Magazine of the Society of the Friends of Natural History in Berlin, en-titled, " Reise über die Gebirgszüge der Alpen zwischen Glaris und Chiavenna," I was undeceived, and found his opinion to be the same with that of M. De Luc, who supposes that horizontal or slightly inclined strata have frequently sunk down at one extremity, and conse-quently *risen* at the other. In the paper just mentioned, Von Buch extends this hypothesis to many tracts in the Alps of Swisserland, and endeavours to explain by it the southern dip of the strata in the north side of the Alps, and the appearances presented by many of the vallies.—*J.*

Rivar-Eid.   In the Vargsund we again find the Nordlandian nature in the formations.   The newer rocks of Alten, the greenstone and the quartz, are lost in the district of Storvig, and consequently before we reach Altensnäss. Aaröe, the island between both capes, is entirely mica-slate, in which the continuous and shining mica surrounds a multitude of nuts of conchoidal quartz, which run along. in the direction of the strata; but there are no garnets to be found in it.   The strata of Aaaröe stretch E. and W. (h. 6.) and dip about 30° towards the north.   Beyond Korsnäss they somewhat change their direction; they stretch E. N. E. and W. S. W. (h. 4.) and dip towards the north-west; as is also probably done by the strata in Seylandt.   They continue so the whole way through the Vargsund, and every where the mica-slate predominates; not only pure, but very well characterised in its subordinate beds.   Such beds of beautiful hornblende appear beyond the Lerrits-fiord, along the steep shore.   The beds become so thick, and they follow one another so rapidly, that the whole formation seems to be nothing but horn-blende, and its deep black is so striking, that we might frequently imagine we saw rocks of coal.   But the hornblende is almost coarse granular, and very shining, and soon therefore betrays its nature.   Its colour has actually, however, excited hopes of coal.   It appears exactly here as it does in the Stiernesund, and there they attempted to burn it.   As this did not however succeed, they thought, as in the trials for alum-slate, that this unripe coal might possess greater maturity below.*   Very thin beds of white and ex-tremely fine granular limestone, which are separated into limestone slates by white mica folia, are very frequent between these black strata; and then others of mica-slate, which are not thicker, appear like stripes over the rocks, and like bands on the black hornblende.   Thus in these sounds the mica-slate has distinctly gained the ascendancy over the gneiss : and as all the strata dip towards the sea, we can have but faint expectations of again finding the older gneiss any where towards the north.

Towards evening we entered Strömmen, (Strömmen Sound) the straits which separate Seylandt from Qualöe.   The whales every where sported around the boat.   They rose and sunk in the water, and appeared again

---

* Amtmann Sommerfeldt, Finmarcks Bescrivelse.   Norsk Topogr. Journal, XXIV, 113.

immediately at a great distance above the waves, like small moveable islands. Their diversified movements contribute very much to relieve the monotony of a voyage in a boat; but in this time of pairing they are somewhat dangerous. These animals frequently mistake a small boat for one of themselves, rush towards it, dive under, and raise the boat aloft, or overset it. Hence no one ventures to sail in the direction of such a fish; and the boat's crew frequently take long circuits when they perceive that they are steering in the exact line of the whale's motion. In the former winter, a whole fishing boat was raised aloft at Harvig in this manner; and as it fell again on the surface of the water, it went immediately to pieces with the violence of the shock, and the fishermen were saved with the greatest difficulty. These animals are therefore left in the Fiords and Sounds to their undisturbed freedom; they are not taken, and only used when one of them is driven dead on the shore. It is not the great species of Spitzbergen; and it is asserted that in these smaller fish the quantity of fat is not sufficient to repay the toil and the labour of the capture. In old times, however, the whale fishery of Finmark was even celebrated. Other mentions that the Norwegians annually visited Finmark on account of the whale fishery, and that he himself killed there sixty whales in two days, of the length of between forty-eight and fifty ells each. Even in later centuries, this fishery was found sufficiently profitable, and so late as 1689, vessels, principally Dutch, used to lie below Finmark. Many traces of this are still to be found on the coast towards the sea. The walls of the church-yard of Harvig are principally built of whale skulls, and on the other islands they are frequently found along the shore.* That the great whale has actually deserted these coasts is not to be wondered at: it is no longer allowed to get down from Spitzbergen; and the activity of the company in Archangel, who allow their crews to winter at Spitzbergen, will at last drive the whale also from that country.†

---

* It should be generally known, that the only accurate figure of the great Greenland whale at present before the public is that of Mr. Scoresby of Whitby, in the first volume of the Memoirs of the Wernerian Natural History Society. All the representations of that enormous animal, from the earliest engravings to that of La Cepede, are incorrect, and give no accurate idea of its figure and proportions.—J.

† Sommerfieldt, 132. All Other's commentators affirm unanimously that morses, and not whales, were caught here, and that thongs were made of their skins. I am persuaded, however,

A narrow and low tongue of land, near to Hammerfest, runs out opposite to Seylandt, which occasions the passage through the Sound to be very narrow, and it is narrowed still more by several cliffs in it. The tide from the ocean is here confined, and precipitated into the Sound with prodigious noise like a waterfall; and the ebb returns with equal fury. The passage of large vessels through this Sound is therefore completely interdicted, and even boats labour against the current with great difficulty. The high pyramid of Tyvefieldt, with its steep precipices, appeared on the other side of the straits: the highest mountain on this part of the island, and to appearance almost like the Niesen above the *Thunersee*. It designates the situation of Hammerfest. We soon afterwards saw the circular bay, the houses on both sides, and vessels in the harbour.

The town, however, promised more in the distance than it realized on our approach. The houses on a small promontory on the south side of the bay appear the commencement, and we expect to see behind the hill the continuation of a greater town. But there is nothing more: the whole town

---

that we may confidently affirm this to be an error. I never heard of morses in West Finland. The climate is also by no means such a one as the morses are fond of; for there are never any ice islands in the neighbourhood of the North Cape.

The morse or walrus (trichechus rosmarus) oft n appear in northern countries in herds of eight or ten thousand on the coasts, and far distant from ice islands. This fact, conjoined with the following observations of Mr. Pennant, seems to leave little doubt of the accuracy of the earlier writers that the morse was formerly caught on the coasts of Western Finmark. " If they (the walrus or morse) are found in the seas of Norway, it is very rare in these days. Leems, p. 316, says, that they sometimes frequent the sea about Finmark; but about the year 980 they seemed to have been so numerous in the northern parts as to become objects of chace and commerce. The famous Octher the Norwegian, a native of Helgeland, in the diocese of Drontheim, incited by a most laudable curiosity and thirst of discovery, sailed to the north of his country, doubled the North Cape, and in three days from his departure arrived at the farthest place frequented by the *horse whale* fishers. From thence he proceeded a voyage of three days more, and perhaps got into the White Sea. On his return he visited England, probably incited by the fame of King Alfred's abilities, and the great encouragement he gave to men of distinguished character in every profession. The traveller, as a proof of the authenticity of his relation, presented the Saxon monarch with some of the teeth of these animals, then a substitute for ivory, and valued at a high price. In his account of his voyage he also added, that their skins were used in ships instead of ropes." Pennant's Arctic Zoology Vol. 2. p. 170.—*J.*

consists only of these houses, with the exception of a single Gaard on the other side of the bay. There are only nine habitations with the clergyman's house; four merchant's, a custom-house, a school-house, and the only mechanic—a shoemaker. The population of Hammerfest does not therefore exceed forty souls, and as it is the most northern, it would also be the smallest town in the world, if a Russian town of the name of Avatcha did not contest with it that pre-eminence. This place has fallen short of the expectations formed still more than Tromsöe. Both places, along with Wardöe, were raised to the rank of towns in 1787. Hammerfest was destined for the central point of the exportation from West Finmark, all vessels were to clear from thence for foreign countries, and this liberty was to be enjoyed by no other harbour in Finmark. It was supposed that Hammerfest would soon draw to itself a great part of the Russian trade, as the sea at the North Cape is not only always open, but even none of the Sounds and Fiords in Finmark are ever frozen, and consequently they are navigable at all seasons; whereas the whole of the White Sea, for four months, is covered with ice, and Archangel completely blocked up. It was supposed that the produce of Archangel would be brought to Hammerfest, and foreign nations, it was hoped, would rather take them off from this place than enter upon the long and dangerous passage to Archangel. All these plans have failed; Hammerfest would hardly take any share in the trade of Archangel, even if Finmark were a Russian province. Trade always takes the simplest courses; and certainly to establish magazines of English and Russian produce in the Island Hvalöe, in a latitude of 70° 40', for the purpose of producing a mutual exchange between places removed at a great distance from one another, merely for the conveniency of being able to navigate these seas a few months longer, is a course remarkable for any thing but simplicity. To induce foreign nations to frequent Hammerfest, the place must offer some more advantages to them than merely an open water. Provisions cannot be obtained here; not even wood for firing; for the inhabitants very reasonably insist on foreigners being prohibited from felling wood in their scanty birchen thickets. The deep snow in winter will only allow them to cut off the tops of the trees; the greater part of the stem, from eight to ten feet, remains rotten under the snow, and can no longer be used. The inhabitants on the other hand can fell the trees before the stems are buried under the snow. A twenty years experience is a sufficient proof that even the trade of West Finmark can never be successfully carried on from this place. The situation and the division of the numerous fishing stations, and fishermen in Finmark, have in like manner divided the merchants who take off the fish and deliver corn in return among the different Fiords; and consequently vessels must take in their lading at different places. Hammerfest will therefore never be more than what it is; perhaps it will even decrease, if any of the merchants at present resident there should again leave the place. The harbour is very small but good. It is formed from the Cape on which the town is situated, and which projects far into the bay. The bay itself, notwithstanding it goes deep into the land, is not sufficiently sheltered from north and north-west winds. In this harbour three or four vessels may lie comfortably and securely in winter; but more will not easily do so: they anchor in from sixteen to eighteen fathoms water.

Hammerfest is almost a whole degree of latitude to the North of Alten. How powerful, however, is the difference of climate and external appearance of the country between the two places! This island produces nothing: nature remains in perpetual torpidity, or suffers under the pressure of a perpetual fog. No trees grow here, and they in vain endeavour to rear a few garden stuffs beside their houses: they will not come up. In the vallies of the island birchen bushes may be seen; they are sufficiently close to one another, but they never become trees. In vain they strive to ascend the declivities of the mountains. At a very small heighth they become shrivelled and disappear, and they cannot reach the higher vallies. Such are the high Alpine mountains above St. Gotthardt; no trace of cultivation or inhabitants. A number of smaller and larger lakes lie scattered among the rocks; and the streams which issue from them are precipitated from one valley into another. There are even lakes at the top of Tyvefieldt, and from the height there a hundred at least are visible, which descend into the vallies, and meet together from all sides. The last birches of any consequence stood here, at an elevation of six hundred and sixty English feet; they were bushes of about three feet in heighth. On the opposite mountains they do not rise higher, nor on the mountains in the interior of the island. If any thing of the birch kind is visible at a greater heighth, it creeps like an herb on the ground, but no where resembles a bush. Their limits do not here exceed eight hundred and fifty English feet; whereas at Skaane-vara; in Alten, they attained a great size at a heighth of one thousand three hundred and eighty English feet. This gives a much lower mean temperature for Hammerfest, scarcely more than 29°. 75′. of Fahrenheit,* more than a degree below the temperature of the Scotch fir. Hammerfest cannot even, therefore, in respect to climate, be compared with the hospital of St. Gotthardt; for to attain the mean temperature of Hammerfest in the Alps we must ascend three or four hundred feet above the pass of St. Gotthardt. If the temperature of Alten is 34°. 25, in this single degree of latitude the temperature has consequently fallen 4°. 5 F. On the coast upwards 1° 36 F. of decrease for a degree of latitude was hitherto an extreme. Such is the difference of climate in the interior of the Fiords, and in places exposed to the open sea. The sun appears to these islands only as a rarity; the summer is without warmth, and they dare hardly hope for a few days possessing some small degree of clearness. In a few moments the north-west wind covers the country with thick clouds from the sea; torrents of rain burst from them, and the clouds run along the ground for whole days. Deeper in the Fiord the showers of rain are slight and passing, and in Alten we there see, during a bright and clear sun, nothing more than a dark stripe of clouds along the horizon towards the north. It is the north wind in particular, and indeed almost solely which incessantly in winter drives the large masses of snow towards Hammerfest. On this account it is true the winter is less severe than at Alten; and we should scarcely see the Mercury freeze here in the open air, a circumstance which happens in Alten. Hammerfest expects clear weather from the south-east, and in winter the severest storms; so severe that it is impossible to remain erect out of the houses. Is it the warm sea air which draws the colder air from the vallies and the Fiord with such impetuosity?

If it were not for the fish in the ocean, who would select such a place for his residence ?

The island on which Hammerfest is situated, *Qualöe* or *Hvalöe* (Whale Island), is by no means small, and may be numbered among the most considerable in Finmark. Although it does not reach the heighth of Seylandt, its mountains do not belong to the most inconsiderable of this country. Tyvefieldt, it is true, is more distinguished from its almost insulated situation and the rapid ascent of its mass, than from its heighth; for the mountain is

----

* But this low mean temperature is again modified by the proportionably very mild winter.

only one thousand two hundred and fifty-one English feet in heighth.* But we see from its summit the ranges of mountains on the east side of the island; a chain deeply covered with snow; and these mountains rise far beyond two thousand feet in heighth. They rise gradually from Hammerfest, and the greatest heighths are quite near to the eastern shore, nearly as at Seylandt. We might therefore be led to suppose the strata of the rocks would in the same manner dip from east to west. This is not however the case at Tyvefieldt, as the strata stretch there E. N. E. and W. S. W. (h. 4—5), and dip very strongly towards the south east. The rocky tongue which shoots out into the Strömen Sound, opposite Seylandt, runs in exactly the same direction; and the snow mountains of Seylandt, towards Langfiord and the *Fieldts* of Jöckulsfiord, run also exactly in the same manner; for we can see this very distinctly from the height of Tyvefieldt. Hence it is possible that this strong dipping towards the south-east may be only an anomaly here, and that the general dipping of the strata on Hvalöe may also be to the west, and on that account the greatest precipices may be situated towards the east.

We find no longer any mica-slate in the neighbourhood of Hammerfest. It is well characterized gneiss; so much so indeed that the mica never appear here continuous, but always scaley, and very thick scaley black and splendent. A great deal of red and small granular felspar lies on it at the bay; but little quartz. Higher up the felspar becomes white and somewhat more rare, and even the mixture is somewhat more small granular. A number of small red garnets are then every where scattered about in it; and that all the way up to the summit of the mountain Even at the very top these garnets are not unfrequent, and entire red streaks *(Flammen)* of them appear amidst the felspar. This gneiss does not on this account become more like mica-slate; for the garnets are not in general so insulated and so purely crystallized as they usually are in mica-slate. We are struck by their colours as much as by their form. Does mica-slate make its appearance farther north or farther south on the island? It would certainly be of im-

---

* Hammerfest, h. 2. Bar. 27. 11. 3. Therm. 12. 5. R. Clear, calm.
Birch-boundary, h. 4. Bar. 27. 3. 7. Therm. 11. 25. R.
Tyvefieldt...... h. 5. Bar. 26. 8. 2. Therm. 10. R.
Hammerfest .... h. 9. Bar. 28. 0. 4. Therm. 9. 5. R.

portance to ascertain this; for till this takes place, it still remains a doubt
whether we ought to consider this gneiss as belonging to the older formation
or to the newer which lies on the mica-slate.

HAMMERFEST, the 24th of July. The harbour has become animated.
There are now eight vessels lying in the bay, four brigs, two yachts, and
two Russians. The former come from Copenhagen and Drontheim, and
they are partly the property of the merchants here. The dried fish is almost
always sent from this place to Copenhagen, and disposed of there or in the
harbours of the Baltic. The Drontheim brig brings a new garrison and provisions to Wardöehuus; and the Russians are occupied in salting in the
inside of their vessels the fresh fish which the Finns are bringing to them
from all quarters. Three other Russians are lying opposite at Söröe. Others
are in Qualsund, some again at Jelmsöe, at Harvig, and in every quarter.
Who would not willingly see them? They are the peculiar benefactors of
Finmark, and the commerce with them is of the greatest consequence to the
country. They come with meal from Archangel, and barter the meal for the
fish caught by the Finns, and which they bring on board the Russians. The
Russian salts the fish, and prepares it in his own manner, according to the
taste of Archangel and Petersburg, a manner quite different from that of the
Norwegians, or of Barcelona, Leghorn, or Naples. The Finns therefore have
not the labour of curing, they are consequently enabled to employ more of their
time in fishing, and can not only gain their whole winters provision of meal,
but also receive it *immediately* from the Russians. If the Finn were to receive
money for his fish he would instantly recollect that the merchant sells his
brandy for money, but not that he could also procure meal and grain for it,
and that he stood in want of these to live throughout the winter; and the
profit derived from the Russians would almost always in a few moments be
consumed in brandy. But this meal, even if it were paid for in money, is
still much cheaper than if received from Copenhagen and Bergen. Even
at Tromsöe the *Vog* of meal from Archangel costs only one and a-half
rix-dollar, whereas the meal of Bergen costs two rix-dollars. It is
true the bread made from the latter is somewhat whiter, which is
probably owing to the mills at Archangel. How often would this people
be exposed to famine, or at least to want, if they had not this facility of acquisition! How often would they be dependent on the good-will of the

merchant! Besides this, if they were to lay up the fish for the store-house of the merchant, that he might send it to Spain, they would not catch the half of what they do at present, and consequently would not be enabled to provide for the half of their necessities; for in summer it is hardly possible to dry the fish. Worms are bred in them in a few days, and they are thus completely lost. The arrival of the Russians has at present every where dispersed the Finns and Norwegians about the fishing places. When they deliver the fish to the Russians, they immediately return that they may avail themselves of the Russian ship again before it leaves the harbour. Each of them at this time catches daily to the value of a dollar, or nearly a whole *Vog* of fish; for the Russians generally exchange a *Vog* of meal for a *Vog* of fish. Such earnings as these cannot be expected in winter. The consequence is that all the coast is at present deserted by the men. This trade is not yet old. The Russians first began in 1742 to explore the Finmark coast, and to catch or purchase fish in the Fiords. This was always considered as a contraband trade, and merely tolerated for the sake of its conveniency. Since 1789, however, when the company to which Finmark was under subjection was broken up, and Finmark relieved from their bondage, the Russians have been expressly allowed to carry on trade here; and their number has annually increased. All the Fiords, Sounds, and islands along the sea are covered with small three-masted Russian vessels, and the number of Russians who frequent this coast in July and August is reckoned at several thousands. In the beginning they did not go far beyond Wardöe: at present they come as far as Tromsöe, and they begin now to trade even there immediately with the fishermen; and notwithstanding the government only permits this trade in Finmark, but does not allow it in Nordland, it has however been found so advantageous and natural, that the prohibition of the government is broken through. It is very probable that the permission of a free trade with the people will be soon extended over the whole of Nordland. The Russians bring more, however, with them than meal; particularly in the Fiords and to the merchants. They supply Finmark with hemp, flax, and tow, with sail-cloth, linen, tar, nails, ironmongery, and even with masts, logs, and deals. They receive in return herrings, hides, cloth, cotton, sugar, coffee, French brandy, eider down, &c. The meal is for the most part the property of the boors along the shore of the White Sea and in the neighbourhood of Archangel. They grind it themselves in their own mills, and pack it up in mats of

birch bark,* each of which packages contains about three *Vogs* in weight.
The meal does not fall through these mats; but it is very much exposed to
the mice, who dig themselves passages through it, and render but too often
a separation of the pure from the impure absolutely necessary. There are
also sometimes complaints of little advantages taken by the sellers, such as
stones among the meal, or sand, which is still worse. This does not however
happen so often, but that Finmark is very glad every year to see the arrival
of the Russians; and it is possible enough that the province could no more
bear to be deprived of the assistance of Russia than fortunately the fishers of
Archangel could bear to be deprived of Finmark. In 1807, Russia issued a
strict prohibition of the exportation of grain on account of the war with
France. The prohibition was generally known in Archangel as well as elsewhere; but Finmark suffered nothing from it. The meal found an easy
way past the Russian custom-house officers, and the province in want of
the meal, according to the general testimony, has received rather more this
summer than it ever did in any former season.

What is thus received is not only sufficient for the wants of the country,
but the merchants also send off a very considerable quantity towards the
south; partly to Drontheim and partly to Copenhagen. The grain vessels
of Dantzic and Archangel cross one another on the Norwegian coast. With
their cargoes of meal the merchants of Hammerfest have been hitherto particularly occupied; and M. Ebbesen, a merchant in Wardöe, sent not long
ago seven thousand vog to Copenhagen, which was brought to him at Wardöe
itself by the Russian boors. This is a singular way of provisioning the
capital. How singular must it appear in the custom-house entries of Copenhagen to read vessels laden with meal from the poor and miserable Finmark! If Finmark, however, is enabled to do this, what might not be expected from the Finns and Norwegians if but a part of the wonderful industry and activity of the Russians could be transferred to them. The
Russians not only *purchase* fish, but they *catch* it themselves, and with very
different success from that of the inhabitants of the country. When the
Norwegian or Finn, at the North Cape or Ingen, succeed in catching any
fish, they proceed towards the land, rest themselves, cook their fish, and
then return to a new capture. But the Russian goes out with his boat

---

* Or lime-tree bark, which is procured in great abundance in Archangel from Wologda.

on Monday into the open sea, and in spite of storms and waves, never returns till Thursday. He sleeps on board his boat, and his comrade must every moment clear out the water of the overpowering waves which break over it. They would not even return at the end of three or four days, if they could preserve their fish longer without being salted. This has never been attempted by the Finmarkers, and it would be as incredible to them as to landsmen, if it did not take place daily before their eyes. A Russian catches at least a hundred *Vogs*, according to the account of very well informed persons, when the Finmarker cannot catch more than four, or at most ten *Vog* in the same place. They affirm, however, that they cannot venture to undertake, with their light and feeble boats, what a Russian boat can stand. This may be very true; but as they do not make the boats themselves, but receive them from others, why do they not order at once Russian boats from Archangel? Perhaps it would not be undeserving the attention of government to purchase such boats in Archangel, and keep them in some depôt in the neighbourhood of this indolent people, to be disposed of to them. There are even some of the Laplanders who are not deficient in a desire to assist themselves; and a Laplander who purchases cod nets for himself would soon aim at the possession of a stronger boat, if he saw any possibility of acquiring it. The Russians generally catch with *lines*; and in this they are highly distinguished. No Norwegian has such long-lines as the Russians. They frequently reach for more than two English miles along the bottom of the sea; and the fisher cannot see from the commencement to the end of them. Such lines, however, are not only kept at the top at both ends, but also in the middle, and they frequently contain from six to seven hundred hooks attached to them, each one fathom and a half from the other. When they are fully set, the Russian does not wait long; he returns immediately to the commencement, and draws out the line. He requires an hour to reach it, and then two or three hours before all the hooks are drawn up. He cannot wait longer, as the sea star (Korstrold) would otherwise destroy the bait of the empty hooks. When the line is drawn, it is again set immediately; and this is incessantly repeated for days, till the fish can no longer be stowed into the boat for want of room,* till they can no longer be kept fresh in warm weather.

---

* The ling, tusk, and cod, commonly called the white fishing, is carried on in Zetland much in the same manner as the Russian fishery described by our author. The regular fishing season

M M

These men are not only active and persevering, but they carry on their operations with a careful selection of the means. They bestow as much attention in catching the bait *(Köder-Madding)* for their hooks, as in catching the fish. In summer they use in the neighbourhood of the North Cape *Smaasey* for that purpose, which they catch in the Fiords, and for which they remain for days in the bays. In spring they prefer *Lodde.* In the district of Wardöehuus, towards Vadsöe, Pasvig, and Peise, whole fleets of them come from the White Sea : they remain for several weeks, for the sole purpose of procuring bait, and they then return to where the proper fishery begins. It may be questioned if ever a Finn, or a Norwegian, in Finmark, went a single mile out of his way for the sake of bait.

It is not to be wondered at, therefore, if the Russians return with rich prizes from the coast of Finmark ; and the Finmarker who complains of the success of the Russians, is exactly like the countryman who complains that the grain of his neighbour's field looks better than his own, without reflecting that to the industry and good sense of that neighbour this success is alone owing. It is in truth like bitter railery, to hear that Wardöe is, as it were, at the distance of a league out at sea, annually besieged by the Russians.

---

commences about the 20th of May, and it terminates on the 12th of August. The boats employed in this fishery are imported from Norway in boards, and are set up in the country. They are light, and have been found to be excellently adapted for the Zetland seas. The stretch of lines which the boats carry varies in different parts of the country. On the west side, some carry one hundred and twenty lines. Each line is about fifty fathoms; so that a boat in this case carries six thousand fathoms of lines, which extend nearly seven miles. The line is about the fifth part of an inch in diameter, and the hooks are fixed to it by small cords, with an interval of five fathoms between each. When the day is favourable, the boats set off for the fishing ground, from ten o'clock A. M. till two o'clock in the afternoon, with no other means of support than a small quantity of hastily-prepared oaten cake, a few gallons of water, and a slender stock of spirits. Having reached the fishing ground they proceed to bait, and set their lines, which, although extending over a great space, are seldom provided with more than three buoys. The boat keeps close to the buoy last floated, and from it the line is hauled in, generally a few hours after it has been set. Eighteen or twenty score of ling have been taken at a single hawl, for it is but seldom that the lines are set twice in the same night. Six or seven score are considered, on an average, to be a good hawl.

Under the most favourable circumstances of the weather and tide, the boats remain at sea from eighteen to thirty hours; and, if a gale of wind comes off the land, they are sometimes out several days. For an account of the Zetland fisheries, see Dr. Edmonstone's interesting work, intitled " A View of the Ancient and Present State of the Zetland Islands."—*J.*

They cover the whole coast with their lines, keep for weeks together on the waves in presence of the inhabitants, and return with full loaded boats, while the people in Wardöe are all the time on the point of starving for hunger. They catch nothing: they have been for years crying out louder and louder that they are ruined by the Russians; that the Russians take the food out of their mouths. " They prevent the fish from coming to us." How? From coming into their houses?—The Russians have never asserted that they are fishing in their own territory; and in all the complaints against them, we never hear that they prevent Finns or Norwegians from following their example. Such an amphibious race, who pursue their only object, the fishery, with such zeal, is not very much calculated, it is true, to adhere very strictly to political or police regulations; and in this respect we ought to consider that the disorders occasioned by them are not much more numerous than they actually are. Even the more recent complaints (1807) against the irregularities of the Russians at Wardöe, derived their origin chiefly, perhaps, to the unfortunate disputes of the royal functionaries at Wardöehuus among themselves; and the Russians served only for a pretext. It is astonishing that Wardöe, the coast, and the whole of Finmark, have not already become a Russian province. This is alone owing to the small fortress of Wardöehuus, and its captain and lieutenant, and a garrison of perhaps twenty men; for this is a firm and definite Danish establishment at the extremity of Finmark, which intimately connects the country with the remainder of the state. The meal trade, on the other hand, draws it towards Archangel, and into the hands of the Russians; and the Russians on the coast would soon turn the scale, gradually extinguish all the old political connexions, and give rise to new ones, if it were not for Wardöehuus; for they would soon erect summer residences on the coast, on account of their fishery: then more secure habitations; without, however, dropping in any degree the connexion with the coast of Archangel. Trade, religion, and jurisdiction, relations, friends, and national spirit, would connect them with their old country. In such a case, no protestations are sufficient to regain the lost territory; no proofs of the evident right to the land, and the former possession, would avail. Military expeditions alone must decide. But Wardöehuus stifles every commencement of this nature before its developement. Will it always, however, remain thus? What the consequences of the abandonment of Wardöehuus would be, is not left to mere conjecture—

they have been already experienced   In the year 1793, the garrison was withdrawn on account of the expence of maintaining it, and Wardöehuus remained empty—but only for two years.   The Russians immediately made their appearance, and the place was again obliged to be garrisoned.   Does not history speak sufficiently plain with respect to the manner in which Denmark lost the boundary of the White Sea, and Candalax, and Cola? and in which it is now losing the three districts of Neiden, Pasvig, and Peise? which are only reckoned as belonging to Norway, in Danish books and maps, but which the opinion of the inhabitants themselves, as well as that of their neighbours, and their political relations, proclaim to belong to another country. When the Norwegians, in former times, proceeded to *Biarmaland* for purposes of trade and plunder, there were then neither Russians nor Swedes in their way.   The former were separated from Finmark by the *Biarmers* ; and the latter had scarcely reached the southern boundaries of the present Lapland, and not so far as West Bothnia.   What limits could Finmark then have but the ocean? Hence the Norwegians always imagined, when they arrived from Archangel over *Gandvicken* (the White Sea) to the opposite shore, they were on the territories of Finmark ; and now they no longer robbed, but they required garrisons.   All this is proved by Schöning *(in Gamle Geographie).* But when the Russians and Tartars destroyed the kingdom of the Biarmers, they drove the inhabitants, Finns, and Karelians, still farther north, as they had already done all the way from Poland ;* and they followed immediately afterwards themselves, and robbed and burnt everywhere.   They not only found the way to the northernmost boundaries of Finmark, but even descended to Nordland, and down to Helgeland ; and their devastations there were so frightful, that Pope John XXII. in order to expel them, gave up to King Magnus Smeck, in the year 1326, the half of a six years' tenth which he levied on Norway and Sweden, that he might be enabled to conquer the land of Canaan .† This was followed by the *Diger* Död (black death) which depopulated Halogaland and Finmark, as well as the southern parts of the country ; and in the midst of this calamity they forgot their widely extended boundaries.   If the Russians had then thought of any thing more serious

---

* For the Finns, *Fenni,* according to Tacitus, formerly lived on the Vistula. *Suhm, Vorrede zu Hammond Missions Historie.*

† Lagerbring III. 302.

than mere plundering; if they had constructed a fort in Nordland or Finmar, as they constructed Nöteborg at the lake of Ladoga (*Orechonitz, Peckensaari*) in the year 1824, the boundaries of Norway would never perhaps have extended to the North Cape. But what was neglected by the Russians was at last done by the Norwegians themselves. They built a small castle on Aaaröe in Altensfiord, which they called Altenshuus : and as this only defended the western part, and not East Finmark, Wordöehuus was at length erected, and Altenhuus abandoned. There is no information with respect to the time when this took place; but Wardoehuus was in existence in the fifteenth century; for as the Russian ambassador, Gregor Istoma, was travelling in 1496 from Moscow by Archangel and through Finmark to Denmark, he found the castle of Barthuus with a garrison of Norwegian soldiers on the outermost cape. The old boundaries to Gandwick could then be claimed; for Finmark was in the possession of Denmark, and the Russians only made their appearance as robbers, who returned again with their booty to their homes, on the other side of Cola and Candalax. The English immediately made their appearance at Archangel. The importance of Archangel came to be known in Russia; new connexions were opened with the interior of the empire, and the Russian plunderers were transformed into fishermen. Russian colonies were founded all the way to beyond Cola, and no where was there a Danish fortification, or a Norwegian dwelling, to prevent possession being taken of the country. Cola itself sprung up. To settle the Danish sovereignty on a secure basis, something more was now necessary than the adventurous youthful journey of Christian IV. in 1602 to Wardhuus and the waters of Cola. He informed his subjects in that quarter that he acknowledged no boundaries but the White Sea; but this declaration had no other effect than occasioning the rather cruel treatment of some English vessels who were peaceably fishing on the desert coast, and who never dreamt of being in forbidden waters here, any more than at Spitsbergen. The Russians remained on the territory demanded back, and hence the declaration was completely fruitless; or rather worse than fruitless, as it occasioned the Russans to construct a fort in Cola, and to send a commandant and governor to that place.* Nothing more could be

---

* Cola was already a sort of fortification even before the journey of Christian. In the Danish state papers on the subject of the Russian boundaries (Büsching's Magazine VII)

done without force; protestations were of no avail. Even at present, the foged of Finmark goes every year to Cola, and protests to the commandant that he is in the exercise of an unlawful authority, and that the Danes preserve their claims to all the territory to the White Sea. This protestation journey was a farce from the beginning, altogether unsuitable to the dignity of any government. The Russian bojars in Cola were not much distinguished for their refinement; and it was not without reason therefore that the fögde were unwilling to lay their protestations before such bojars: they were obliged to endeavour to soften the unpleasantness of the impression with presents of brandy, furs, and spices. This was so successful, that the journey of the foged was almost looked on at Cola in the light of a tribute paid by Finmark. The protestation itself has hardly found the way to Petersburg. The Danish claims to the important place of Cola, and the coast so thickly inhabited by the Russians all the way to Candalax, appear now so singular in Finmark, that they are no longer believed in, and the last protestations, at least, (1803 and 1806) were made in a spirit very different from that which was originally intended. The question was no longer concerning Cola. Nature every where asserts, at last, her rights. Such protestations belong to cabinets, and not to frontier forts; and it would be better if they were never made at all. Far from the state of Denmark having any hope of ever extending its boundaries to the White Sea, they may rather expect Russia to seize an opportunity some day of depriving them of the whole of Finmark, if the Danish policy do not take means of preventing it. The Norwegians will then withdraw from it, and the uniformly mild and benevolent government of Denmark will no longer spread happiness and tranquillity among these remote wildernesses. The Quäns, however, will remain, the Russians will settle every where, and the province will become more populous. The Flensburgers will send less brandy, but in the place of it, the Russians will arrive with meal, and it is possible that the Finns will learn to apply the superfluity which they can no longer consume in brandy in the building of houses, and the purchase of fishing materials. Both these wants may be supplied by Archangel in remarkable perfection, which Norway can only do at present with difficulty and labour.

---

there is a Danish complaint, that Cola was surrounded with palisadoes in the year 1582, and was made by that means an Ostrog. A Bojar was always sent there after that period.

KIELVIG on Mageröe, the 27th of July. I was conducted in the night by three Finns towards Maasöe through the fog, with a light wind, and with short and almost inconceivably rapid strokes of the oar. The Norwegian sinks his oar deep in the water, and drives the boat forward with all his force; the Finn makes up by rapidity what is wanting in strength. Each of them believes his own mode the most efficacious; that of the Finn is however the most prepossessing; for we generally judge of the internal motion from that of the exterior, and as the march of ideas of the Finns is rapid like the strokes of the oars, in this respect, we may rather compare them with Italians than with Goths; and their violent and perpetual disputation corresponds tolerably well with such an impression. The vivacity of these men, and their astonishing toughness and durability, carried us through the Hava Sound to Maasöe, a space of five Finmark miles, equal to about fifty English miles, if not more, in less than twelve hours. A thick fog every where on the water prevented any thing like a prospect. We only saw passing through the Hava Sound, close to the main-land, the steep rocks of about two hundred feet in heighth, appearing through the fog, the stratification of which was easily recognizable, E. S. E. and W. N. W. (h. 7.) and which dipped strongly into the land towards the south. The strata may have been mica-slate. About ten o'clock in the morning we landed on the small island of Maasöe among the trading-houses, but opposite in a deep bay, which is connected by a narrow Eid with the harbour of Maasöe. By this the island is divided into two parts; and yet the *Eid* is only a few hundred paces in length, and scarcely fifteen feet above the level of the sea. It is singular enough to see the stripes of inundation run up on the one side parallel with the circular head of the bay, and surrounded with small bivalve shells and stones, as if this Fiäre had been but a few moments before washed by the waves. It was the same opposite, down towards the harbour of Maasöe. And yet the most remote tradition preserves no account of this *Eid* having ever been overflowed by the tide, or that it even reached the half of that heighth. If such tides were to be apprehended, the houses of Maasöe could not possibly stand. These are documents which connect the great geological phenomena with the more recent history of the earth, but which we stand in need of much more experience than we yet possess to explain satisfactorily.*

---

* Maasöe 70° 59′ 64″. In hac insula, quæ olim no duabus, imo tribus cestabat insulis, lucu-lentissima habentur signa *decrescentiæ seu imminutionis maris*. Hoc loco, decrescentia maris

Along the harbour of Maasöe reside one merchant, the clergyman, the schoolmaster, and the vassal; the two first in houses, and the others in earthen huts; the church is a log-building. Sea and sky, *Fieldt*, fog and rain, are here one and the same. The sun seldom or never penetrates through the clouds, and then appears but for a moment above the perpetually swelling waves, the high coast of Mageröe, and the singular rock of Stappen towards the North Cape. They seem like spirits, which no sooner appear through the fog than they disappear. There are a few scanty herbs among the rocks, but no trace of bushes; nothing to suggest the idea of trees. What a residence! A stranger is carried off in his first year by the scurvy, and if from his youth, his strength, and precaution, he overcomes the pestiferous influence of the climate, in a few years, however, his health is for ever destroyed, even should he return to more southern regions, or into the Fiords. The man of talents sinks under such a pressure. Yet here a clergyman resides, to whom courage and strength are so necessary in his vocation; and clergymen have been seen here for six, eight, and twelve years in their office, till scurvy and despondency brought them to their grave! It is only a residence for fishers and Russians. The spirit is oppressed beneath these fogs. Are priests or public officers ever sent into the Pontine Marshes? Or have the Portuguese ever had the conscience to allow their countrymen to remain even one year on St. Thomas's without being relieved? The clergyman on Kielvig lately died of the scurvy; another came, and he also was carried off in the space of a few weeks; a third was sent, and in a few weeks he also followed the others. They were at last compelled from such necessity to transfer the residence of the clergyman into the Porsangerfiord towards Kistrand. The sun again made his appearance there, and the earth was diversified with herbs and birch-bushes, and it was possible to receive shelter from the scorbutic winter. The flock of Kielvig have suffered nothing from it; the clergyman easily finds his way to them out of the Fiord, and can dedicate his time and labour to them, without necessarily sacrificing his own health. Are they waiting in Maasöe and Loppen for the experience afforded by Kielvig before making a similar transfer of their clergymen's seats? But let them only look at the lists of

---

a me dimensa reperta fuit in linea perpendiculari a superficie maris, pedum Vienn. 110, *says Father Hell in Ephem. Vind Anno.* 1791. **319.**

their clergymen, and see how many of them return with impaired health, how many with loss of strength and invincible depression of spirits, occasioned by the perpetual disappointment of their expectations to be relieved from this banishment!—Let them inquire into the melancholy history of the former clergyman of Loppen; and then let him who sends the clergymen into these desert islands, and does not translate them to better situations in the course of a year, clear his conscience of it if he can.

The Prästegieldt of Maasöe includes part of Mageröe to the North Cape, Jelmsöe and Ingen, and then several miles of the main-land. Two hundred and seventy-five Norwegians live on these islands, and two hundred and seventy-two Finns on the main-land. Part of the Laplanders of Kautokejno also come in summer, who descend between Porsanger and Altensfiord, to the very farthest point; but still Maasöe remains in every respect one of the smallest and most wretched Prästegieldts in Norway.

The island may be about six hundred feet in heighth, in the near environs of the harbour. The rocks are divided as if into plates, like roof-slate; but yet it is gneiss; the felspar appears even small granular, partly red and partly white, and the mica lies in separate folia on one another, in the manner of gneiss. The tabular figure of some of the strata is occasioned by the straight slatey juncture of the ingredients. The strata every where stretch very distinctly S. S. E. and S. and N N. W. and N. (h. 11. and 12.) and dip towards the west; and this is perhaps pretty general in all the surrounding islands, if we dare trust to the form of the rocks and that of the land.

Towards evening I was conveyed over the Fiord, which is about nine English miles in breadth, to the Sound of Mageröe, by Norwegians. The violent current out of the Sound was against us; we run into *Finnbugt*, about the middle of the Sound, and on the island of Mageröe itself, to wait for the return of a favourable tide. The Norwegians live there in earthen huts, which being covered over with grass, bear a resemblance to small hillocks; dwellings like those of the Tungusians, or like the *Gammers* of the Finns. The interior, however, looks more like a house. When we squeeze ourselves through the three feet high door, which is made to shut of itself, we go through a dark passage to the various compartments of the hut; a similar door opens into the dwelling-room; and this apartment differs in nothing from the usual dwelling-place of the peasants at Bergen. It is constructed of logs, quadrangular up to the roof, which is a quadrangular pyramid, with a square opening in the middle, that at night is closed with a blown up fish bladder, and through which the light enters, and the smoke issues out during the day. The furniture consists of a table, and a bench behind it; the bed of the master of the house, and a cupboard or press, and chests are ranged around. The children and servants sleep on the outside of the room, or beside the cow. The kitchen is a large chimney in the corner of the room. This is actually the most convenient manner of laying out a house in climates like these, where not a twig for firing is grown. The thick earthen wall makes a cellar of the hut, in which the temperature does not come in contact with the external temperature for weeks. Whether it storms or snows without, whether it is winter or summer, cannot be felt in one of these earthen huts; but in a common northern log-house, every external change is felt in a few hours in the inside. The air penetrates through doors and windows, and finds its way over the whole house. It is singular, that the richer class, the *Storkarle (great fellows)*, as they are called by the Laplanders, or the *Lords*, as they are called in the canton of Schweitz, or the people of *condition* as they call themselves, do not adopt this mode of constructing houses of earth, and pass the summer in the larger log-house, and the winter between earthen walls. For nothing prevents them from ornamenting the inside as well and comfortably as the taste of the inhabitants can wish; and though in such a dwelling there is little light, and almost no prospect, during four months of continual night little of either can be expected.

The rocks of the bay were *gneiss*. They did not, however, at first sight resemble it; for they were strikingly black. This proceeds from a number of insulated beds of mica, and a little quartz, in very fine granular mixture. The mica folia are so fine, that we have often difficulty in recognizing them. The gneiss in general is very fine slatey, and therefore unpleasant, on account of the indistinctness of the ingredients. In the upper part, towards the summit, clay-slate was seen reposing on it very distinctly. All the strata stretch N. N. E. and S. S. W. and dip nearly 50° towards the west, as is the case nearly throughout the whole Mageröe Sound.

They conveyed me in the morning over to Kielvig very slowly: they call this passage a mile; but we set off about four o'clock, and landed there only at twelve. These are miles truly suitable to a dreary and desert region. The weather was admirable. The Sound opened more and more; I saw

Sverholt, the sharp cape between Porsanger and Laxefiord, and at last, at a distance of between fifty and sixty English miles, Kynrodden or the *Nordkyn*, the last point of the main-land of Europe, extending far into the sea. It seemed as if we were approaching the end of the world. The rocks on Mageröe seemed to grow steeper and steeper; we proceeded through between them, and the small island of Altesula, and at last landed at a place where the shore seemed totally inaccessible. There, however, lay Kielvig in a bay, consisting of the church, the deserted habitation of the clergyman, and four or five houses belonging to the merchant and his servants. No more people live here, and they could scarcely do so ; for we go over the whole ground in a few minutes on which a house can possibly stand. It is a narrow space between the waves and the rocks, covered with perpetual snow. Who could have the heart and the courage first to build here? The elements seem in perpetual agitation. The high waves and the storms from the north and north-east make their way without obstruction, and with great violence into the very interior of the bay; and a small island above, a rock at the distance of a few hundred paces from the land, affords an insecure protection to vessels. From the west, the wind sweeps with such violence down the fissures of the rocks, that the vessels are frequently compelled with the greatest precipitation to weigh anchor, lest they should be dashed to pieces among the rocks. They can only anchor safely in calm weather. But where is there any calm at the North Cape?

And yet Kielvig is much frequented. The bay lies at the entrance of the Sound of Mageröe, and in the way of all who come from the east to Finmark or proceed downwards to Nordland. The great Porsangerfiord ends here ; and even the vessels from Archangel to England sometimes touch here before proceeding round the North Cape. Moreover, several of the fishing places are quite near, and it is convenient for the fishers to get their fish disposed of so easily. That such a multitude is alone perhaps a sufficient cause for the inhabiting such an inhospitable spot is proved by the experience of the former year. The fishermen lay several weeks at the mouth of the bay, and at Helleness, about two miles farther into the sea, and they brought the merchant about five hundred *Vogs* almost every day. The fish frequent this station in May or June, when they are in pursuit of the *Lodde*. The lodde appeared in too great numbers in the present spring; they came so early as February ; and they were not then followed by the large fish. According to the fishermen, the water was then too cold for them : this would be singular ! Is it then so much warmer in the open sea, and in the regions of the North Pole ? Kielvig is at present, as well as Rebvog, a factory of the house of Knudson in Drontheim.

KIELVIG, the 31st of July. The ascent of the rocks here resembles a stair ; and yet they can only be climbed where a small stream descends from the valley above. This valley is nearly five hundred feet above the sea; it is an *Eid*, a high isthmus, which fully separates the tongue of land of Kielvig, which projects towards the east, from the rest of the island. The mountains on both sides rise to a considerable heighth ; and notwithstanding the valley sinks immediately towards Kamöefiord, the mountains do not follow it, but run steeply and perpendicularly into Kamöefiord. The highest rocks to the west of the houses of Kielvig, and which seem to hang immediately over the place, have been employed for astronomical purposes ; and there is now a signal post on them, visible at a great distance, which was erected in 1796 by Bützon, the astronomer of Copenhagen. His measurements have not been made public ; but the barometer gave eight hundred and forty-seven English feet for this heighth, and for the southern summit nine hundred and fifty-two English feet. These were the highest mountains in the whole neighbourhood ; and they command an extensive prospect over the Fiords and the sea towards the east. The steep, high, and almost insulated Cape of Sverholt, on the other side of the Porsangerfiord, appears to lie deep beneath our feet. It is not so high by a great deal ; and the *Porsangersfielde* do not reach this heighth till after they run a great way in-land. Beyond Sverholt, the land runs out to an infinite distance into the sea, at nearly the same heighth, till it at last falls at the extreme rock, the Nordkyn, *(Kynrodden)* suddenly and abruptly into the sea. The mountains on the tongue of land to the east of Kielvig are, however, still higher. The barometer gave them one thousand one hundred and two English feet.* There

---

* 27th. h. 12. Kielvig Bar. 28. 3. 1.........Therm. 7. 25. R. clear. Faint east wind.
h. 2. Kielvig......8. 5...Light clouds, sunshine.
h. 3. Signa post..27. 4. 8.........Therm. 6.
h. 4. Heaps of stones in the south 27. 3. 6. Therm. 5. 5.
h. 8. Highest Fieldt, east, from Kielvig27. 1. 4.
h. 9. Kielvig.....28. 2. 4 ....Therm. 6. clear, north-west wind.

I saw the *North Cape* above Kamöefiord and Hulfiord, at the distance of about fourteen English miles. The high chain of the tongue of land between these two Fiords would hide the view of the cape, if they were to continue at an equal heighth through the interior of the island; but they decline, and allow us distinctly to see how the precipitous row of rocks of the North Cape, covered at present with snow at top, shoot into the sea. They are higher than the rocks of Kielvig: perhaps they may be one thousand two hundred Paris feet; and they consist of a long row of pyramidal points, such as we might expect on a cape exposed to the fury of the whole ocean. The rocks which surround the Fiord here are very powerful. But how dreary and desolate is the interior among the mountains! All is lifeless, or merely a commencement of life. In the lower parts, large spots of snow are still lying; and the heights are covered with huge heaps of stones, without the smallest vestge of grass, or any other vegetation, with the exception of sometimes a little white moss. It appears like a new earth sprung from the deluge. Nature never resumes her influence here, and we gladly flee from such dreary spectacles of dessolation.

Early on the morning of the 28th I ascended the mountains towards the interior of the island for several miles. I never saw a range of mountains so broken and indented. I got several times to the top of the mountain, and then came down again on an *Eid*, which divides the whole island. If the surface of the sea were but raised a few hundred feet higher, we should see here a whole archipelago of small islands, in the place of one large island. The first of these *Eids* is about two English miles from Kielvig; the valley, however, is about three hundred feet above the sea, and terminates with such frightfully perpendicular precipices towards Breivig in the sound of Mageröe, that we can hardly approach the brink without giddiness. At the distance of about five English miles farther, we come to Honningvogseid, deeply buried among the mountains. The land is there not even twenty feet above the level of the sea; and a small lake, which continues along the valley, occupies the greatest part of the space of an English mile, which the breadth of the Eid perhaps amounts to. It is so low that the fishermen actually consider this isthmus as a sound which separates the island of Kielvig from the greater island of Mageröe. They frequently drag their boats with no great difficulty from Honningvog overland to Skibsfiord, the interior of Kamöefiord, and save in this way from twenty to twenty-five English miles of a passage, which is frequently very disagreeable, round the eastern point of Kielvig, and exposed to every storm. Beyond this *Eid* the sea appears incapable of encroaching farther upon the interior, Huge blocks are towered up above one another, to the size of rocks and entire mountains, and their number is endless. They hardly seem to hang together; their forms are most singular, and they almost always environ small lakes in the bottom, which only find an outlet towards the Eid through narrow crevices. There are nearly a hundred of these cauldron-like environs, like a row of small *Crateres*. The whole is finally closed by a black, perpendicular, and inaccessible wall of rocks. It is hardly possible to imagine a stronger picture of horror and devastation, even in a country which bears no trace of the living powers of nature. Towards the northern side there was, however, a possibility of reaching the summit of the range of rocks over the blocks. The blocks are heaped up more and more towards the height, and the whole upper part resembles a ruin. It seems as if the whole mountains had been tumbled over one another, and as if the fragments had fallen down in wild confusion towards the isthmus. From the point of the rocks there is an extensive prospect over a great part of the island ; but I could not see the North Cape. The mountains in that direction, and towards the west, are some hundred feet higher, at most about one thousand three hundred, or one thousand four hundred Paris feet ; for Honningvogseid was one thousand one hundred and sixty English feet above the sea.*

The strange rocky appearance of this island at last began in some degree to relax: a green valley now opened down towards Skibsfiord, with small lakes and dwellings in it; and the mountains on the farthest side rose in an undulating and connected form to their greatest heighth. There was every where large and wide spreading masses of snow: the influence of the summer had been very small there ; and if the snow ever leaves the summits it can only be for a few weeks. These mountains approach the perpetual limit of snow. They were not, however, fully one thousand four hundred Paris feet above the level of the sea, and not so high as many mountains at Alten, on which the birch grows with freshness and vigour. A few birches were to be

---

* h. 2. Honningvog 28. 1. 8. . . . . Therm.   8. 5. R. . . . . . Strong north wind.
  h. 5. Honningvogs Nordfieldt towards Skibsfiord. . . . . . . 27. 1. 3. 10. . . Calm, clear.

seen also here on the declivities of both sides of Honningvogseid; but in what form! They had not even the appearance of bushes. They rose but a few feet above the ground, and had not strength enough, however low, to spread their branches along the ground: small sapless twigs, which from their leaves alone we can discover to be birches. I could trace those feeble remains for nearly four hundred Paris feet downwards, when they wholly disappeared. Hence they place the temperature of the coasts of Mageröe, Kielvig, and Sarnäss, about half a degree R. below the mean temperature of Hammerfest: at about—1°. 5 R. or 28°. 63 Fahr. If the limit of perpetual snow is one thousand six hundred Paris feet in these regions above the highest boundary of the birches, it may be perhaps somewhat more than two thousand Paris feet above Mageröe. There is no mountain or rock, however, on this island of such a heighth, or in any of the islands in the same latitude; and therefore we cannot expect to find here perpetual snow, and much less can we expect to find glaciers.

The composition of the rocks of this island, and all their geological relations, are in the highest degree remarkable; and this is the case at our very first step on land at Kielvig. They are the more deserving of an accurate consideration, as they not only determine the constitution of the most northern points of Europe, but actually furnish us with explanations of the order of northern formations, which we can hardly find either so full or so distinct in southern regions, or in the interior of mountains.

That the formation of Kielvig was not gneiss, was evident at a great distance. For on all the cliffs and small rocks around the shore, the external folia appear sharp and indented, and follow one another like the leaves of a book. They are distinct clay-slate rocks. If we examine them more narrowly, we find the slates composed of fine and shining folia, which are very distinct in the light of the sun; but we do not find here the glistening and the uniformity of composition of mass of the clay-slate. Between the folia there are always lying a number of small brown crystals, small prisms, which their minuteness prevents us from ascertaining with accuracy, but which from some of the larger crystals may perhaps be *hollowspar*, *macle*, or *chiastolite*. The cross fracture of the slates is fine earthy and somewhat splintery. The whole exterior of these rocks, the fine slatey structure, the earthiness of the cross fracture, pronounce with sufficient distinctness that we are entering on *clay-slate;* but when the light of the sun is reflected from the shining plates we instantaneously give up all idea of clay-slate. It is nothing else however. The *Bützow* signal post on the summit of the rocks is immediately above it; and there even in single pieces the nature of the clay-slate can scarcely admit of a doubt. The basis is continuous, and the scales of mica, although still very abundant in it, are only, however, scattered on the continuous base. Beds of massive brown rock crystals, in which delicate fissures are frequently coated with chlorite, and less frequently imbedded crystals of felspar appear. Large folia of talc are also not unfrequent, and small greenish grey splintery cones resembling the serpentine stone. All these strata, from the surface of the sea upwards, stretch uniformly N. N. E. and S. S. W. (h. 2) as at Finn-bugt, in the Mageröe Sound, and they dip under a very great angle towards the north-west into the interior of the island and the mountains. This is the case on the west side of the house of Kielvig. When we ascend Kielvigs-Eid, however, we soon find above in the valley *small granular granite* instead of the clay-slate; containing in it black insulated mica folia, and a great deal of hornblende; and this granite by no means lies below the clay-slate; the line of separation of both formations may be followed for a considerable length; and it is undoubtedly and clearly seen that the clay-slate is conti-nued beneath the granite. Hence it is soon lost in the interior of the rocks and towards Kamöefiord, and also in the mountains of Kielvigs-Eid towards Honningvogs-Eid, and so much so that clay-slate in general only seems to form the external border from the Sound of Mageröe to the last eastern point of the island. The granite which lies on it changes itself frequently into straight slatey gneiss, and in this we often find large and beautiful garnets. Such is in fact the whole tongue of land from Kielvig to the last cape in the ocean; but the opposite side westwards towards Honningsvogs-Eid is dif-ferent. There the granite soon becomes a fine granular green stone; and the latter obtains at last so much the ascendancy that it becomes evident that granite and gneiss are not independent, but only feeble repetitions of older formations. This might almost have been already conjectured from their in-ternal composition; for the granite not unfrequently contains *diallage* in the mixture, and but little quartz, and it also receives a foreign appearance from a number of long thin crystals of an iron-grey metallic-shining fossil, perhaps titanic iron ore. The diallage increas s; and the quartz and mica continue to decrease; till at last the granite becomes green-stone without any visible separation. The change in the nature of the stone was then betrayed by the exterior; for the clove-brown *diallage* is weathered on the surface of the blocks, and looks *tile-red*, and frequently like garnet. This is very striking. The felspar does not remain white, but becomes grey; and the mixture becomes so firm, that it is with difficulty we can break off small pieces. Iron pyrites are also not wanting, as is always the case in diallage and hornblende stones. On the other side of Honningvogs-Eid, all the way to the highest mountains of the island, the green-stone becomes at last coarse granular, and the ingredients of felspar and diallage are beautifully discernable.* It now bears an entire resemblance to the rock of the *Zobtenberge* in Silesia, of *Prato* in Tuscany, and of Mount *Musinet* at Turin. The brown diallage is distinctly foliated, with a single cleavage and small conchoidal in the cross fracture, and then alone glistening. We often imagine we recognize the crystalline forms of the imbedded fossil, viz. a broad four-sided prism accuminated with four plains, which are set on the edges like stilbite. The felspar is more easily disintegrated by the weather than diallage, although

---

* Diallage occurs associated sometimes with felspar as in the rock of Mageröe, or in other instances along with jade, a mineral very nearly allied to felspar, or with jade and felspar, as in Corsica, Italy, and Switzerland. These rocks are well known to the Italians under the denomination *Gabbro*. Von Buch proposes to retain this name in geognosy, and to consider these compounds as varieties of the same species of mountain rock. Gabbro has hitherto been found principally in primitive country. I believe that it occurs also among the transition and even the Floetz rock of Scotland, where it has been confounded with greenstone or serpentine. In primitive country it is certainly newer than mica-slate, and probably rather newer than clay-slate. It is frequently associated with serpentine, and there is an uninterrupted transition from the one rock into the other. Indeed, the transition is so complete that we cannot refrain from inferring that serpentine is but a compact, and less distinctly crystallized gabbro.

The Corsican varieties of gabbro have been long known to artists under the name *Verde di Corsica:* most beautiful masses of these abound in the beautiful Laurentine Chapel at Florence.

Certain varieties of this rock were known to the ancients, and they appear to have brought them along with porphyry and sienite from Upper Egypt. Mr. Hawkins collected specimens of gabbro in the Island of Cyprus, and it would seem that the Cyprian copper mines, so much celebrated by the ancients, were situated in Gabbro. It is a frequent rock in Tuscany—it forms hills in Silesia along with serpentine, and in Moravia it lies under transition clay-slate. At Crems, in Lower Austria, it rises through rocks of the coal formation, and it forms hills that rest upon clay-slate, near Bergen in Norway. M. von Buch in the text considers the diallage rock or gabbro as identical with greenstone, an opinion which I believe he has now relinquished.—*J.*

the latter sooner loses its colour. Hence the surface of all these blocks of rock are so rough. The diallage projects, and beside it we see the cavities in which the disintegrated felspar was situated. This coarse granular diallage-stone continues for a great way into the interior of the island, and may form a considerable part of it. The stratification is only recognizable in the pure granular green-stone, but it becomes then very distinct, and stretches westwards from Honningvogs-Eid N.E. and S.W. (h. 3.), and dips 60° towards the north-west.

Clay-slate may therefore hardly make its appearance in other places than the neighbourhood of the sound of Mageröe, and probably not towards the North Cape ; but the diallage-stone also no longer appears there. The island of Stappen *(the mother with her daughters* of the English), so well known on account of the capture of the Puffin *(alca artica)* consists of gneiss ; as also the nearest rocks of the steep North Cape, on which we can land in the neighbourhood of Tuenäss. This gneiss is more striped than slatey, and the ingredients are altogether fine granularly connected with one another, but still distinctly enough. The mica is black, in very fine folia, which lie single and insulated. The felspar is in great abundance, pale flesh-red and white, and almost transparent ; and the completely small grey quartz-grains are also easily distinguishable. This gneiss certainly is not placed above the diallage-stone ; for its extent is too great for that purpose. But as both stones are connected on these points, this is difficult to decide in an island of so great extent, and so desert in the interior. Not far from Kielvig a very small bay lies between the perpendicular rocks, which is called Little-Kielvig. The stone is there actually transformed from clay-slate into mica-slate ; for there is no longer any basis here ; the whole is a collection of an infinite multitude of shining folia lying upon one another, and not such folia as appear on greywacke slate, but fresh and scaley, above one another, as they usually are in gneiss. Pretty thick beds of pot-stone *(Grydesten)* frequently lie between ; it is greenish white, coarse, and very frequently splintery, translucent, and entirely similar to the *jade*, if the hardness were only greater. But the stone is scarcely semi-hard ; small white talcy folia are frequently scattered in it. Such beds are not however to be found in the clay-slate to the west of Kielvig in the bay of Mäet.

When we compare all these appearances with one another, we find the order of the formations which determine the constitution of the south-eastern part of Mageroe, till towards the centre of the island, to be nearly as follows : First, the older *gneiss* of these islands, and on the whole western coast of Mageröe. In the Sound at Finnbugt the clay-slate lay distinctly on it. Then comes *mica-slate* in *Little Kielvig*; then the clay-slate of the rocks and mountains in the *Mäet* at *Kielvig*. Then again *gneiss* on the top of the eastern mountains of Kielvig, then fine granular *granite;* both far from extensive. Next comes very small, and almost fine granular *diallage-stone*. *Lastly*, the coarse granular *diallage-stone* to the interior of the highest mountains of Mageröe. This formation consequently follows the clay-slate, and may be separated from it by a small repetition of gneiss and granite; and this coarse granular stone is not the older; it lies much more on the fine granular. Hence we might expect the former even in Alten eastwards, in the mountains between Alten and Porsangerfiord, and perhaps at the source of the Pors-Elv, the stream of the saw-mill.

These outermost points afford us a general result for mineralogy, that diallage-stones belong to the remotest members of the primitive formation, and nearly touch on the transition formation ; and *this is not contradicted* by *Silesia, Prato, Genoa, and Cuba* *.

The interior of Mageröe, however barren and unfriendly, is not, however, altogether unoccupied. About five or six hundred rein-deer run about almost wild among the mountains : they are allowed full freedom in winter, and in summer only are driven together by the mountain Laplanders, who use their milk. These rein-deer and ermine are the only wild quadrupeds of the island; for the bears and wolves, the furious and dreaded enemies of Finland, have never been able to penetrate here. The sounds are too broad to allow them to swim over. Every person, and more especially the Norwegian, keeps his cows and sheep in the neighbourhood of his dwelling. They have no doubt difficulty enough to provide a sufficient store of provender for them in winter; for even if they were disposed to attempt mea-

---

* A formation, somewhat similar to that described in the text, occurs in the Saxon Erzgebirge, and I believe also in Scotland. In the Saxon Erzgebirge we observe the oldest gneiss covered with clay-slate, which contain beds of flinty-slate, green-stone, lime-stone, porphyry : over these rest, in a conformable position, newer granite, which alternates with sienite, gneiss, and porphyry.—*J.*

dow cultivation here, where could they find room between the rocks? They
fall upon another, and certainly very singular method, of supplying themselves. They are acquainted beforehand with places beneath the rocks, of
perhaps a few paces in circumference, on which grass grows in summer,
though it does not attain any heighth. It does not, however, cease growing
beneath the snow. In the middle of winter they dig through the snow, and
draw up the grass, which is then high, beautiful, and fresh, with a pick-axe.
This labour is not always, however, without danger. As these bountiful
spots lie generally close under the deepest rocks, they are consequently exposed to the descent of avalanches *(sneeskred)*.

In the foregoing winter (1806) a Lapland mountaineer at Sarnäss sent
his two sons, who were children, to one of these grassy spots, at about seven
or eight English miles distance. They scratched up the snow, filled their
nets with grass, and hastened back ; but in descending from the *Fieldt* they
were both overtaken by and buried under an avalanche. Their dog, which
had run on before them, returned, found the avalanche, and kept scratching
so long in it, that at last, with his assistance, one of the boys was enabled to
get out. He immediately sought for his brother, but not in the right place.
The instinct of the dog succeeded better ; he found out the place, and incessantly dug at it till he at last also uncovered this unfortunate boy, who
was lying on his belly unable to assist himself. Even the cattle know how
to find out what grows beneath the snow ; not only the rein-deer but also
sheep. M. Bang, in Kielvig, forced his sheep to remain out of the house in the
winter of last year ; for his provender was exhausted at home. They scratched
through the snow, like rein-deer, from twelve to fifteen feet in depth, and
in spring they were fatter than they had ever before been seen. What can
warm the ground in winter in a zone, the mean temperature of which is
beneath the freezing point ? This phenomenon seems common to all Finmark, and not confined to Mageröe. The stream which runs into the bay
at Hammerfest, flows throughout the whole winter ; and the inhabitants of
Hammerfest procure their potable water in winter from it. The small degree of heat of the summer is exhausted in the first cold months of winter
and cannot possibly influence the surface of the earth in the middle of
winter. In well-secured cellars it never freezes either in Kielvig, Hammerfest or Alten. The temperature of the cellars of Kielvig cannot there-

fore be the mean temperature; and the cause which raises them above is an addition from the interior of the earth, originating in some source with which we are unacquainted. How very different from the regions of Siberia or North America, where we are assured that the earth is never thawed for a depth of more than a foot, and frequently only for a few inches!

KIELVIG, the 3rd of August. In the course of a few days the *Fieldt* was covered with vivid flowers, and all the snow had disappeared. The spring was transformed into summer. The thermometer stood for several days at 65° F., and here it cannot rise much higher; for whenever the sun ceases to have any power, which happens but too frequently, the heat immediately falls to 54° F. at midday, and to from 45° to 50° at night. Hence thunder storms are so rare here: years elapse without any thing of the kind being heard, though, when they do happen, they are sometimes powerful enough. M. Bang witnessed a very severe thunder storm two years ago, in the month of August, from the north-west; and consequently from the sea. The winters are less dreaded here on account of the cold than on account of the storms; for their fury exceeds all description. They sweep down the Fieldt with the utmost impetuosity from the west and north-west. All is in agitation; no sound can be raised, no human voice is audible for the raging of the elements. The inhabitants endeavour, in sullen expectation, to secure themselves with double cloathing and furs against the cold, and are compelled to satisfy the cravings of hunger with what little they can find around them; for no fire burns, and the rocking house is perpetually on the point of falling about their ears:—a dreadful situation, which frequently continues for days. These storms generally make their appearance when the sun begins again to rise above the horizon; but it is very remarkable that they always diminish with the dawn of evening, and are less violent throughout the night. At break of day they return again with their former fury. They may be more terrible at Kielvig than at other places on this coast, but these powerful hurricanes in winter are, however, peculiar to the whole coast of Finmark. Father Hell relates that he at one time could not observe the thermometer before the window, because no person durst venture out of doors for fear of being swept into the sea, though at a pretty considerable distance from it. This air pours down from the north-west from the neighbourhood of the pole, and probably, therefore, flows up into the Northern Ocean from the equator. The extreme cold in Wardöehuus, in 1769, as observed by Father Hell, was—14° R. or 0°.5 F. in January, and — 12° R. or 5° F. in February, the former during a north-west wind, and the latter with a south wind. This was no unusual year, and yet the cold was not greater than we frequently enough observe it in Germany, and even at Paris. We may therefore conclude that we are high in our estimate when we state the mean temperature of January, in the extreme islands of Finmark, so low as — 9° R. or 11° 75 F.; for Uleoborg, at the end of the Bothnian Gulph, in January, is — 10°.83 R. or 7°.64 F., and yet the mercury is every year there on the point of freezing.

———

REBVOG in the PORSANGERFIORD, the 5th of August. The three Norwegians who were to carry me over the Porsangerfiord imagined they might avail themselves of the north-east wind, notwithstanding the high waves which it drove in from the sea. This succeeded very well till we came to the middle of the sound, but there a wind sprung up from the south-east. The new waves which came down the Fiord struggled for ascendancy with the former, which were higher, and this gave rise to whirlpools, in which the one set of waves rose high above the other, and, by their collision, produced a most furious shock. It is impossible for a feeble boat to combat with the wind and with such dreadful whirlpools. We were obliged to put back again for the Sound: we followed the course of the wind. run through Altesula, and re-landed on Mageröe, in the deep and secure bay of Sarnäss. Nothing stood there but *gammers*, earthen huts, like hillocks, covered with grass and flowers. This, however, was the habitation of the vassal and his children; but the interior betrayed the owner rather for a Norwegian, not altogether of the poorer sort, than for a miserable Finn. All the houses on these islands should be similarly constructed; for they can neither be carried off nor shaken by the storm: its force is lost on the thick and round walls, and its raving is scarcely heard within. Here, also, there was no want of light, notwithstanding the windows were placed in a wall of more than three yards in thickness. In winter these windows are closed up, and they remain in the bowels of the earth quite separated, and fully secured from this perpetual commotion and agitation of the winter both on land and sea. The construction of houses is an art which has yet made but very little progress, even in the southern part of Norway.

The rocks of the bay of Sarnäss are not indeed so steep as at Kielvig, but they are still singularly indented: they run in small rows in the direction of the strata, and gradually rise to the greatest heighth, with narrow vallies between, in which there are generally small lakes. The strata stretch here, as on the island of Altesula, N. E. and S.S.W. (h. 2.) and dip strongly towards the north-west. In both places we find clay-slate, with numerous mica folia, and beds of white quartz in it. The fine granular diallage-stone appears first on the height.

The Fiord is only half the breadth from Sarnäss to the main-land that it is to Kielvig. We rowed over, early in the morning of the fifth, in complete security till we were close to the land. Then we were overtaken by a storm from the west, accompanied with heavy rain, which compelled us to wait at the extreme point of the desert Porsangernäss for greater tranquillity in the elements. A wretched *gamme* of stone has been built there, scarcely two feet in heighth at the entrance, and not four in the middle ; but it was a welcome receptacle for us from the heavy rain, and a boat like our's. We can only learn properly to estimate the importance of a roof in desert and dreary regions.

Porsangernäss is every where conspicuous from its white colour. At a distance we might be led to suppose it was covered with snow; and this idea 1 long continued to entertain when I saw the rocks from the mountains of Kielvig. It is nothing, however, but pure white quartz, which lies in huge and thick beds between distinct and characteristic mica-slate. The strata rise out of the ground as sharp as knives, and it is with the greatest difficulty that we can make our way over them. They stretch E.N.E. and E. and W.S.W. and W. (h. 5.6) and dip strongly towards the north. The quartz is so thinly stratified, that it falls to pieces in magnificently large plates of but a few inches in thickness, and several feet in length, like marble tables. The gamme is for the most part constructed of it, and it might have almost been raised out of a single plate. If these admirable stones were found on a southern coast, they would not lie long useless. The mica-slate which encloses this quartz contains a huge multitude of small red garnets, like the mica-slate on the southern mountains of the country of *Glatz*, and also a multitude of nuts called *tytter*, which are firmer than the continuous mica of the base, and always rise above the surface of the strata. They consist of a fine granular mixture of white talcy folia of red coarse garnet and white felspar, and they not unfrequently give a very singular appearance to the whole.

The garnet in the mica-slate on Sverholt, the extreme cape of the opposite side of the Porsangerfiord to the east, are equally frequent*.—The newer rocks on Mageröe are consequently only limited to that island, and are no longer to be found on the main-land—It is singular enough that the very last island towards the north should be precisely the most remarkable and various in its composition. Even the islands to the west of Mageröe, and which are almost in an equally high latitude, appear to contain none of all these stones; at least mica-slate with garnets is very peculiar to the island of Söröe, and predominates in an especial manner in the Finnefiord on the farthest side of the island, where the garnets are stuck around in the rocks as large as hazel nuts. The quartz of Porsangerness continues into the Fiord, and almost to Rebvog, between four and five English miles to the south of the Näss. The strata are always equally thin, but they gradually change their inclination, and instead of dipping towards the north, dip gently towards the south. Small birch bushes also gradually make their appearance between the rocks. Every mile we proceed up the Fiord discloses a better vegetation with the sinking of the mountains. The country round Rebvog is still however singularly dreary and naked, and on seeing the large beautiful and new house in the bay, one of the best in Finmark, we should in fact be astonished how it came to be erected in such a wilderness, if we did not at the same time observe several Danish brigs and three or four Russian ships in the harbour. Rebvog is an excellent and secure bay, and quite near one of the best fishing stations; and several vessels not only yearly sail from this place to Spain, but the exportation from it has also become one of the most considerable of all the trading places in Finmark. We are not to be surprised, therefore, if we find here as well at so many other places polished and cultivated men. Still, however, we cannot help feeling an unusual degree of pleasure when we find a few miles from the North Cape Ariosto and Dante Moliere, Racine, Milton, and the flower of the Danish poets. The influence of great men is thus extended over the most remote spaces, and their spirit comprehends and finally diffuses itself over the whole globe.

---

* Sommerfeldt. Norsktop. Journal xxiv. iii.

The climate of Rebvog may nearly coincide with that of Hammerfest, notwithstanding the place lies somewhat more to the north; but Hammerfest is not so deep in the Fiord, and not so far from the open sea. At Rebvog the leaves first begin to appear on the birch bushes towards the end of June or beginning of July. This, as was remarked by the attentive and intelligent Wahlenberg, is seven full weeks later than at Upsal, and a week later than the birches of Utsjockis, near the banks of the Tana-Elv, where Scotch firs begin again to grow,* and a week also later than on the greatest height of Fillefieldt, or in the valley between Fogstuen and Jerkin on the Dovrefieldt.

QUALSUND, the 7th of August. I again left Rebvog in the night, with two young and mettlesome Finns, and with a guide over the mountains, a *Wappus*, or a *Loots* as the Norwegians very characteristically call those persons who are acquainted with the mountains, and can find their way among them. Such a journey is undoubtedly equally dangerous in eternal fogs as the passage among shallows, cliffs, and skiärs. My *Wappus* was not a little proud of his *Loots* knowledge, notwithstanding we may conceive that it requires no great art in clear weather to find a way over an isthmus of fifteen or sixteen miles in breadth. The Finns threw out their hooks the whole night while they rowed. They no doubt caught a few tusks; but nearly the half of the fish which they took fell back again into the sea, which was not to be wondered at, for if the hooks of Bergen among the Finns are not found advantageous, they, by means of a singular practice, quite common to all of them, alter these hooks in such a manner that the error instead of being diminished is thereby much increased. They bend the shorter arm almost entirely down to the right angle, and then fasten above the longer arm a heavy piece of tin in a new direction towards the barb. It becomes therefore altogether impossible for the fish to swallow the whole of this at once. Instead of penetrating their intestines, the point of the hook can at most enter their head or the jaws, and hence the greatest number of the fish either slip from the hook, or it brings along with it the part of the head on which it was fastened. Example has as yet had no effect upon the Finns. They adhere with the greater pertinacity

---

* Wahlenberg Kemi Lappmarck Topog. 18.

P P

to this ridiculous custom, because it is peculiar to themselves, and the opposite of what is practised by the hated Norwegians. Clergymen alone have the power to combat such prejudices. If the Finns could first find examples among their own countrymen, their eyes would at last be opened to the improvement. At Mageroe the cod nets have never enjoyed the reputation of driving back the fish ; and on that account they are actually used there by several Norwegians. This practice has gone even through the Finns to Köllefiord, and there are in fact several Finns who fish with nets in Kielvig, and the Prästegieldt of Köllefiord in a great and almost incredible advance.

Early in the morning we crossed Smörfiord, and at last entered into Olderfiord beneath high mountains covered with bushes, about ten English miles from Rebvog. We landed close by several Finnish houses—sea Finns, who live by fishing, and keep few or no rein-deer. We found only women. All the men in the Russian season lie out at sea, and do not return to their homes in the Fiords for many weeks. These women could hardly be more slightly accommodated. The *gamme* or Lapland hut in which they shut themselves in is not more than eight feet in diameter, and like a baker's oven, of about four feet in heighth. It is constructed of branches. The branches are covered on the outside with grass, but so carelessly put on that the wind penetrates through every part. A quadrangular opening in the middle serves both for window and chimney. There they sat closely crowded together, mother, daughter-in-law, and daughters, occupied with embroidering the collars of the coats of the men, or in weaving woollen bands. The limited space is accurately divided out to each ; the daughters do not come over to the more distinguished side of the mother, nor the mother, except accidentally, to that of the daughter. The fire, or small hearth in the middle, separates the herile and the servile sides. By means of this order established we can obtain a place in a gamme occupied by a large grown up family without being in the way of one another; a remarkable example of what may be effected by method. Nobody would believe, without seeing it, that so many beings could be occupied in such an earthen hovel, without being a hindrance to one another.

The sea Finns are not obliged to be nomades like the Fieldt Finns; for the few rein-deer which they may possess they must give up to the care of others. They are therefore fully enabled to build more durable habitations, *gammes*, or even houses like those of the Norwegians. But this, however, is not done by them; and it not only prevents them from making any further advances, but destroys also the sources of the prosperity of the country. They think they cannot do without at least three habitations. The winter abode is deeper in the Fiord in the vicinity of the woods, and so close to them, that the birches which are felled are almost at the very door of the gamme. In summer they draw nearer to the sea for the sake of the fishing stations. They frequently change also their residence in harvest for the sake of a fresh pasturage for their cows. Their dwellings, the gammes, are only calculated therefore for a few months duration. All the property is moved about with the master of the house up and down the Fiords, and at most there is at the winter place a small house of stakes, in which they preserve their boats and necessary winter stores. With such a wandering mode of life, property becomes a burden; for it is too great a load to remove from place to place, and it also requires a more roomy and careful erection of the gamme. Hence we may account for the bad economy of these sea Finns. A more durable house, a substantial gamme, would excite new wants in them, and with these wants industry and activity. Of this we have actually the most fortunate examples. There are several sea Finns in Altensfiord, in Näverfiord, and at Korsness, who no longer change their residence. Their gammes have a much greater number of conveniences, and they have built houses besides the gammes they live in, for the sake of storing up what is necessary for their sole sustenance. These men have become prosperous, and their property round the gammes is at present as well cultivated as we might expect among Norwegians or Quäns.

There are other sea Finns, who are not contented with a change of residences for summer and winter; they again remove to another winter station when they have destroyed the woods, or when the trees are not near enough to them. The spot which they leave is lost for centuries; for the place of the wood which is so cut down is never supplied. The Finn only uses the head of the tree, because that alone appears above the snow; the stem remains in the ground, rots, and prevents the growth of new trees. Hence the woods of the Fiords are disappearing every day; and this is an irreparable loss. For it is evident that a wood is an object of great importance in regions where woods do not thrive remarkably, and where there are so many individuals in the outward and treeless islands in want of them.   Such a division of property as we find among the Norwegians and Quäns, if it were possible in districts so thinly inhabited, would perhaps prove a remedy for this evil, and fix the sea Finns in definite habitations.

We ascended the Olderfiord, and at the end of the Fiord we crossed over hills of several hundred feet in heighth.   There we could overlook almost the whole of the large valley which traverses the isthmus, connects Refs Bottn with Smörfiord, and the latter again with Olderfiord.   The valley is like a plain, extensive and large, and of very inconsiderable elevation ; and the ground is every where covered with an excellent birch wood.   The mountains decline gently from the north into this valley, and the birches ascend a great way up the declivity.   This is such a joyful and animated prospect as we never again find in our course up this Fiord.   The wood might certainly be a great assistance to those who live near the sea if the smallest land-carriage were not impossible in this country for want of men and horses.

We turned towards the south, and ascended the valley through the bushes. We kept mounting for several hours before the form of the birches betrayed a greater elevation.   The lateral valley then became broader and marshy, and consequently the way over the wet ground was attended with difficulty. These marshes at a distance look more like meadows than lakes; for the water is wholly covered with black heath, and innumerable little clusters of the small dwarf birches (*Krampe-Birk*, Betula nana) which rise one or two feet above the surface ; and mountain brambles (*Rubus chamaemorus*).   When we proceed over such marshes, we must look for the spots nearest to the greatest number of these clusters, and spring from cluster to cluster among the birch bushes : certainly a sufficiently difficult and fatiguing manner of proceeding when the way runs for whole miles over the marsh. When we at last reached the top of the valley we saw ourselves on a widely extended table-land.   Insulated and long extended rocky hills rose above it like islands, and between them in the plain the marsh gradually proceeded downwards.   The birches on the rocks now became small bushes of a few feet in heighth, and crept along the ground ; and at the greatest heighth over which we passed, in the middle of the way between Kistrand and Reppefiord, they seemed on the point of disappearing altogether.

This height was eight hundred and two English feet above the sea.* Eastwards, towards Kistrand, the mountains were much higher, and they rose there far beyond the limit of birches, perhaps to between one thousand four hundred and one thousand six hundred Paris feet. They were near enough, however, to enable us to perceive that the small bushes actually ascended higher on them than the heighth of those hills towards Reppefiord, at most about two hundred feet. Hence the boundary of birch vegetation may be placed at one thousand and ninety-six English feet above the sea. We may from this form a judgment of the climate of both Kistrand and Reppefiord, and the amelioration which takes place in the course of the passage up the Fiords. For the mean temperature of these fiords would nearly rise to—0°. 18 R or 31°. 6 F. and consequently be above the boundary of firs. This is extremely probable; for a few miles deeper into the Porsangerfiord we actually find Scotch firs, and at the end of the Fiord, in the Porsanger-Bottn, there is a fir-wood like that of Alten, only not so extensive for the surface over which the wood can spread is not so large. In Kistrand we may live nearly as at Talvig, at least the difference is of small consequence; but how different it is from the climate of Kielvig, only twenty-eight English miles distant! The difference is as great as from Helgeland to the farthest parts of Nordland. The limit of snow on the *Fieldts* of Reppefiord would commence at nearly two thousand eight hundred and seventy-five English feet above the sea; but such heights are to be found in the chain of mountains between the Fiords of Alten and Porsanger, only on a few points to the west of Altensfiord and the valley of the Alten-Elv.

We saw at a distance, on our way, the gamme of a Fieldt Laplander on the mountain, with the rein-deer around it. They come from Kautokejno, and to the number of three or four families roam over these hills in summer all the way to the sea. This is no great number for the extent of country; but the want of rein-deer moss on these mountains necessarily limits the multitude who use it. The marshes are too large, and the islands which rise out of them too rocky and bare. The rein-deer find scarcely any thing for their sustenance in the whole of the Table-land between the

---

* 6th August, h. 8 Olderfiord Bar. 28. 0. 3. Th. 10. 8 R. clear, calm, south wind.
    h. 6 p. m. greatest heighth 27. 1. 3 .... 8.....clear, south wind.
  7th August, h. 7 Qualsund.......... 27. 10. 6........ ....clear.

mountains of Kistrand and Qualsund, which is about ten English miles in
breadth ; and therefore the Laplanders can only remain for a few weeks on
single mountains, and are obliged to proceed farther north towards the
mountains, above the Fiords which belong to Maasöe Persecuted by the
innumerable swarms of gnats which hover over us in clouds, and which
incessantly disturb, torment, sting, and smart us, if we are not incessantly
occupied in defending ourselves from them, we at last descended into the
large and green valley of the Reppefiords Elv. The valley, when we
entered it, declined from the south, and we saw at a distance, before
entering it, its beautiful wood of large birches, alders, and aspens. Near
the place where we descended the valley immediately inclines towards
the west, and soon afterwards enters the Reppefiord. The river dashes
down through the trees, and birches lying in the water tore up by the roots,
others with a feeble hold hanging over the banks, others lying in heaps at the
side of the stream, were the strongest proofs of the violence of this water, and
the devastation which it creates in its descent from the mountains.   It is also
one of the greatest rivers of all those which fall into this Fiord, and, with the
exception of the great Altens-Elv, has not perhaps its equal in all West Finmark. Hence it is also called *Laxe-Elv*, a river in which salmon ascends,
from which three families settled on the plain at the mouth of the river
derive a rich support.   This was evident from their *gammes* ; for they bore
no resemblance to the little earthen huts we saw in Olderfiord, but looked
like a little colony.   The solid dwelling *gammes* were situated in a pretty
extensive circle and hedged round ; farther on there were stables for the
cattle, then storehouses of logs covered over ; and several pieces of ground
about the place were carefully converted into meadows. The cows stood before the enclosure with their heads turned towards one another in the midst of
smoke, into which they rush of their own accord. This smoke is made for the
purpose of securing them from the sting of the gnats in the night.   It has a
singular appearance. Perhaps there are few circumstances more characteristic
of nature in Lapland than these groups of cows in the smoke, surrounded
by gammes and meadows; and the birch-wood and mountains in the distance.
   The formations from the Olderfiord onwards over the mountains are like
the mountains themselves, in nowise distinguished.   In the whole way we
find the same mica-slate which commenced at the Porsangerfiord, with a
gentle inclination towards the south, and without the smallest trace of

gneiss. The mica-slate is never, however, very distinct; it is too fine slatey, and the mica is not fresh. It is situated on the boundary of the clay-slate.

It is but four English miles and a half from Reppefiord to Qualsund: a hospitable and well-contrived place. The respectable house on the height, the brigs in the water, with a large gate at the entrance through green meadows up to the house, the church, and several other houses deeper in the bay, form altogether a surprising and agreeable prospect And though there are no trees in the country, there is, however, a great deal of variety in the green valley at the side of the place, and in the surrounding mountains. Hence we may partly reckon Qualsund among the most agreeable districts in Finmark, though it does not equal the noble situation of Alten.

The merchants are here the true princes of the country. We might divide the land with as much certainty, according to the circles of their influence, as it is at present divided into Prästegieldts. If the clergymen influence the minds of the inhabitants, in return, their temporal felicity is almost always in the hands of the merchants. The Finns and Norwegians will place their all at stake for the sake of drinking brandy at the merchant's till they fall down. They drink much more than the value of the fish which they bring with them, and the debt is entered in a book, which they carry back, but the contents of which they never compare with their circumstances. The debt at last exceeds the value of their property: they must assign over to the merchant, if he desire it, their house, and the whole of their little possessions, and they esteem themselves fortunate when they are allowed to remain on the spot as farmers. Hence an iniquitous and avaricious merchant is a pest to the country ; but fortunately there are upon the whole but few of such a description in this province ; though if many of them resembled M. Clerke, the active merchant in Qualsund, the most happy and beneficial consequences might every where result from their influence. M. Clerke has put Quäns into several of the Finnish residences which have fallen into his hands, and this has been attended with the most favourable consequences. The industrious Quän soon accomplishes what the drunken Finn never could have done ; and if this does not at last excite the attention of the sea Finns, they will in all probability be wholly driven from these coasts. On this subject I once heard a Finn (in Naeverfiord) complaining with a degree of national pride not a little comic. He felt the future consequences of the

settlement of the Quäns the whole way up to Hammerfest, and he complained of the injustice of not restoring to Finns the places obtained from
Finns, the indigenous nation. Such a view of the subject may easily be
pardoned to a Finnish Laplander, and we may even hear it with some degree
of pleasure; but in the mouth of a rational Norwegian, it is somewhat the
same as lamenting, that in the free States of America, instead of the indigenous, wandering, and scalping Iroquois and Chippiways, many millions
of foreign husbandmen are living on the produce of the soil, and that many
thousand spots are now inhabited by emigrants in regions formerly peopled
by the wild beasts of the forest and the rattle-snake. So long as the Finns
shall be possessed of their rage for brandy, nothing can be expected from
them that has the least tendency to improvement. If a Quän therefore lives
where a Finn formerly lived, the place is occupied by a being of a superior
character, and upon the whole much more humanized. We must judge
of men from essentials, and not from forms. That the mind of a Finn is
not capable of every degree of cultivation, as well as that of the Finlander
in Finland, or the Sclavonic Russian or Pole, who would think of denying?
But then this cannot take place till brandy becomes a rarity in the country.
The merchants themselves acknowledge that upon an average not less than
from twenty-seven to thirty rix-dollars is annually consumed by the Finn in
brandy; this is much more than a whole cask, and also more than half
of the annual earnings of a Sea-Finn. They do not drink for the purpose of
lightening the severity of their labour, or for the purpose of keeping themselves warm in winter on sea, for they very seldom have brandy in the boat
with them in their voyages. They do not drink to assist in the digestion of
their meals of fish and fat fish-livers; for they seldom have any brandy in
their gammes, and neither Norwegian nor Finn drinks it with fish-livers.
All is consumed at the merchant's, and before his door, and the Finn would
be himself astonished if he returned from the merchant's without becoming
raving mad with the liquor, and afterwards lying for several hours senseless
and dead drunk before the door. The scenes which take place when the
Finns are assembled on particular accasions, such as fairs or court-days, may
be easily imagined. Particular edicts have been issued for the purpose of
prohibiting the merchants under a heavy penalty from furnishing brandy to
Finns before the expiring of the first court-day; but the cases notwithstanding are very frequent, when *Sorenscriver* and *Foged* have been obliged to

return without doing any thing, and to fix a new court-day, because, although the Finns made their appearance, they were lying senseless along the ground, like so many cattle. They do not drink so immensely with impunity. The brandy at last deprives them of their appetite, and they become feeble, powerless, and worn down, and are at last unable to perform the most necessary operations. This is so striking, that one would imagine it would be a warning lesson to them. But they cannot be convinced. The charms of brandy are too powerful. With the greatest self-complacency, on account of the unanswerable nature of their argument, they assert that brandy is equally strong and equally nourishing as bread, for like bread it is prepared from grain. Thus all the little intellect which the mind of a Finn may be supposed to possess, every spring of activity, and every incentive to improvement, are destroyed and eradicated.

It is thought, and at first sight it appears natural enough, that the wretchedness occasioned by this rage for brandy, and by which Finmark is so severely oppressed, would first be diminished, and at last disappear, if all access to this poison was cut off from the Finns; and that the surest way to effect this would necessarily be to prevent the merchants from selling brandy to them, or preventing any brandy from being carried to Finmark. Hence they are convinced that a great deal of the miserable condition of the Finn attaches to the conscience of the merchant, and they rail at, and complain of, the immorality of the distillers of Flensburg, who have the shamelessness to send every year such uncommon quantities of brandy into Finmark. The poor Flensburgers! Very probably they may never have entertained a suspicion of the mischiefs occasioned by their exports, and how heavily it lies upon their conscience. They probably have not the slightest knowledge of the place where their brandy is carried to, except that it goes northwards; for the Nordlands in the Danish states is as indefinite an expression as *le Nord* to the French. If we are to carry responsibility so far, and make the Flensburgers accountable for the damage done by their brandy in the world, why not also include the peasant of Schleswick, who sells them the rye and wheat of which it is made? Why not the coppersmith who manufactures the vat? the seaman who transports the cask to the Finns? How are we to expect from any manufacturer that he is to sit down and weigh the good and bad consequences with which his manufacture may be attended? He who ventures to estimate all the consequences of an action *not immoral in itself*, and to condemn or approve it according to his own view of the matter, takes a bold and audacious grasp of the wheels of fate, which none ever yet attempted with impunity, but the few powerful minds whose strength has enabled them to take a free, certain, and unobstructed view of the movements of the world. A manufacturer seldom extends his views so far. Limited to the world of the creation of his manufacture, the nature, quantity, and market of his produce, are his highest principle. How can it be otherwise? Is the preparer of verdigris, arsenic, or opium, to be deterred by the consideration of the mischief to which his produce may give occasion? Or is he not warranted in placing so much confidence in the prudence and morality of other men, as to suppose that they will as well as himself avoid the immediate pernicious influence of such dangerous articles? The reproaches against the merchants for exporting brandy to Finmark are therefore highly unjust. Can the exporter judge of the quantity he may safely send without injury to the people, where he is to stop, and whom it is to injure? Is the nature of the trade in general capable of any such consideration? The whole responsibility, if it can possibly attach to the seller, must be limited to the retail dealers in Finmark; for these persons see the immediate effects of this destructive beverage. There indeed many cases may occur where we cannot justify them. That an edict should be necessary to prevent the merchants from selling brandy on Sunday to the Finns before divine service is over no doubt does not reflect any great credit on the merchants, but it is what we ought to expect; for why should Finmark alone escape the influence of the demon of self-interest? That the traders follow the Finns with brandy almost to their very houses, that they should stimulate the people to intoxication, and that they should contrive and multiply the opportunities for that purpose, is neither a worthy nor an honourable course of proceeding; but it is a course which frequently takes place. But can we reproach the merchant for allowing free operation to the rage of the Finns for brandy when he comes to his shop? Is he who comes only to Finmark for the sake of gain to set up for the moralist of the Finns? Can he determine the quantity which each may bear without being injured by it? And is he not to purchase the fish or the skins of the Finn, because, perhaps, he may give more brandy for them than is sufficient to lay him, if taken at once, senseless on the floor? Certainly not: in that case the Finn alone is accountable for the mischief he does to himself.

No doubt, many a philanthropist will here break out in complaints and wishes (and how often do we not hear them?) that this people had never been incited to have any connexion with trade, and that they might have continued to live happily and unnoticed in their former innocence. Would to God, that neither the Norwegians nor the brandy, had ever found their way to these Fiords! This might be all very well, if a troglodite life of this sort were either the happiness or destiny of a people; if men were not to keep pace with nature in their career. And what then is the felicity which these Laplanders enjoyed? It does not rest on their own conviction; for every Finnish Laplander, who never before knew brandy, would immediately after becoming acquainted with it, esteem that an infinitely happier condition which allowed him a facility of enjoying his brandy. If we are to call this an imaginary felicity degrading to humanity, then the former careless, infantine happiness which this people enjoyed in the visionary innocence of their natural state was in no respect more dignified or becoming a human being. But is a virtue worth much which the possessor is unconscious of, which is merely the result of habit, or rather the consequence of an impossibility to do evil? No, if strangers should have only introduced vices among the Finns, they must also have given them a capability of exercising genuine virtue, and in this respect alone contributed to ennoble their nature. Man can alone rise in the scale of being by the collision of mind with mind ; and we ought to rejoice when we see people, who formerly lived insulated and alone, carried along by the world in its progress. In the desert, the child never becomes a man ; and in a limited space, where there is only room for a few ideas, no nation can ever be formed.

The Finns therefore will never improve so long as they are domineered by their rage for brandy. But neither moral considerations nor royal edicts can put a stop to the importation of brandy. The latter cannot, because the Finmark trade as at present constituted would be destroyed, and because it is utterly impossible to stop the course of so overpowering a torrent as the want of brandy in Finmark. Edicts which must remain unexecuted and without effect are prejudicial to the character of a government. The evil must be attacked in its source, if we wish to remove it. This is sufficiently evident. We shall neither allay the thirst of a person labouring under a burning fever, nor improve his condition, if we obstinately refuse to give him any thing which he can drink. In like manner, the nature of the Laplander or even of the Norwegian of Nordland, will not be changed if it were even possible to deprive them entirely of brandy. For what gives rise to this inordinate rage for brandy? Has it its cause in a peculiar organization of this people? This is by no means probable, if we consider that this rage is shared with them in an equal degree by negroes, panting under the line, by the Iroquois in temperate zones; and again, by the Esquimaux in the coldest of all the inhabited regions of the earth. On the other hand, the gentle Hindoos never drink, neither do the industrious Chinese; and the Russians, who are elsewhere addicted to an excessive use of strong drinks, are wonderfully moderate in Finmark. Hence the ruinous propensity is not produced by external causes; the ground lies in the constitution of the inward man. It lies in the low degree of cultivation, and the thoughtlessness of these nations. The Russian in Finmark is bent on returning with a rich cargo of fish; he resolves to dispose of his goods, and when he reaches home, to convey the fish again to a new market. He is always occupied, and the chain of his occupations is always present to his mind: he knows that the success of his first beginnings determines the result of his final operations. He has gained the incalculable advantage of feeling an interest in his existence, and the knowledge of a definite object why that existence should be spun out. It is not so with the Laplander, with the negro, the Iroquois, or the Esquimaux. The present moment is alone prized by them, and what lies hid in futurity has little concern for them. They can never fall in the world, because they have never risen. The consideration of the destruction of their domestic and civil prosperity through brandy can consequently never influence them; for their domestic ties are extremely feeble, and civil relations they have none. The brandy on the other hand gives them the feeling of the moment and of their existence, and on that account it is such a favourite with them. Give them an object to occupy their thoughts; for till then they will never cease to drink. Why is the Quän not so great a drinker? Because agriculture consists of a series of occupations which exercises his attention, and which makes him look with an anxious eye at the beginning of the year towards the end of it. The pernicious consequences of brandy are therefore easily perceived by him, for he soon feels how much they frustrate his veiws. Why again is the Norwegian at his fishing stations in Lofodden so prodigiously addicted to drinking? Because fishing with him is merely an insulated, unconnected operation, as well as with the Finn, and not, as in the case of the Russian, connected with a number of others in distant perspective. Were it possible for the example of the Quän to produce at last some effect on the Laplanders, and induce them seriously to divide their time between their sea pursuits and the cultivation of their land, brandy would soon become a much greater rarity in Finmark. We must not despair that this will one day take place. A young Finn is susceptible of ambition, when he is stimulated in a manner conformable to his powers; and if many of them possess a disposition to please, and the bustling good-nature which is observable in the few that a traveller has an opportunity of seeing, they also possess other qualities, by means of which they may be influenced. However much they hate the Norwegians, they appear to repose great confidence in clergymen, in merchants, and royal functionaries, in the *Storkarlen*; for they are not held in contempt by them, as by the other Norwegians; and the people are accustomed to receive more favours than ill-treatment from them. The older Finns again have, with the rage for brandy, visibly sunk in character. With little plan or reflection in their transactions, selfishness seems to have gained a complete ascendancy over them. A young Finn will do a kind action, because it may prove acceptable, but an elderly one will scarcely move without some evident advantage to be gained by it; and he is little disturbed by his conscience when he violates his duty, as soon as he imagines he can do so without fear of punishment. This is not because he takes a pleasure in giving pain to others, but because the sensual enjoyment of the moment is the only advantage he can conceive, and the only object of his thoughts. But let the minds of the young Finns be gained over, let their actions be first directed in smaller circles to lower objects, and then gradually to higher, and the people will at last be raised to men and to citizens; and we shall finally be convinced that nature never exclusively destined one people to serve, and another to command.

ALTENGAARD, the 9th of August. It was very gloomy when I left Qualsund: the fog descended to a heighth of about three hundred feet on the mountains: it had rained the whole night through, and it now appeared as if every thing would dissolve into fog. We proceeded up the Vargsund, and with a west wind crossed the Strömen Sund, the course to Hammerfest The fogs began to ascend higher and higher; and in the course of the hours which I passed with the industrious Finn family in Näverfiord, the sun broke out, and the clouds disappeared. Above the Vargsund hangs the border of the perpetual cloud canopy which is spread over the islands towards the ocean; and from this place, as we ascend deeper into the Fiord, the sun and clear weather are no longer reckoned among the number of rare blessings. On the extreme point of Korsness we actually saw the sun rise again quite clear from the north; and in a few hours afterwards we endeavoured to defend ourselves from its rays. This calm, and the clearness of the air, brought up the whales: the whole Fiord appeared full of their spouts; and under these circumstances it was deemed a wise and necessary precaution by my Finns to keep as near to land as possible. We proceeded therefore round the east side of Aaröe, and we thereby added considerably to the length of our voyage. But our hope of obtaining a wind from the sea, the *Havkulje*, at the end of the island, was disappointed. Notwithstanding the air in warm and clear days always rushes into the Fiords, this day other causes were operating. The clouds appeared in thick masses in the valley of Alten, and similar masses were also driven about from the north-west. These soon betrayed their nature. Lightning and thunder were powerfully and incessantly discharged from the clouds above Alten, and the whole was at last quickly driven towards Porsanger. The other storm reached us on the Fiord itself. It resembled a thunder-storm in a tropical climate. In a few minutes we were driven the last four English miles from Altenness to Kongshavn: the rain flowed in thick and aggregated drops, and hail of the size of peas fell with a hissing noise into the water. Who would have expected such weather in Finmark? They were not balls of snow, but firm and solid icicles, which were very sensibly felt on the legs and arms as they fell. And what was singular, they were not round, but in the form of pears, with the point upwards, and with concentrical shells round the thicker half. The drop was not merely frozen in the clouds, but also in the descent, where formerly there was no ice temperature. The freezing water had sunk down on the firm icy substances, and only strengthened the undermost half. Is this freezing in such deep regions not a consequence of the rapid evaporation in the strongly heated air through which these drops had to fall?

Thus I again reached Altengaard, where I soon forgot the disagreeable impression of the weather, in the circle of so amiable and respectable a family as the one which now inhabits that place.

ALTENGAARD, the 17th of August. We crossed over to Talvig on the fifteenth In three hours we were in the beautiful bay beneath the high rocks, and at the edge of the green declivity which environs in such an animated manner the whole bay. I crossed through meadows for about an English mile to Stor-Vand, a large lake which runs deep into the mountains. It brought the *Klönthal* to my mind; it is on the same solitary and grand scale. The surface of the water on the opposite side advances close beneath a high rocky mountain, from which a waterfall dashes down precipitately quite in the Alpine manner. Snow lay on the summits around. The scattered boats and fishers in the bays of the lake appeared lost amidst the great and mighty masses. Such a view, such grandeur, such an impression in Lapland!

As I ascended the following morning amidst noisy waterfalls over the nearest rocks above the houses of Talvig, one thousand one hundred and forty English feet above the Fiord, snow mountains made their appearance above for a great extent, and yet near enough to allow access to them over the points of the Fieldt. It is extremely entertaining to climb great and rapidly ascending heights in these climates. As in the ascent of Montblanc we gradually rise beyond all the points which seemed immeasurable from the valley, so in like manner the Lapland vegetation with which we are familiar in the vallies gradually disappears under our feet. The Scotch fir soon leaves us; then the birches become shrivelled; now they wholly disappear; and between the bushes of mountain willows *(fieldt-weider)*, and dwarf birches, the innumerable clusters of berry-bearing herbs have room to spread—bleaberries *(blaabaer, vaccinium myrtillus)* on the dry heights, and mountain-brambles *(rubus chamaemorus)* on the marshy ground. We at last rise above them: the bleaberries no longer bear: they appear singly, with few leaves, and no longer in a bushy form. At last they disappear, and they are soon followed by the mountain willows. The dwarf birch alone braves the heighth and the cold; but at last it also yields before reaching the limit of perpetual snow; and there is a broad border before reaching this limit, on which, besides mosses, a few plants only subsist with great difficulty. Even the rein-deer moss, which vies in the woods with the bleaberry in luxuriance of growth, is very unfrequent on such heights. On the top of the mountains, which is almost a table-land, there is no ice, it is true, nor glaciers; but the snow never leaves these heights, and a few single

points and spots above the level are alone clear of snow for a few weeks.
Here the Laplanders seldom or never come with their rein-deer, except in
their way to the vallies. It is a melancholy prospect; nothing in life is any
longer to be seen, except perhaps occasionally an eagle in his flight over the
mountains from one Fiord to another.

But on the small mountain-caps above the level I had an extensive view
over the mountains and fiords. All that we have such difficulty in over-
looking in this indented and intersected country lay now extended beneath,
and could be taken in at a single glance—the singular rays of *Fieldt*-points
of Quaenanger appearing like indented quartz: in the openings of the rocks
the still higher chain of Lyngen behind. Southwards, towards the Swedish
boundaries, the mountains seemed an interminable plain, with merely a
few long mountains on it destitute of character, which were visibly greatly
inferior in heighth to the mountains of Talvig. Eastwards, towards Porsanger,
and above Refsbottn, still higher mountains appear; but singly, and only
in a blue distance, and scarcely visible. The mountains of Talvig are alone
commanded towards the north, where the long-extended snow-chain of the
Jöckulsfiord enters deep into the ocean above Stiernöe and Seylandt. I
saw here distinctly the manner in which the ice is then separated in clefts
from the powerful mass of snow, and precipitated into the Fiords. On
Seylandt also the ice beneath the snow was not to be mistaken; the last
remains of the great northern Kiölen mountains. We can see here also
clearly and distinctly the manner in which these mountains split towards the
Fiords of Lyngen and Alten, and that there is no mountain in the direction
of East Finmark at all to be compared to them. To the south of these
heights the level scarcely seems a mountain range; but to the northwards
there is nothing but a long range of Alps and Glaciers.

I stood long on *Akka-Solki* lost in this prospect. The mountain runs in
an almost insulated manner between two vallies, about ten English miles to
the south-west of Talvig, and is only commanded by a few other surrounding
heights. It was three thousand three hundred and ninety-two English feet
above the Fiord. The summit, a huge ruin of millions of blocks lying above
one another, had only lost its snow a few days before; but there still lay a
broad snowy mantle along the declivity, which never leaves it, notwithstanding it is here exposed to the action of the sun, the rain, and the wind;
—it was a commencement of the perpetual snow region. Stor-Vandsfieldt,

the highest mountain of this district is about four English miles still farther to the south-east, and it is separated from Akka-Solki by Stor-Vand, and lies between the latter and Kaafiord. This summit is rather more than one hundred and fifty feet higher, and may be about three thousand five hundred and fifty English feet above the sea. There the snow actually lay to the very top, and the summit, as viewed from Alten, is never altogether free of snow, from which place it is very distinctly visible. If there were a level of any extent on this height we should no longer find particular spots clear of snow, and glaciers would be generated towards the Fiords. The lower limit of snow above the mountains of Talvig in the latitude of 70°. may therefore be somewhat about three thousand five hundred and fourteen English feet, or five hundred and fifty toises above the sea. The heighth of the mountain table-land, between Talvig and Quävanger, can hardly at an average be estimated above two thousand nine hundred and eighty-two English feet; and hence there are no glaciers above the Jöckulsfiord, and on Seylandt, although here and there may be found perpetual snow.

The appearance and limits of the different plants, bushes, and trees, on the declivities of the mountains, are still more striking and grateful to the view as we descend; for it is a return from wilderness to cultivation. And although these limits can never perhaps be actually laid down with accuracy, it is visible that they do not oscillate here more than a few hundred feet. Hence the following table, which is the result of the data of the barometer, will be found to err very little in the heighths laid down as the limits at which the different productions disappear.

Limit of snow above Talvig in 70° lat. - - - - 3514 English feet.
Betula nana (Krampe-bir, Dwarf-birch) - - - 2742
Salix myrsinites (Whortle leaved willow) - - - 2150
Salix lanata (Downy willow) goes higher; it rises
     above the *Betula nana*, and approaches near to
     the perpetual limit of snow.
Vaccinium Myrtillus (Blaaber, Whortle-berry, or
     Blaeberry) - - - - - - - - - - - 2031
Betula alba, Birch-tree - - - - - - - - - 1579

The Scotch firs are strangers at Talvig; they are not found together in woods or thickets, but here and there among the birches. The high and

R R

perpendicular rocks round the bay prevent the lower parts for several hours in the morning from receiving the rays of the sun, and the spring sun does not rise above these rocks. Hence Talvig is actually colder than we might expect it to be from its situation, and the difference when compared with Alten is very striking. This cause has no influence, however, on the higher regions above the range of rocks. Hence we may also take the limit of firs at Talvig at nearly seven hundred Paris feet *.

   © That enterprising traveller and acute observer, Wahlenberg, in a work lately published, gives us a very interesting view of the distribution of plants and animals as we ascend from the Gulph of Bothnia to the summits of the Alps of Lapland. The following extract from that work cannot fail to gratify the reader.

   " 1. On approaching the Lapland Alps (Fjall) we first arrive at the line where the spruce-fir, (*pinus abies*,) ceases to grow. This tree had previously assumed an unusual appearance; that of a tall slender pole, covered from the ground with short, drooping, dark branches; a gloomy object in these desolate forests! The *rubus articus* had already, before we arrived at this point, ceased to bring its fruit to maturity. With the spruce we lost the *rosa cennemomum*, *convallaria bifolia*, &c. and the borders of the lakes are stripped of their ornaments of *arundo phragmites*, *lysimachia thyrsifolia*, *galium boreale*, and *carex globularis*. Here is the true station of *tussilago nivea.*—The last beaver-houses are seen in the rivulets, and no pike nor perch are to be found in the lakes higher up. The boundary of the spruce-fir is three thousand two hundred feet below the line of perpetual snow, and the mean temperature is about $37\frac{1}{2}°$ of Farenheit.*

   " 2. Scotch firs (*pinus sylvestris*) are still found, but not near so tall as in the lower country. Their stems here are low, and their branches widely extended. Here we see the last of *ledum palustre*, *salix pentandra*, *veronica serpyllifolia*, &c. The bogs have already a very sterile appearance. Near the outermost boundary of the Scotch fir grows *phaca alpina*. Higher up are hardly any bears to be met with, and the berries of *vaccinium myrtillus*, (the whortle or blacberry, or billberry) do not ripen well. *Salmo laveretus* (the gwiniad), and the *S. thymallus* (the grayling), soon after disappear from the lakes. The upper limit of this zone, where the Scotch firs cease, is two thousand eight hundred feet below the line of perpetual snow, and the mean temperature about $36\frac{1}{2}°$ of Faht. A little below this point, or about three thousand feet before we come to perpetual snow, barley ceases to ripen; but small farms, the occupiers of which live by grazing and fishing, are met with as far as four hundred feet higher; and so far also potatoes and turnips grow large enough to be worth cultivating.

   " 3. Beyond this the dwarf and stunted forests consist only of birch. Its short thick stem, and stiff, widely spreading, knotty branches, seem prepared to resist the strong winds from the Alps. Its lively light green hue is delightful to the eye, but evinces a weakness of vegetation. These birch forests soon become so low that they may be entirely commanded from the smallest

   * The snow line in these regions is about four thousand two hundred feet above the sea.

The first rocks above Talvig, from which the waterfalls descend, still consist of clay-slate, which frequently resembles talc-slate.   It is glimmer-

---

eminence. Their uppermost boundary, where the tallest of the trees are not equal to the heighth of a man, is two thousand feet below the line of perpetual snow. This zone is therefore much wider than the preceding. Long before its termination, *alnus incana, prunus padus,* and *populus tremula,* were no more to be seen. A little before the birch ceases we miss the *sorbus aucuparia,* which for some time had not presented us with any fruit ; the *rubus arcticus,* already likewise barren, *erica vulgaris,* &c. Where the birch forest becomes thinner, the reflection of the heat from the sides of the mountains is strongest. Here in many spots we find the vegetation of *sonchus alpinus,* and *aconitum lycoctonum,* remarkably luxuriant. The drier spots now become covered with *lichen rangiferinus. Tussilago frigida,* and *pedicularis sceptrum-carolinum* have their place to the utmost boundary of the birch. Thus far only charr *(salmo alpinus)* is found in the lakes, and higher up all fishing ceases.

" 4. All mountains above this limit are called Fyall Alps. Near rivulets, and on the margins of bogs only, is found a little brush-wood, consisting of *salix glauca,* whose grey hue affords but little ornament to the landscape. The lower country is covered with the dark looking *betula nana,* which still retains its upright posture. A few juniper bushes, and some plants of *salix hastata,* are found scattered about. Every hill is covered with *arbutus alpina,* and variegated with *andromeda cœrulea,* and *trientalis europea.* The more boggy ground is decorated with *andromeda polifolia* in its greatest beauty, and *pedicularis lapponica.* On the sides of the mountains, where the reflected heat has the greatest power, grow *veronica alpina, viola biflora, pteris crispa,* and *angelica archangelica.* This zone extends within one thousand four hundred feet of the line of perpetual snow. The Glutton *(Mustela gulo)* goes no higher than this. The berries of *rubus chamœmorus* ripen here, but not at a greater degree of elevation.

" 5. Now no more brush-wood is to be seen. The white *salix lanata* is not above two feet high even above the rivulets, and *salix myrsinites* is of still more humble growth. *Betula nana* occupies the drier situations, but creeps entirely upon the ground. The hills are clothed with the rather brown than green *azalea procumbens,* and *azalea lapponica,* which give this zone its most peculiar feature. Verdant spots between the precipices, where the sun has the greatest power, produce *lychnis apetala, erigeron uniflorum,* and *ophrys alpina.* In boggy places, *aira alpina, carex ustulata,* and *vaccinium uliginosum,* are observable. The only berries, however, which ripen at this degree of elevation, are those of *empetrum nigrum;* but these are twice as large as those that grow in the woodlands, and better flavoured. The upper boundary of the zone is eight hundred feet below the line of perpetual snow. The Laplanders scarcely ever fix their tents higher up, as the pasture for their rein-deer ceases a very little way above this point. The mean temperature is about 34º. of Faht.

" 6. Next come the snowy Alps, where are patches of snow that never melt. The bare places between still produce a few dark shrubby plants, such as *empetrum nigrum,* but destitute of berries, *andromeda tetragona,* and *hypnoides,* as well as *diapensia lapponica.* Green preci-

ing, and still very thick slatey ; in the cross fracture coarse splintery and earthy. This is the same rock as that of Kongshavnsfieldt, and we should probably soon also find here the quartzy sand-stone of that mountain. All the strata dip westwards into the mountains ; a circumstance which also betrayed the direction of the rocks and other precipices towards the east. In the course of the mountain westwards this rock gradually changes into the micaceous clay-slate of Kielvig on Mageröe, and at the end of about two English miles along the mountain ridge we find a summit of high rocks, which rise above the clay-slate, consisting of a beautiful small granular mixture of leek-green diallage, which is somewhat more finely granular than that of Honning-vog, and more resembling that of Prato in Tuscany : inter-mixed we observed grey-coloured long crystals of felspar, and grass-green epidote in very small aggregated crystals, like those we see in the gneiss of Mont Blanc ; and lastly a great bed of iron pyrites and a number of iron grey metallic grains. Before this epidote was frequent in the quartz beds in

---

pices exposed to the sun are decorated with the vivid azure tints of *gentiana tenella,* and *nivalis,* and *campanula uniflora,* accompanied by the yellow *draba alpina.* Colder and marshy situa-tions, where there is no reflected heat, produce *pedicularis hirsuta* and *flammea,* with *dryas octopetala.* This zone extends to two hundred feet below the limit of perpetual and almost uninterrupted snow.

" 7 Beyond it perpetual snow begins to cover the greatest part of the ground, and we soon arrive at a point where only a few dark spots are here and there to be seen. This takes place on the Alps of Quickjock, at the elevation of four thousand one hundred feet above the level of the sea ; but nearer the highest ridge, and particularly on the Norway side of that ridge, at three thousand one hundred feet. Some few plants, with succulent leaves, are thinly scattered over the spongy brown surface of the ground, where the reflected heat is strongest, quite up to the line of uninterrupted snow. These are *saxifraga stellaris, rivularis,* and *oppositifolia ; ranunculus nivalis* and *glacialis, rumex digynus, juncus curvatus* and *silene acaulis.* The mean temperature at the boundary of perpetual snow is about $32\frac{4}{5}°$. Faht.

" 8. Above the line of perpetual snow the cold is occasionally so much diminished that a few plants of *ranunculus glacialis,* and other similar ones, may now and then be found in the clefts of some dark rock rising through the snow. This happens even to the heighth of five hundred feet above that line. Farther up the snow is very rarely moistened. Yet some um-bilicated lichens (gyrophoræ), &c. still occur in the crevices of perpendicular rocks, even to the heighth of two thousand feet above the line of perpetual snow. These are the utmost limits of vegetation, where the mean temperature seems to be 30°. of Faht. The snow bunting *emberiza nivalis* is the only living being that visits this elevated spot."—*J.*

the clay-slate. The position of this diallage rock is here as well as at Kielvig distinctly above the clay-slate. Higher up the mountains this formation again disappears, and gives place to those slatey mountain rocks, which, without being entirely clay-slate, must however be numbered among them. A thick bed of snow-white small and fine marble also appears on the declivities amongst it, which is surrounded by small collections of water that are here extremely frequent.

A deep valley, the Utsvadal, here interrupts the farther ascent of the mountains. It runs parallel with the range of mountains, and descends towards the Longfiord. The farther declivity is an immense mural precipice, altogether inaccessible, and if there were not a narrow isthmus between the commencement of this valley and Stor-Vand towards the ascent of Akka-Solki, the mountains would be altogether inaccessible on this side.—With this valley and this range of rocks the nature of the formation immediately changes. Limits are here set to the clay-slate. The mica-slate makes its appearance with all its characteristics of composition and beds, and we might frequently be induced to believe we were ascending the *Nuffenen*, between the *Levantine* and *Valais*. The mica is continuous, very shining, very thin slatey, and interspersed with a number of small garnets. Small beds of white dolomite frequently lie between the strata, also white quartz beds as at Porsangerness, and very often black beds like those which occur so frequently in the Swiss Alps, consisting of very small mica folia, thickly heaped together, with some intermixed hornblende. This is the rock of the mountains between Quänanger and Altensfiord, apparently not once, as usual elsewhere, interrupted by strata of gneiss. Such is the composition of the summit of Akka-Solki, three thousand three hundred and ninety-two English feet above the sea.*

---

h. 8. a. m. Talvig-Prästegaard   28. 0. 9.   Therm. 8. R. clear N. E.
h. 11. Talvigsfoss-Sörfieldt.... 27. 0. 0.   ...... 10. 6. still trees. ·
    12. rocks south-eastwards in the same mountain ridges 26. 7. 3.   11. 2. birches cease.
h. 4. Akka-Solki............. 24. 11. 1.   ...... 8. 8. strong west wind, clear.
h. 6. Mountain willows cease.... 26. 0. 6.   ...... 9. 4.
b. 10. p. m. Talvigsfoss-Sör-fieldt 26. 11. 8.   .... . 8. slightly clouded, particularly in
        the north-east.
h. 11. p. m. Talvig-Prastegaard.. 28. 0. 8.   ...... 5. 6. calm, cloudy almost 2 degrees R.
        warmer on the Fieldt.
    Talvig sea shore..... 28. 1. 7.

The clay-slate and its subordinate rocks are wrapped around the higher mica-slate, in the form of a mantle, nearly to the heighth of one thousand six hundred feet. The mica-slate is the fundamental and most abundant rock of the mountain, and this circumstance renders it still more probable that this mountain ridge is not a subordinate arm, but truly a divided portion of the principal mountain range.

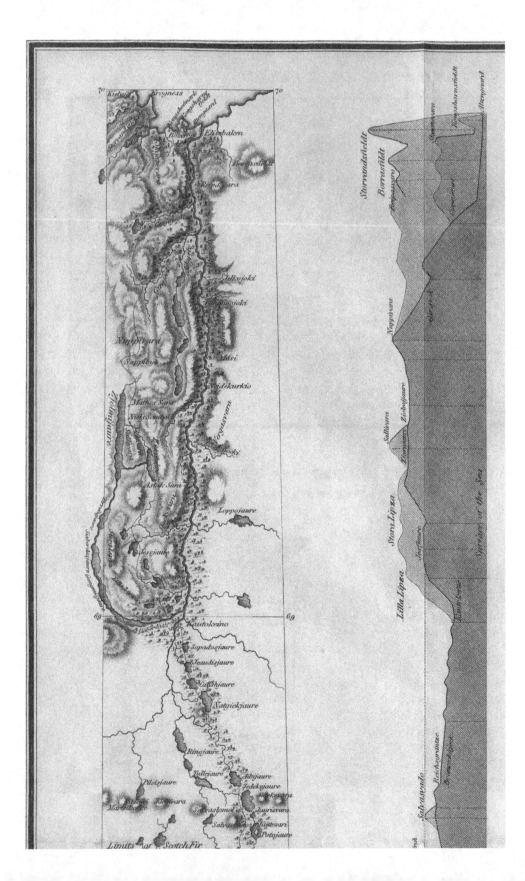

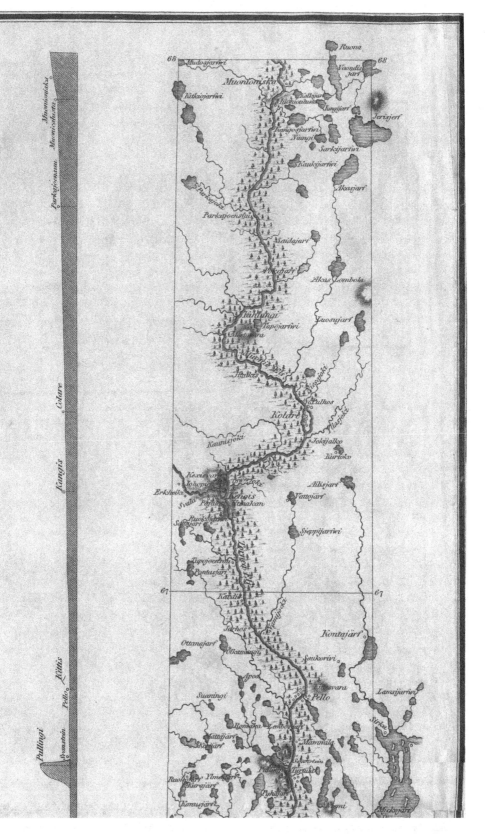

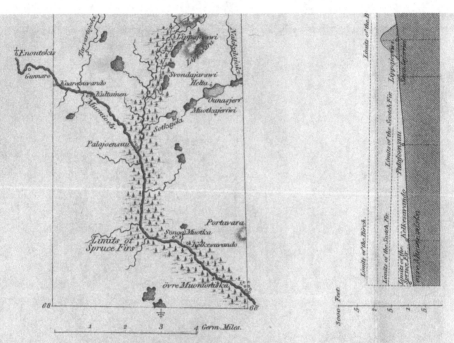

1   2   3   4 *Germ. Miles.*

3000 *Feet.*

I. *PLAN and SECTION of the ROAD between the*

FROZEN OCEAN and the BOTHNIAN GULPH.

II. THE COUNTRY round ALTEN in F

Published July 12ᵗʰ 1823 by Hen

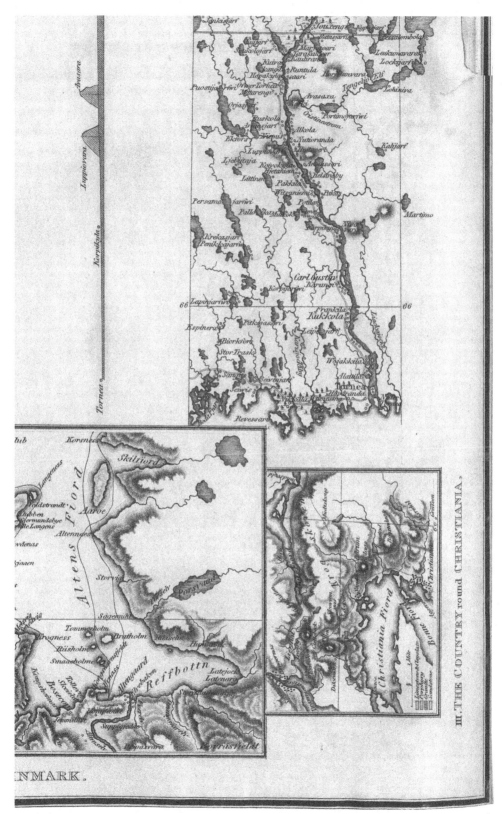

Senksengo Kupsoari
Patessjarvi Cossiembola

Saajari Mar..saari
Sorkolojari Kirokkalioar Leukunvaara
Kkahiranda Lockjarvi

Kuivas Runtula Harsu..vara Eli
Kanga Kuukuytassaari Tengelio
Hepakytlag..saari Lohisiva

Puostija Tviri Omer Torhea
Mitarengi Avasaxa
Orjagsri Oistinestrom
Portimojarivi

Kuskola Alkola
Ivasjari
Elsiors Venus Nutioranda
Kalijari
Lupp..vlota Klistijari
Ljebkoira Koivokolia Antarssari
Hietanien Helsingby
Littinen Pukslu
Witaniemi Pihan
Persam jariwi Pallas Pala
Palt..mala Martino
Karpen..la N..la
Krekagiari
Penik-biparin

66   Lapinjariwi   66
Korpijarsri
Espinerde Pitknasssri
Leipiesu..en
Bierksiore Franksla
Stor Trasks Kukskola Liackopsuo..a
Sinana Wojakskila
Deorbjari Aldaba
Sewris Ilasiranda
Kitalo Nibesama Tornea
Revessaru

Korsnesso Skilfiord
..lub Langeness
Aaroe
Holdstrandt
Klubben
..ermundshoe
..le Langens
..wdenas Altenndes Storvig
..giaen
Po..hace
Sagernill..
Tommergolm
Krogness Bratholm
Rusholm ..siescli
Smaasholme Huguark
Sagegaard Altengaard Latejocl
Pasecy Scbebolsn Lotmarv
Kingiskeboma ..gnass
..mitlita Sappisedde
Oupasvara Torristields

Altens Fiord
Reffbottn

Christiania Fiord

Bonne

CHAPTER VIII.

JOURNEY FROM ALTEN TO TORNEO.

Departure from Alten.—Valley of Alten.—Dwellings of the Laplanders on the Mountains.—Rein-deer Flocks, and Huts of the Laplanders.—Zjolmijaure.—Names of the Laplanders.—Breaking up of a Lapland Family.—Diet of the Laplanders.—Rein-deer Milk.—Siaberdasjock.—Kautokejno.—Spreading of the Finlanders in Lapland. —Salmon Fishery in the Tana.—Course of the Kiolen Mountains.—Mica-slate of Nuppi Vara.—Granite of Kautokejno.—No Mountains between the White Sea and the Bothnian Gulph.—Departure from Kautokejno —The Kloker, the Interpreter of the Clergyman.—Fisheries in the Inland Lakes.—Boundaries between the Kingdom. —-Disputes and Wars before they were settled.—Entry into Sweden.—Re-appearance of the Scotch fir.—Difference of the manner in which the Rein-deer Moss spreads in Sweden and Norway.—Lippajärfwi.—Palajoen Suu.—Striking of Salmon in the River Muonio.—Boundaries of the Norway Spruce.—Muonioniska.—The Laplanders and Finlanders are different People.—Granite on the Boundaries between the Kingdoms.—Gneiss at Palajoen Suu and Muonioniska.—Water-fall of Gianpaika. —View of the River Muonio.—Colare Kengis.—Rappa River, Red Granite at Kengis.—Iron-work of Kengis.—Lapland Iron-stone Mountains.—Tärändo-Elv.— Departure from Kengis.—Pello.—Pullingi at Svanstein.—Makarenji.—Excellent Road to Torneo.—Cultivation of the Land.—Gneiss at Korpikylä.—Clay-slate at Wojakkala.—Transition Formation at Torneo.—Spreading of the Finlanders in recent Times.—A Country is not depopulated by emigration.—Lists of Exports from Wester-Bothnia and Lapland.—View of Torneo.—Architecture of the Town.—Manner of Living.

KAUTOKEJNO, the 11th of September, 1807. The two rein-deer, with their driver, *Mathes Michelsön Sara*, had agreeably to engagement come down from the Fieldts. These animals were loaded with the most necessary requisites for our journey, and with them, two Laplanders, a woman, and a child, I left Altengaard as I would leave a home, on the evening of the third, and a few hours afterwards I reached Bosecop. This remote country, besides the attractions which it has received from nature, the grand and interesting style of the environs, the variety of new phenomena which strongly recommend it to our notice, possessed a superior charm for me in the highly distinguished and agreeable society which are here collected. Their repeated and incessant acts of kindness and benevolence continued for so many months towards a stranger whom they could never expect to see again, with the polish and the attraction of their conversation, could not fail to produce such an impression on my mind. Although strict justice, wisdom, and knowledge, are qualities which we ought not to look upon as extraordinary in any governor of a province, I felt a particular pleasure in the consideration that even the head of the most remote province of the Danish dominions possessed these qualities in so eminent a degree.* At the last habitation, about two miles beyond Bosecop, I took my leave of them, when I began to think, for the first time, that I was three degrees beyond the Polar Circle, among wilds and deserts.

We soon entered the wood: the rocks of Skaana Vara appeared nearer and nearer, narrowed the valley, and formed perpendicular precipices along its sides. All traces of habitation disappeared. The high and majestic Scotch firs stood thickly around, with excellent stems, and the small marshes in the wood were surrounded with alders and aspens. On entering deeper into the valley the view became suddenly frightful. The trees lay in heaps above one another, torn up by the roots almost in every direction for large spaces, and the few solitary stems which remained erect were quite lost among them: an image of the alarming nature of the storms in winter. Most of the trees lay with their heads down the valley. The storm had swept down from the south, and when compressed between narrow ranges of rocks, the firs are not always able to withstand it.

At the approach of evening the Laplanders took the rein-deer up several cliffs which were covered with rein-deer moss like snow, and there they tethered them. We passed the night ourselves contentedly under the trees by the side of a clear blazing fire.

These *Fieldt* or mountain Laplanders require time for their operations. I lost several admirable hours of the morning, before the woman had bathed her child in warm water, and then till the man had again loaded the rein-deer. We reached in half an hour a lateral valley, and a stream which pours

---

* Hillmar Krogh of Drontheim has been governor (*Amtman*) of Finmark since the Spring of 1807.

down it, called the Gurjajock. There we left the great valley of Alten, and began to ascend the new valley towards the west, which rises pretty rapidly for the space of five English miles at least. Hitherto we had still seen traces of the cows and horses which the inhabitants of Alten allow here to run about almost wild in summer; but the last vestige of cultivation at last forsook us. The Scotch firs became smaller and more scanty, and the birches became more frequent; and as we lay down at mid-day on the banks of a small lake we found ourselves beyond the region of firs. This lake, *Gurjajaure,* was actually eight hundred and ninety-eight English feet above the sea, and consequently above the heighth which the observations on Skaane Vara had given as the boundary of their growth. Our ascent became now less rapid; the vallies began to widen; and the mountains to become marshy levels. On the long extended rockless mountains the birch bushes grew scantily and dwarfish, their growth being probably prevented by the storms; the ground was also less covered than usual with rein-deer moss. All nature was here bare and dismal. Several leagues farther, towards the south-west, at the termination of these dreary levels, the Northern Ocean suddenly appeared in the distance, for the last time, like a ray of light, piercing through the darkness. I never saw it again. It was a part of Refs Bottn to the right of the source of the Alten stream above the valley of Alten. We now descended a flat and broad valley, and prepared our night-quarters on a sort of island in the *Carajock,* a small stream, which appears to be of some importance in spring, but which was then almost dry. It probably unites in its course to the eastwards with the *Aiby-Elv* which is laid down on the maps, before it flows into the stream of Alten. We durst not have ventured much farther if we wished to avail ourselves of birches for our nightly fire. The small birches became visibly shrivelled, and were thinly scattered over the plain, so that it was evident, without the protection of vallies and cliffs, they could not have possibly stood. Our island in the Carajock was one thousand five hundred and thirty one English feet above the sea. We might have ascended about nineteen or twenty English miles beyond the wood; we travelled but slowly; for a rein-deer is like a gazelle, destined by nature to run and not to carry. Notwithstanding a horse could, with the greatest ease, have carried more than double of the load with which these animals were burdened, they became fatigued however in a very short time, and we were obliged to halt, and allow them time to recruit their strength with the moss, which they greedily devoured. During the night we tied them with a long thong to some bush or piece of rock, round which the ground was thickly covered with excellent moss. They slept or lay very little, but continued to eat the whole night through.

On the following morning, the fifth, we ascended an entirely flat, parched, and dismal valley for about five English miles, till we reached the height of Nuppi Vara, which is, according to the barometer, two thousand six hundred and fifty-five English feet above the sea. This was the greatest height of this table-land; for we commanded from it a prospect of many leagues in every direction. The snowy chain of Lyngen appeared again in a long range towards the Fiord, notwithstanding it was at least forty-six English miles distant; and we could now see very distinctly how these cones became lower and flatter where the Fiord terminates, and the chain continues to run along the mainland. The Fiord is a fosse *(graben)*; the chain the wall above it. At the foot of Nuppi Vara a long marshy level runs towards Quaenangerfiord, containing a number of small marshy lakes, a desert and dreary prospect. Every thing is here solitary and dismal. The snow had long disappeared; but nature still remained dead and torpid. The dwarf birch (*betula nana*), the true companion of these mountains, could only support itself here with weak and powerless branches; the mountain brambles (*rubus chamaemorus*) in vain endeavoured to put forth fruit: they could only bear leaves; and here and there could alone be seen a spring-flower endeavouring, with great difficulty, to blow in harvest. A few solitary bushes of mountain-willow seem to make their appearance here, more in defiance of the inhospitable climate than as a covering to the earth.

The barking of dogs below announced the vicinity of a herd and the hut of a Lapland family. We made all the haste we could towards it; for the rising storm and rain from the south-west seriously admonished us to seek shelter for the night. We soon found the hut or *gamme* at the foot of the hill, and on the bank of the Great Marsh. They received us, but not in a friendly manner. The Laplanders are not Arabs. Where the spruce and Scotch firs, and where birches will not succeed, the nature of man seems equally defective. He sinks in the struggle with necessity and the climate. The finer feelings of the Laplanders are to be developed by brandy; and, as in eastern countries, a visit is announced by presents, the glass alone here softens their hostile dispositions. Then, indeed, the first place in the bottom of the tent, opposite to the narrow door, is conceded to the stranger. We lie in the circumference of a room containing at most eight feet in diameter; the fire or smoke of the hearth in the middle prevents the draft from the door; and hence this back space is the place of the master or mistress of the herd. The children sit next them, and the servants next to the door. When a stranger demands entrance he is commanded by Lapland politeness to keep himself on his legs in the inside of the door, and sometimes even before a half-opened door. The master of the house then asks him the cause of his arrival, and also the news of the country; and if he is pleased with the account, he at last invites the stranger to approach nearer. He then becomes a member of the family; a place in the house is allotted to him, and he is entertained with rein-deer milk and flesh. The Arab invites into his hut, and asks no questions.

It was well for us that we passed this night under a roof. The storm raged furiously, and the rain struck like sand against the roof of the *gamme*. It was not a little wonderful that the feeble hut could withstand such a hurricane. It is built of stakes, which are united together in the middle in form of a cone. Several other cross stakes hold them together below. Over this frame there is nothing spread but a piece of coarse linen, generally sail-cloth, in such a manner, however, that a quadrangular opening at the top remains uncovered for the smoke to issue out of. A great part of this covering lies also loose on the ground, and serves to protect their milk and other household concerns against wind and weather, and to cover over their stores; and then these articles, and the covering over them, form altogether a sort of mound, which prevents the entrance and draft of the external air into the gamme from beneath. Another large and loose piece of sail-cloth is drawn round this outward covering on the side from whence the wind blows. This side is therefore always protected with a double covering. The inside seats consist of soft rein-deer skins and white woollen covers. The quality of this skin and cover also determines here the rank of the place and the person who is to occupy it *. This is certainly a slight habitation;

---

* In *Knud Leem, on the Laplanders*, all the sorts of gammes are circumstantially described, and drawn with great accuracy. We have also a faithful enough representation of them in the *View of the District of Quickjock in Baron Hermelin Ritningar tel Bescrifningen öfwer Swerige.* Stockholm, 1806.

and it is almost inconceivable how a large and frequently numerous family can find room in such a narrow space for many months together. But all the members of the family are seldom assembled together at the same time; the herd of rein-deer demands their presence and their attention even during the night,and such stormy and dreadful nights as the one we passed here in *Nuppibye.* Men and boys, wives and daughters, take the post of watching by turns twice or thrice a-day; and each goes out with several dogs, which belong in property to that individual, whose commands alone they will obey. The former guards in the mean while their return with their hungry dogs. Hence it not unfrequently happens that eight or twelve dogs march over the heads of the persons sleeping in the gamme in quest of comfortable spots for themselves to rest in. They certainly stand in need of rest, for all the time they are out with their master, watching the flock, they are in continual motion. The welfare and the security of the flock rests wholly on them. By them alone are they kept together in their destined situation, or, when necessary, driven to others. The wolves, the dreadful enemies of the Laplanders, are by them driven away from the rein-deers. The timid animal runs frightened up and down the wilderness when the wolf approaches; the dogs then by their barking and snarling keep the flock together, and by this means the wolf will not easily venture an attack. If the rein-deer is to the Laplander what his field is to the husbandman, the dog is to the Laplander what the plough is to the other. When he returns wearied to his gamme, he always willingly shares his rein-deer flesh and his soup with his dog, which he will hardly do with either father or mother.

It is an unusual, a new, and a pleasing spectacle to see, in the evening, the herd assembled round the gamme to be milked. On all the hills around, every thing is in an instant full of life and motion. The busy dogs are every where barking, and bringing the mass nearer and nearer; and the rein-deer bound and run, stand still, and bound again in an indescribable variety of movements. When the feeding animal, frightened by the dog, raises his head and displays aloft his large and proud antlers, what a beautiful and majestic sight! And when he courses over the ground, how fleet and light are his movements! We never hear the foot on the earth, and nothing but the incessant crackling of his knee-joints, as if produced by a repetition of electric shocks:* a singular noise, and from the number of rein-deer by whom it is at once produced, it is heard at a great distance. When all the three or four hundred at last reach the gamme, they stand still, or repose themselves, or frisk about in confidence among one another, play with their antlers against each other, or in groups surround a patch of moss. When the maids run about with their milk vessels from deer to deer, the brother or servant throws a bark halter round the antlers of the animal which she points out to them, and draws it towards her: the animal struggles, and is unwillingto follow the halter, and the maid laughs at and enjoys the great labour of her brother, and wantonly allows it to get loose that he may again catch it for her. The father and mother have quietly brought their's together, and filled many a vessel, and now begin to scold them for their wanton behaviour, which has scared the whole flock. Who would not then think on Laban, on Leah, Rachel, and Jacob? When the herd at last stretches itself to the number of so many hundreds at once, round about the gamme, we imagine we are beholding a whole encampment, and the commanding mind, which presides over the whole, in the middle.

They were already returning from their summer pasturage on the high mountains along the sea, to the woods which surround the church of Kautokejno. Numbers had already gone before them, and numbers were still to follow. They descend, in fact, always more and more from the mountains the farther they advance into the country; for towards the interior there are no longer any ranges of mountains, and mountains are visible only above the Fiords: the highest are precisely where they are straitened between two Fiords. The view from Nuppi Vara, towards the south, stretches therefore over an endless level, on which Sallivara, Dasko-Vara, Stora-Lipza, &c. seem more like hills than mountains. On the other hand, towards Talvig, and above Kaafiord, the whole mass of mountains suddenly rose, covered with furze for the whole length of their course, as if mountains first began there. The rein-deer feed there in summer at a heighth of between two thousand and two thousand eight hundred Paris feet, and one thousand six hundred feet above the sea. The winter *gammes* of the Laplanders at Kantokejno are not above seven

---

* The incessant crackling of the rein-deer in their course is thus very sensibly and judiciously compared by M. Schmidt — *Reise durch Schweden & zu den südlichen nomadischen Lappen.* Hamburg, 1804.

hundred feet above the sea.   That the Laplanders, the nearer they approach
to the sea, should also be obliged to drive their flocks higher up the mountains, is a singular peculiarity of these mountains.

We left Nuppibye on the evening of the fifth, and reached, about mid-day,
the border of the long and narrow Zjolmijaure, and the gamme on the brink of
the lake.   It was between four and five English miles from the gamme below
Nuppivara.   The herd belonged to Mathes Sara, my guide, who was to exchange the exhausted rein-deer with fresh ones at this place.   The herd,
however, was at a great distance, and could not be expected before the evening.
We entered the gamme.   The grown-up son was within, but he did not rise
up or welcome us, and nobody would have suspected that he had not seen
his father for a number of days.   Distrust had completely blinded him.   In
the evening he went out to the herd with the younger brother, and the
daughters returned.   Why did not the herd also come?   Why were they not
to be milked at the gamme as usual?—The women thought the distance too
great, and it would be too difficult to drive them to the gamme that day.   The
son took the rein-deer that had been employed on the journey along with
him, but he did not send back fresh ones.   The night passed away.   In the
morning still there were no rein-deer.   " I must seek them myself," said
Mathes Sara.   The women told him the place where the herd was feeding.
He ran about the whole day, and returned breathless and worn out with
fatigue in the evening, without having seen a single rein-deer.   His wife and
children had given him a false direction, and while he was seeking the herd
in one quarter, it was driven to one directly the contrary way.   It did not
come home this evening any more than the former, and was nowhere to be
seen in the neighbourhood of the gamme.   Still less appeared the rein-deer
which were stipulated for on the following morning.   The will of Mathes, it
seems, was not the will of his family.   They did not hold the stranger in sufficient estimation to consent to let him have rein-deer for the prosecution of
his journey; and the bargain with the master of the house had no power
over them.   Mathes's exhortations and his threats were equally powerless.
Certainly there was here no patriarchal authority of the father over his children: to cause the father purposely to wander among the desert mountains,
and in the wilderness, was no display of submissive respect.   But what breaks
through all the resolutions of the Laplanders brought us at last also the
rein-deer which we were in such anxious expectation of.   The mother could

not withstand the impression of the brandy. She was moved by a feeling of gratitude, whispered a word in the ear of her daughter when she returned home late in the evening, and in a few minutes the electric-like crackling of the cattle, and the barking of the dogs, announced the anxiously-expected arrival of the herd. And yet we had in vain two whole days been seeking for them. Here the mother evidently had the management of matters: it was the same thing also in Nuppibye, where the feminine authority might be styled, perhaps, more hard and severe, for the movements of the mistress of the house there were by no means of a mild and gracious nature. The wife of *Torbern Kaafiard*, my other Laplander, who was daughter of Michel Sara, had also a decisive power over him. Yet how necessary it is to observe a foreign nation long and attentively before venturing to pronounce respecting its manners and customs. The internal state and condition of these families could hardly bear an application to the whole community. For how is it possible to separate the idea of a patriarchal authority of the father from that of a nation of Nomades.

Zjolmijaure* lies nearly two thousand two hundred and thirty-six English feet above the level of the sea; its naked banks still bear no trace of birches. The Laplander can procure no firing except the dwarf birch *(Betula nana)*, or mountain willow, both of which, it is true, grow very well here as shrubs of the heighth of two or three feet. They run along the banks of the small streams, and wherever they can find any moisture; and a small stream is frequently wholly concealed by them. We can scarcely, however, warm ourselves at a fire made of such materials: the leaves alone give out a flame; the moist wood goes off in such thick clouds of smoke, that even the Laplanders themselves rush out of their gamme to draw fresh breath. This prevents the people from residing here in the winter. They are compelled to return to the woods. Even on the mountains above Talvig, and above Langfiord, where the dwarf birches almost entirely disappear, the Laplanders have frequently an insufficiency of wood to cook their flesh and their broth, and on that account the gamme is then often at a great distance from the place where the

---

* Cholmijaure, according to the English pronunciation. The above is agreeable to Leem's Orthography of the Lapland words, who very properly censures Samovich, and would wish to see the Hungarian Orthography applied to Lapland words, on account of the connexion between the two languages.—*Kiöbenhavns Selsk, Schrift.* Tom. XI.

flocks are feeding. The summer on such heights cannot be of long duration; it is as if we were living above the cloister of the *Great Bernhardt.*
We never again experienced a fine day on this range of mountains. On the
sixth of September there was a violent storm in the night from the north-west.
In the morning, not only the mountains, but the plains along the lake were
covered with snow. It is true it did not remain along the banks, but on the
height it was seen the whole of that and the following day. The sun could
now no longer draw out flowers and herbs.

We first put ourselves again in motion about mid-day of the eighth. The fog
lay deep, and the thermometer stood at only four degrees and a half of Fahrenheit, above the freezing point. Mathes was of opinion that there was some
risk in venturing ourselves in such weather through these wastes; for the fog
prevents the view of the distant hills, which are the guides through the country, and traces of paths on the ground are no where to be found. But it succeeded. The fog ascended about two hundred feet, and allowed us just a
sufficiency of prospect to enable us not to lose the proper direction. We
went between four and five miles down the banks of the lake. There we
found the gamme of the rich Aslack Niels Sombals. He received us in
a friendly manner, introduced us himself into the gamme, put the kettle on
the fire, and cooked a rich and abundant supply of rein-deer flesh for Mathes.
He mixed milk and meal with the broth of the flesh, and handed it to
Mathes. The daughter brought me some milk, which she had brought from
the distant flock in a tin flask, and she insisted with friendly earnestness that
I should completely empty it. Without a doubt, the nature of Laplanders
varies as well as that of other beings. Why should kindness and benevolence
be strangers to this people alone?

Mathes conducted me through a lateral valley down towards the lake of Zjarajaure, which was narrowly confined between the steep clay-slate rocks. It
seems it abounds in fish, which are not only caught by the Laplanders, but also
very much by the Finns of Kautokejno. They remain for several weeks in
summer in a gamme not far from the outlet of the lake, catch the fish, dry it,
and return with it to Kautokejno, where it serves them for a winter store.
The high and bare rocks by the side of it give an indescribable dreary and
dismal character to the water. They at last prevented us from following its
banks; and we were obliged to ascend a height of about three hundred feet to
the westward. Here we saw ourselves in an instant surrounded with rein-

deer. As far as the eye could reach all was in motion, and far and near the barking of dogs was incessantly heard. " That is the herd of Aslack Niels Sara, my brother's son," said Mathes, with a tone of self-complacency " He is a rich man: he possesses well on for a thousand rein-deer. He has every day rein-deer flesh, and he possesses clothes in superfluity. We must pass the night in his gamme, for we can no where be better off." When we got to the gamme, Niels came out. " My dear Mathes," said he; " I cannot receive you. A few hours ago two Lapland strangers arrived here, who have taken up all my spare room." So we were obliged to go on. After we had been half an hour on our way, Mathes said to me, with a tone that indicated the state of his feelings: " It was not well done in Niels to refuse us a place in his gamme." " But how could he help it, when all his spare room was already taken up by strangers?" " It is all very well," answered he with keenness; " but where there is room in the heart it is soon found in the gamme." *

We went to Aslack, Aslackson Sara, the brother of Mathes, on the height at the end of the Zjarajaure *(Charajaure)*, and met there with a friendly reception. It appeared almost as if the whole country was peopled by the family of *Sara*. And this family was also bound together by a name peculiar to themselves. One could scarcely believe on a first view how important this phenomenon is, and how characteristic it is of this people. Wahlenberg, in his excellent description of Kemi Lappmarck (Stockholm, 1804), has already remarked that all the Laplanders, whether engaged in fishing, or in roaming with their rein-deer among the mountains, preserve their pedigree by particular names. In the *Forsamling* of Cuare, the great family of *Morotaja* lives; another great family, *Kua*, the *Sagats*, the *Musta*, the *Valle*, *Sari*, *Padar*, &c. To these names they add the baptismal name of John *(Jounes* in Laplandish); Andrew *(Anda)*; Oluf *(Volla)*; Peter *(Pietar)*; Nicholas, or Niels *(Nikke)*, or Aslack, a very common name on the mountains of Kautokejno. To this is also added the name of the father; as in the name of the master of our gamme, Aslack Aslackson Sara. In Norway, Sweden, and also in Denmark, there are, properly speaking, no family names; and though they are here and there to be found, they owe their origin to a very recent period. If the father, for instance, is called *Oluf Nielson*, the son is perhaps called *Hans*

---

* *Hvor er Hierterum, da er Huusrum*, says a Danish Proverb, which the Germans may well envy the Danes.

T T

Olafsen, and his son again, *Carl Hanssön*. The grandson has no longer any thing of the name of the grandfather, and relations are not distinguishable from their names. If several persons bear the same name, they are distinguished from their places of residence; as *Ole Steensrud*, and *Ole Biölset*, because the former lives at Steensrud, and the other at Biölset. Hence we find the names of all the Danish towns very frequently employed as family names. This was also the case in Germany, as well as in France and Italy, till the power and the consequence of the rising towns produced a middle class between the master and the slave, between the prince and the vassal. It still prevails in Russia and Poland, and wherever one class oppresses the other, and governs their inclinations. This, it is true, is not now the case in Norway; but in this, as well as in many other things, the Norwegians follow the old custom. Are the family names of the Laplanders to be considered, therefore, as an old memorial of their perpetual liberty? If this is the case, we ought to hold them in honour.—Or is this an accidental custom of this people, which we do not find to prevail among other free nations? And can we recognize the connexion between nations by such undoubted primitive customs? Family names, it appears, are also common among the Samöiedes; as also among the tribes related to the Laplanders, the Burätes, the Ostiacks, and the Baschkirs.

Aslack Aslacksön Sara was thinking of leaving the hills of the Zjarajaure, and drawing nearer to Kautokejno. We had scarcely set our feet out of the door of the gamme in the morning, when in less than half an hour the house was entirely destroyed, and the rein-deer laden with all the utensils, and in motion to the new place of destination. They were bound together in rows of five with thongs, like the beasts of burden on St. Gotthardt, and they were led by the mother and daughter over the mountains, while the father went before to prepare the new dwelling, and the other children conducted the free herd to their place of pasture. The flock amounted to about four hundred head. We had yet seen none under three hundred. With this number a family is said to be in moderate prosperity. It can be maintained on it. They can afford to kill as many rein-deer as are necessary for food and clothing, shoes, and boots, and to sell besides a few rein-deer skins, hides, and horns, to the merchants for meal or brandy, or woollen stuffs. On the other hand, a family lives very miserably on a hundred of these animals, and can hardly keep from starving. Hence, if they are brought down so low, they must give up the free pastoral life on the mountains, and draw towards the sea, and endeavour as sea Laplanders to gain from that element what they can no longer find among the mountains. But their desires are always fixed on the mountains, and every sea Laplander eagerly exchanges his hut and his earnings for the herd of the *Fieldt*-Laplander. The charms of a free life among the mountains, and of independence, may have less effect on the producing of this inclination than the actual good living of the Fieldt or mountain Laplanders, which the sea Laplander cannot even procure on holidays. Every day I have seen rein-deer flesh cooked in all these gammes for the whole family, and generally of young fawns, in large iron kettles. Each person certainly received more than a pound for his share. When the flesh was cooked, it was immediately torn asunder by the master of the house with his fingers, and divided out among the family; and the eagerness with which each person received his allowance, and the rapidity with which they strove as for a wager to tear it with teeth and fingers, are almost incredible. In the meantime the broth remains in the kettle, and is boiled up with thick rein-deer milk, with rye or oatmeal, and sometimes, though seldom, with a little salt. This broth is then distributed, and devoured with the same hungry avidity. The sea Laplander, on the other hand, has only fish, or fish livers, with train-oil, and never has either the means or opportunity of preparing such costly soups. The former not only relishes his flesh, but finds in it a strong nourishment. In fact, how few boors in Norway or Sweden, or even in Germany, can compare their meals, in point of nutrition, with this. In winter, the food of the Laplanders is more multifarious. They then catch an incredible number of Ptarmigians *(Ryper, Tetrao Lagopus)*, wood grouse *(Truren)* and a number of other wild birds, partly to eat, and partly to sell. They not unfrequently also shoot a bear, which they eat like the Norwegian peasants. They have then also no want of rein-deer flesh; for the frozen pieces may be long preserved. They can even preserve the precious milk in winter, although they can then derive none immediately from the rein-deer. They expose it in harvest to the frost, and preserve the frozen pieces like cheese. When melted after a lapse of several months, this milk still tastes fresh and deliciously. When a stranger then enters the gamme, whom they wish to see, the frozen piece of milk is immediately set to the fire; the guest receives a spoon, with which he skims off the softened exterior in proportion as it melts. When he has enough the rest is preserved in the cold for other guests. Such pieces are not unfrequently brought by the Laplanders down to Alten, and then disposed of to advantage; for the inhabitants of Alten eagerly purchase this milk. They use it like cow-milk, and can mix a good deal of water with it without injuring its quality. In its pure state it is even too fat for domestic purposes. Notwithstanding, even in the middle of summer, each reindeer yields but little milk, it would be quite impossible, especially for any length of time, to consume the whole quantity at once. In October the milk season generally ceases, and recommences about the end of June or beginning of July. The rein-deer calve about the middle of May. The Laplanders call a doe or female rein-deer a vaija, when it has calved in the third year. It is allowed to suckle the fawn for six weeks, which is then slaughtered, or allowed to provide its own nourishment; and they can then have milk for three or four months. A moderate *vaija* about the end of July yields the quarter of a Swedish kanne * per day †. With a herd of a thousand head the quantity procured from all the vaïjas would be very considerable, and perhaps sufficient to maintain a whole family on milk alone. But their prosperity necessarily requires the possession of considerable flocks, that they may always be able to slaughter a deer when the wants of the family require it, without thereby injuring the flourishing condition of the herd. A great part of the Swedish Laplanders in *Kemi Lappmark*, and especially in the *Församling* of *Enare*, live in quite a different manner. They live there for the most part by fishing, and have but seldom a few rein-deer; on the other hand they generally possess eight or ten sheep, but no cows. In summer they scarcely eat any thing but fish from the fresh water lakes, and drink with great eagerness the water in which the fish has been boiled. In winter they must put up with dried fish (*Sick, Salmo Lavaretus*), and with soups (*välling*), of water, fir bark, and rein-deer tallow. They peel off, in summer, the innermost bark of the firs, divide it in long stripes, and hang them in their dwellings to dry for winter stores. When used, these stripes of bark are minced in small pieces along with the rein-deer tallow, boiled together for several hours with water, till in consistency they form a thick broth, and then eaten. A little ewe milk, and a few mountain bramble-berries (*Hiortron, Rubus Chamaemorus*), contribute very little to the

---

* The Kanne contains about eight-tenths of an English wine gallon.—*T.*

† *Erich Grape,* (formerly a clergyman at Enontekis, and since 1807 at Nedercalix, near Torneo), *Beskrifning ofver Enontekis Sockn in Nya Handlingar af Svensk. Vidensk. Selsk.* XXIV.

improvement of this wretched diet.* Well may they account the rein-deer Laplander happy, and envy his situation.

We remained but for the space of two English miles with the caravan of Aslack Sara. Our way lay down a long flat mountain-valley, covered along the declivities with dwarf birches and mountain willows, and at the bottom with marshes, or with innumerable small lakes. There are no distinguished objects to be found here; no where any rocks. We approached one of the highest hills of this quarter, the Stora Lipza, and on reaching the foot of it we ascended the eastern declivity of the valley, and then we found ourselves on a seemingly boundless plain as level as water, perhaps the only one of the kind in the whole way from the Northern Ocean. In the middle of the plain, and all the way to the Lipza, there was a pretty considerable lake, the Jessjaure, and a hut shone over from a peninsula which stretched a great way into it. It was one of the numerous dwellings which the Finns of Kauto-kejno have built along the banks of the lake. These active and industrious men wander about in summer from one lake to another among the mountains, catch their fish and dry them, and then repair to other districts, till they are driven back by winter to their houses at Kautokejno. They must also derive from the mountains the stores for their winter subsistence. It is singular enough, that where there is scarcely a twig for firing, where wolves and bears only make their appearance on account of the rein-deer, where almost the whole vegetable world is torpid, two classes of people, entirely different, find the means of their subsistence in such a different manner. The rein-deer have drawn nomades into these mountains, and the innumerable lakes those who lived in permanent habitations.

We might have proceeded about another five English miles over the plain of Jessjaure, when we at last began to descend very gently towards Kauto-kejno. We followed the small Lipzajock which rose here on the height, and by it we were about evening guided to the finest birches which the climate allows to grow again in this place. We hailed them gladly as old absent friends, and encamped ourselves for the night in their beneficent neighbourhood.

According to the barometer we were here one thousand six hundred and fifty English feet above the level of the sea, and consequently still at a very

---

* Wahlenberg Kemi Lappmarks Topographie, 48.

considerable heighth. We could therefore have hardly expected to find birches much sooner; they commenced here in fact somewhat higher than at Alten. They might be nearly the same here as at the Carajock, which is somewhat beneath the country where we passed the night, and about one thousand four hundred and fifty-two English feet above the sea. The absolute limit of birches would therefore at Kautokejno far exceed one thousand seven hundred English feet, and perhaps approach to one thousand eight hundred English feet; that is to say, small bushes with their branches creeping along the ground. Hence this country has gained something over Alten in point of climate, nearly as much as to correspond with its greater distance of a full degree of latitude from the pole.

We followed along the foot of a small mountain chain from the Storalipza, which we reached at Jessjaure. The flat summit over our nightly places of repose was called by Mathes Sara the Little Lipza *(Lilla Lipza)*; it was, in fact, a continuation of the greater, and both appeared more like hills than mountains. Their declivities are neither steep nor rocky; and they are certainly not more than six hundred and forty English feet above the plain, though somewhat more than two thousand one hundred and thirty English feet above the sea. They are, however, the greatest mountains of that quarter; for Doskovara hardly exceeds them in heighth, or the mountain range between the plain of Jessjaure and the stream of Alten. These mountains were certainly not striking externally. They became less and less so the deeper we descended the valley; for they were soon covered with birches to the very tops, and with a lively green. About four or five English miles from Lipza they had altogether sunk down to the line of vegetation. The birches below had acquired the character of woods, and they covered in the most agreeable manner the small hills on the banks of the lakes. Between these run high rein-deer moss and mountain willows. Many places bore a resemblance to the high mountains of *Jura.* Beneath the Gutisjaure the streams were no longer precipitated over cliffs: the Muddijock creeps in innumerable serpentine windings, as in a flat country, and impedes not a little by its high and marshy banks the passage over so small and inconsiderable a stream. At last, in the afternoon, we reached the banks of the frightful Siaberdasjock. We had already been told at the Zjarazaure of the difficulty of passing the river at this season of the year. A Laplander there told us that he had had the greatest difficulty in wading through it at the outlet of the Zjolmijaure; and how much more difficult must it be so many miles farther down! I was much struck with this account. Zjolmijaure then runs in the middle between Kautokejno and Alten, yet not towards the sea, but towards Kautokejno. What a singular course! The descent of the country, however, seemed to confirm it; and we were also told by the Quans in Kautokejno of the number of lakes which the Siaberdasjock must run through from its source to that place. We saw clearly when we came to it that the stream must have already run a great way: its breadth, its depth, and its great rapidity, were a sufficient proof of it; and it was not without great difficulty that we found a ford. It is the principal source of the great stream of Alten, and we may well compare it with the *Mulde* of Freyberg, or the *Flöhe*. The distance is, however, undoubtedly greater from Nuppevara, whose southern waters descend in this way towards Kautokejno than the distance of the other feeders of the stream of Alten; it is even often the double of the distance from a number of the sources which are situated towards the frontiers of the kingdom. It is not unimportant to remark this; for this singular separation of the waters between Alten and Kautokejno also shews us where we are to look for the farther continuation of the great Kiölen mountains, which are traversed in their whole breadth by the Alten-elv below Masi.

Kautokejno is not far from the banks of the Siaberdasjock. Every trace of the mountains has here vanished: an open flat table-land spreads itself out, and wide prospects are opened over marshes covered with vegetation, *myrer*, and over innumerable little lakes. We are soon surprised with the unusual and new prospect of green meadows, with hay-cocks in one quarter, and cows pasturing in another; and at last we see little houses ranged like a street on both sides of the way, as in a village. On the opposite side of the river the neat church is situated, which is also surrounded with houses.

Thankfully did I welcome the house of the clergyman. With what pleasure we again look upon a table or a chair after having been for some time deprived of them! This house is small and miserable, and contains very few conveniences. But what keen enjoyment may the little that we find here have given to travellers! In this season of the year, however, the clergyman does not live here himself. He is left by his congregation at the end of winter, who direct their course towards the sea. He generally follows them, and passes the summer up among the Fiords. I saw the present animated preacher, M. Lund, in Alten. Notwithstanding he has been accustomed to more southern climates, he found even the winter of Kautokejno not without its pleasures. The gammes of the Laplanders lie close around him, and it is easy to procure from them the produce of their flocks. The hunting with large snow shoes *(skier)* is productive of much pleasure, and diversity of occupation, and supplies the dexterous hunter with a rich booty in rein-deer, or wild fowl. In winter the merchants leave Torneo for the fairs of Talvig, Quänanger, and Utsjocki, on the sea-coast, and generally go through Kautokejno; or the merchants of Finmark proceed through it on their way to Torneo with fish and hides, from whence they return with butter, medicines, and other productions of southern latitudes. There are also in winter several factories belonging to the merchants of Alten, for the purpose of carrying on trade with the Laplanders, who then live in its vicinity. In summer only are these regions dreary and waste. In winter they become animated.

The small houses on the left side of the river, to the number of thirty, perhaps, which were all shut up, were not destined to be inhabited. They were the storehouses of the Laplanders, in which are preserved their clothes, furs, and winter utensils. They are constructed of birchen stakes, in the manner of the northern storehouses (*staboe*), as we universally see them. The logs of these houses rest upon piles, which run one or two feet above the ground, and between the log and the pile they place overspreading horizontal pieces of deal or stones. The mice and rats cannot long keep hold of the under horizontal plane of the board, and therefore can never make their way into the house. The assemblage of these little magazines is a social bond, for which the Laplanders are indebted to the introduction of the Christian religion. They are uniformly in the neighbourhood of the church, and probably were unknown to the Laplanders before. By receiving churches they acquired a point of union among themselves. Here they have in reality a fixed property, but the only one, perhaps, which they possess. In these houses, however, they do not lay up their riches, for they have not yet renounced the singular inclination to conceal or bury their valuable articles, and especially coins, below stones in holes and caverns among the mountains. The place is known to the proprietor alone, but to none of his family. When he is dying he will seldom discover his treasure to his heir, and from the interminable nature of the wildernesses among the mountains, it is seldom found. It is affirmed that Sweden and Denmark have lost considerable sums in this way in Lapland; and the same thing would still perhaps take place, if the great want of the precious metals in both kingdoms did not oppose a bar to the gratifying of this desire. The Laplanders have almost every where become acquainted with Danish and Swedish bank-notes; they receive payment for their rein-deer, horns, and hides in paper. But paper is not well adapted for concealment under stones, or in holes and ditches. What is not worn to pieces in the hands of the Laplanders is soon probably converted into brandy.

The fixed and constantly settled inhabitants of Kautokejno consist of eight families of Quäns, colonists from Finland, and the same who fish during summer in the lakes among the mountains for more than twenty English miles around. In this way they gain their principal subsistence even for winter; and they are even enabled to purchase something in addition. They do not neglect grazing, however. Each of them keeps some cows and sheep, which are partly fed on the beautiful meadows which surround Kautokejno, and the hay collected from them, and partly on rein-deer moss, which they mix up with the hay. The cows are very fond of it when it is dried, and yield good and rich milk from it. The houses of these Quäns are certainly slight, but still they are houses with *perten (schwarz stuben)*, with bed-places, stables, kitchens, and storehouses. Such a situation disposes them for a still greater cultivation, and raises them high above the Laplanders. They would certainly yield in nothing to their brethren in Torneo and in Finland, if they had to struggle less with the climate. They have frequently attempted agriculture, I was told by the very intelligent sexton, but it would not succeed; and they were equally unsuccessful in their attempts to cultivate garden-stuffs. This may in a great measure be produced by the numerous lakes and marshes in the bottom of the valley, which frequently keep the atmosphere in summer several degrees lower than at the top of the hills; and in harvest the influence of the sun, which is then more efficacious than even the temperature of the air, is obstructed by the low fogs.

These Quäns appear to have been settled here before the emigration to Alten in 1708. Were they settled here, when Charles the XI. built the Church of Kautokejno, in the year 1660? Probably not; for Tuneld relates, that in the year 1696, only three Finland families *(Nybyggare)* were living

in Lapland. Kautokejno belonged then still to Sweden. The king sent a
clergyman here, and honoured the church with a small library of Finnish
prayer-books and sermons, which are still preserved, and included in this pastorate all the Laplanders, not only the whole way to Alten, but also along
the course of the Tana river to the northern boundaries of Sweden, where
Utsjocki is at present situated. A church or place of worship *(Bethaus)* was
soon erected on the Afjuvara, about half way between Kautokejno and the
Carasjock, for the sake of the far distant Laplanders who lived on the Tana-
Elv (Tenojocki), and the Carasjock. But even this church was found at last
insufficient; and it was found necessary on the Swedish side to erect another
church still where the Utsjocki flows into the Tana river at a place which the
Norwegians then called and still continue to call Aritzbye. The Swedes,
however, transferred the name of the river to the church. The Lector Von
Westen, the apostle of the Laplanders, says, indeed, that the Swedes built
the Church of Utsjocki, as they saw that no concern was taken in Denmark
about the salvation of the Laplanders,* and he gives us very clearly to understand that they founded their new church on the Danish territories. It is
probable, however, that the Laplanders had very small influence on the
determination of the Swedes, and that they were much more influenced by
the emigration of the Finlanders in 1708, who proceeded down the Tana
river. For this reason probably Utsjocki was not added to Enare, which it
lies near to, but was provided for as a chapel of ease, or annexation of
Kautokejno, which lies at a considerable distance from it. This union continued for the space of forty years, with some opposition it is true on the part
of the Norwegians, who claimed the whole course of the Tana, till at last
the treaty in 1751, fixing the boundaries between both kingdoms, by
laying down a clear line, put an end to these disputes.† Denmark renounced her claims on Utsjocki, and on the other hand Sweden gave up the
whole of the pastorate of Kautokejno to Norway, a tract of country containing fully three thousand eight hundred square English miles, though not
more than one human habitation for every seven square miles. Immediately
after this agreement, in 1747, Sweden removed the then clergyman of Kautokejno, who was afterwards to be appointed from Norway, to Utsjocki, separated

---

* Hammond, Missions historie, 664.

† See the whole of the treaty in *Buschings Magazin*, II. 198.

Enare from Kusamo, and annexed it to Utsjocki as being much nearer.* Some changes also took place on the part of Norway. The church of Masi for the mountain Laplanders, who were reckoned to belong to Norway, became now superfluous, and a great burden could now be removed from the clergyman of Talvig who had the charge of them in winter. The congregation of Masi was joined to the church of Kautokejno, from which the way was neither long nor difficult; and the increase of the congregation by this annexation is still felt, for the church is at present too small, and unable to hold the congregation. The new Finlandish colonies were by this agreement divided between both churches. Those who dwelt on the Carasjock, and deeper down towards the mouth of the Tana, became Norwegians, while those who settled on the right bank of the stream did not change the Swedish authority. For the convenience of the former, the Norwegians transferred the church of Afjuvara to the Carasjock, and Afjuvara fell into that state of ruin and waste, which from its situation among the barren mountains seems to have been destined to it by nature. On the other hand, the Finlanders of Carasjock live in the midst of a wood of Scotch firs, along a stream abounding in fish and surrounded by rich meadows. They were therefore enabled to build houses and other conveniences as at Elvebacken in Alten, and we here see again, as well as there, almost a complete village. In the year 1807 a new and large church was built for them. Their countrymen, at the mouth of the Tana, seem to succeed very well also there: they have built houses for themselves, notwithstanding they are above the region of firs; they have enclosed meadows, and feed cows and sheep. But grain did not succeed better with them than with the inhabitants of Utsjocki. It seems as if grain would not thrive where the Scotch firs will no longer grow. Wahlenberg names two Norwegian places on the Tana inhabited by Finlanders; Seida, three Swedish miles above the mouth of the Tana, with three families; and Polmack, four miles still higher up, with six or seven prosperous families. Seida lies on the usual way from Alten to Wardöehuus; for to avoid the dangerous passage by water along an open coast sheltered by no islands, the royal functionaries are in the habit of proceeding to Wardöe by entering from Kielvig the Laxefiord, then going over Hops-eid, and entering the Tanafiord, and at last ascending the Tana to Seida. From this place there

---

* Wahlenberg, Kemi Lapp. 49, 42.

is a distance of about nine or ten English miles over land to the Varangerfiord, up which they ascend with great ease to the island of Wardöe,
which lies before it.   These places now belong to the congregation of Tana,
with which they form an annexation to the Prästegieldt of Köllefiord.

Thus, to the remotest boundaries of the country, the Finlanders or Quäns
live in fixed habitations: they have not merely penetrated into the country of
the Laplanders; they have actually begun to surround it.   They will at last
proceed from the mouth of the Tana to cover the coasts of the Fiords, as in
Altenfiord.   The unfortunate race of Laplanders is driven higher and higher
up the mountains, cut off from communication, and in some measure destroyed for want of subsistence.   This is the fate of every people who set
themselves against cultivation, and are surrounded by a people making a
rapid progress in civilization.   The cultivation of the wastes of Baraba in
Siberia, and the wonderful growth of Kentucky and Tenessee in America,
have in our times driven out nations of nomade shepherds, and extinguished
almost the very names of many of them.

Though the boundaries of the two neighbouring kingdoms are now determined with geometrical accuracy, and though each of the countries knows
very correctly from what Laplander or Quän it has a right to demand contributions, yet this has not been sufficient to remove every cause of dispute.
The important salmon fisheries of the Tana very frequently set the fishers
who live near one another at variance, and these quarrels are transferred to
their governments.   The large fat and very excellent salmon of the Tana
was formerly an object of exportation from Finmark to Holland; and we
often hear that the Dutch would not look on any other salmon after the usual
cargo arrived from the Tana.   This trade has almost entirely ceased: there is
now seldom more than fifty barrels of salmon exported from Goldholm, the
trading place on the Tanafiord; for the inhabitants there now require the
salmon for their own subsistence.   They place their contrivances (Stängsel)
wherever they imagine the greatest number of fish will go, without concerning themselves about political boundaries.   The Swedes catch their
stores on the Norwegian side, and the Norwegians frequently place their
stakes (Stangen) on the Swedish side of the river.   If the person who is
situated high up the river catches fewer, he complains that the person
below him prevents the salmon from coming up.   If the Norwegian has
met with great success one year in a particular place, the Swede will en-

deavour to get the start of him the following year by placing his contrivances there. The Norwegians in the same way usurp the Swedish
stations. The whole breadth of the river is frequently covered, and the
salmon, according to the accounts of those higher up the river, are all caught
at once. On this subject the complaints have become very loud; they have
already reached the central boards of audience (*Central-Bchörden*) of Stockholm and Copenhagen; and commissioners have more than once been dispatched to put an end to their differences. This might easily be done if two
kingdoms were not concerned in the dispute. The excellent salmon regulations at Alten, by which each salmon fisher is shown where he is to set his
net, and how far he is to go into the river, might easily be applied to the Tana.
But to effect this the Swedish and Danish commissioners must be agreed.
The former reside in Torneo, and their business brings them only in winter
to Utsjocki. The Danish magistrate can seldom fall in with them, and he
cannot always commit the business to a person capable of bringing it to a
conclusion. The affair is of some importance for the province. If the
salmon fishery has drawn the Quäns down the Tana, the security of their
employment cannot be a matter of indifference to them.

Kautokejno lies eight hundred and thirty-four English feet above the
surface of the sea, and actually at the southern or eastern declivity of the
Kiölen mountains. For though the waters flow towards the Frozen Ocean
and not towards Sweden, we can however reach the sea in Sweden without
going over the smallest mountain; but we cannot reach Alten without traversing the mountains in their entire breadth. We cannot too often repeat
this, so long as we find in books that these mountains run for the whole
length of their course along the Norwegian and Swedish boundaries, and so
long as we are told that they divide themselves at the source of the Tana
into two arms, the one of which is lost at Vadsöe, and the other runs in the
direction of Finland, between Russia and Swedish Lapland. And yet the
maps of Baron Hermelin have pointed out the course so beautifully and accurately! There is not the slightest trace of the Kiölen mountains on the other
side of the Tana. Kautokejno, Masi, the Carasjock, and then the Tana and
the Tanafiord, are the southern and eastern boundaries of the range of mountains. To the north of this line there is still a continued elevation of more
than a thousand feet, with summits on it upwards of two thousand feet in
heighth. But eastwards, towards Vadsöe, the country becomes a plain, and

though insulated mountains still occur, they are scattered without order, and bear no trace of the connexion of a range of mountains. Towards Finland the separators of the waters of both oceans sink so low, that on the level, mountainless, and marshy plains of the Jvalojocki in Enare, not only birches thrive and Scotch firs, but even spruce firs stretch up towards the Frozen Ocean.* The Kiölen mountains split between the Fiords of Finmark; and with Sverholt between Porsanger and Laxefiord, with the Nordkyn between Laxe and Tanafiord, they sink into the ocean.

It is singular undoubtedly that the river of Alten *(Alatajock)* should run through the midst of the mountains, but however singular, it is not the less certain; the river dashes over perpetual falls all the way from Masi; the valley through which it runs becomes narrowed, and the mountains rise like perpendicular rocks above the water. The bed of the river becomes at last a fissure, and no person has yet followed it into the straits, through which it makes its way at the water-fall of Pursoronka, about fifteen English miles above Alten.

The valley of the Rhone divides in the same manner a chain of mountains in its whole breadth; and the whole of Jura is also traversed by the Rhone below Geneva, and by the Isere at the Pass *des Echelles.*

Even the rocks and the internal contents of the mountains point out this course. From Kautokejno eastwards, and towards Kola and Finland, we no longer find the rocks of which the mountains of Alten are composed, and gneiss and granite, which in the whole way from the Frozen Ocean could not break through the newer formations, become now predominant for whole tracts of country.

At our first nights stations in the wood in the Altens-Elv our rein-deer pastured on a rock, the white colour of which appeared even through the bright rein-deer moss. It was still the same quartz and quartzy sand-stone, only with a few white mica folia rarely intervening, which I had already seen in the neighbourhood of Kongshofmark, and at the Skaanevara. I did not expect it would be otherwise; for Akka Solki, above Talvig, seemed to prove that newer rocks resting on older filled the interior of Altensfiord. I was not therefore surprised when, at our second nightly station, the micaslate made its appearance with continuous mica, and with a number of small

---

* Wahlenberg.

garnets; and containing a number of grey quartz beds. The strata lay almost horizontally, and dipped only a very little towards the west. But this mica-slate did not continue. On the declivities towards Nuppivara it soon became shining clay-slate, entirely similar to that of Kielvig; and farther on the clay-slate again appeared with indistinct marks, and with the effects produced on it by the weather, which rendered it so disagreeable in appearance at Kongshavnsfieldt and at Talvig. No rocks rise out of the flat summits of these mountains: we must investigate their contents from the few blocks which are not wholly covered; so that we can seldom find places which are not rendered wholly undistinguishable by the weather. High ranges of rocks, precipices, and deep valleys, in general appear only where the strata are very much inclined; for if they are nearly horizontal, the hills are separated from each other not by two valleys, but by comparatively inconsiderable hollows scooped out by the streams, and the internal structure of the hills remains concealed from us. If we then ascend the mountains, we go continually from older to newer rocks; and we shall not again find, as at Montblanc, and in many other places in Switzerland, and even at Akkasolki, on the summit which we have attained with so much difficulty, the same stratum which we could have equally well observed in the line in which it stretches in the bottom of the valley. The formations of the regions of Nuppivara, or along the Zjolmijaure may therefore, without hazarding much, be deemed newer than those which appear on the Carasjock, or in the height of the Gurja valley. Here also we no where again see pure continuous mica; the micaceous, quartzy clay-slate wholly usurps its place, and not unfrequently there appear between the strata thick beds of a small granular rock composed of black diallage, and white felspar with much iron pyrites. In the rocks of the Zjarajaure the clay-slate becomes black, and quite similar to drawing slate, and that increases in no small degree the melancholy impession of the lake. Farther on, towards the mountains of Lipza and Jessjaure, when a block on the plain betrays its contents, it is almost always that small granular mixture of diallage and felspar which constitutes the formation of Bratholm, and of the rocks at the saw-mill at Alten. There is no where a trace of gneiss, nothing even to indicate it. Finally, along the Muddejock, and down towards the Siaberdas-jock, the mica-slate again appeared; or rather hornblende slate; for the small rocks are black from the small granular hornblende, and the mica-slate appears subordinate. Farther on, the marshes towards Kautokejno prevented all discovery of the stones beneath them; but at Kautokejno itself, on the small hills in its vicinity, we not only perceive again the gneiss to which we have so long been strangers, but also the granite which is so rare throughout the north. How much we are struck with these white fresh and shining masses after being so long without seeing them! Yellowish white mother of pearl felspar, with grey transparent shining diamond quartz in small granular mixture, and insulated black mica folia between. We must, however, look a great way in every direction before we again find this stone above the surface. The land does not rise sufficiently: rocks are altogether out of the question.

Hence granite is the base of the Kiölen mountains in Finmark; and primitive clay-slate, diallage-stone, and more rarely mica-slate, are the masses with which they are elevated. But even these stones no longer appear beyond the Tana, and consequently the Kiölen mountains also do not appear. Mica-slate also appears here and there in the gneiss, with lime-stone beds in it; as on the way from Seida to Vadsöe; but this mica-slate may be also older than that which appears without gneiss at Talvig and Alten.

This remnant of these mountains between the Altens and Tana-Elv is not upon the whole higher than what lies between Kautokejno and Alten, and in point of heighth can by no means be compared with the mountains which run between Nordland and Finmark. I looked over this part of the mountain range from Akkasolki. There was no elevation visible from them which had attained the region of perpetual snow, except perhaps the solitary Vorie Duder, several miles above the source of the Porsanger-Elv, which is the highest eminence of these districts, and probably approaches near to three thousand six hundred and twenty English feet in heighth. Rastekaise, the highest between the Carasjock and the Laxefiord, is not clear of snow till August, and spots of snow still continue to lie on its declivities. It may be at least three thousand two hundred English feet in heighth. There is no higher mountains either eastwards or southwards till we come to Ural. Throughout all Finland there is probably no height which exceeds the vegetation of birches; and therefore there is no mountain above two thousand one hundred and thirty English feet in heighth. There is none even towards the Russian frontiers and the White Sea. Small mountains of a few miles in extent are scattered over the country here without any visible connexion; and what is exceedingly remarkable, they are not more frequent even at the separation of the waters between the Bothnian Gulph and the White Sea than more near to the mouths of the rivers. Such a small mountain is Peldoive, between the Tana and the Enareträsk, only a few miles distant from the great range of mountains. Hence it may here rise above the birch region, and may actually be more than two thousand one hundred and thirty English feet in heighth.

LIPPAJAERFWI, the 13th of September. The sexton in Kautokejno and two Finlanders accompanied me over to Sweden, and I soon found how much I gained in not travelling with rein-deer. Notwithstanding the shortness of our day's journey, the animals were generally so wearied in the evening that we were obliged to drag them on by force; and at times, when we were most disposed to proceed with expedition on our journey, we were obliged to search for moss-fields on their account. The Quäns, on the other hand, were determined and jovial men, who knew no difficulty in the road, and who were in the same good humour in the evening as at their first outset in the morning. The sexton is an important personage among them, for he translates phrase by phrase to the congregation the Danish sermons of the clergyman into the Finnish or Laplandish languages. Although his translations may frequently fail to convey the sense of what is delivered, which we may easily believe, yet this occupation necessarily gives him the habit of connecting propositions together, which must have some influence on the rest of his conduct, and from which a traveller with no small pleasure feels the beneficial consequences. It would no doubt be more advantageous for the Laplanders, if the preacher did not stand in need of this troublesome assistance; but the attempts of the Danish government to accomplish this have been always hitherto unsuccessful. An attempt was made in the beginning of the former century to send young Laplanders to Drontheim to be there educated, and afterwards formed into Lapland preachers at Copenhagen. In 1754 a particular seminary was established in Drontheim, in which young persons were instructed in the language of Lapland, to be afterwards employed as missionaries in Finmark. Both these attempts have proved unsuccessful. In 1774 they fell upon a plan still less likely to succeed. The Laplanders were to learn Danish; their youth were to receive their education through the medium of that language, and in this manner it was hoped that the Danish would be every where introduced, and become the prevailing tongue —a Herculean undertaking, which no government has yet been able to

execute, and probably never will; for of all its peculiarities, a nation reasonably considers its language as the most sacred and inalienable; as nothing
more loudly and immediately excites the feeling of freedom and independence.
If a language has ever sunk under the pressure of a conquering nation, this
wonder has not been effected by regulations of police. The people are
either wholly extinguished, and their language along with them, as was the
case with the *Venedi (Wenden)* in Lüneburg, and in many other parts of
Germany, or the ruling language is so often introduced in all the civil and
domestic relations which connect two active nations together, that nothing
remains to be communicated in the old speech of their fathers but the
memorials of former times. But in such a case the language so introduced
generally gains as much from that which is supplanted as the latter does
from it. The language of the Norwegians and the French became a peculiar
Norman dialect, as the Norwegians adopted the French usages and customs
along with their own*. The Anglo-Saxon, French, and Danish, became
English, as the English were formed of the Norman conquerors, the Anglo-
Saxons, and the Danes. In this way it is true the Danish language has not
made any great progress among the Laplanders; and *Fieldt,* or mountain
Laplanders, who are too remote from the sea to come much in contact with
the Norwegians, very seldom yet understand a word of Danish. But how
much greater is their confidence, where the preacher can address them in
their own language! What was not the violent and proud Thomas Von
Westen enabled to effect from this circumstance? And in like manner the
mild and philosophical Simon Kildal at Stegen? With the very best inten-
tions, the preacher can do very little good through an interpreter; and so
long as the preacher himself is not in a condition to rectify the singular
ideas of the Laplanders, they have merely with the adoption of the Christian
religion exchanged an old superstition for a new one. They are frequently
praised as good Christians, on account of their eagerness to receive the
sacrament in the church as often as possible; but they merely do so, as it
appears, because they look upon the sacrament as a sort of charm to preserve
them from the attempts of evil spirits. It is not long since they always
took a cloth into church with them, into which they spit out the sacramental
bread, which they then wrapt carefully up, and divided at home into

* Gibbon, Chap. LVI. Note 17.

innumerable small crumbs. Every beast of their herd received so long as any remained one of these crumbs, and the Laplanders were convinced that by this means the flock would be secure from all injury. It would certainly be easier to persuade young people, especially in Nordland, to undertake the study of the Lapland language, if they had a certainty of being promoted to good situations afterwards, than to do any good by preaching in the Danish language to the Laplanders. Would any clergyman ever repent his having learned the Lapland language, and thereby communicated instruction to the Laplanders, even though in a short time he should be removed far from that people into the southern part of the country?

We embarked in a boat at Kautokejno, and ascended the Kautokejno-Elv for about an English mile to the first waterfalls which retard the navigation. Hitherto the river resembled a landlake: it is broad, and almost motionless, and yet it produces excellent trout, which may well be compared with those of the Bartholomæus Lake, in Berchtesgaden. We landed among small birch bushes, and proceeded over almost imperceptible hills, and over rein-deer moss, till we came to Sopadosjaure, when we again entered the boat, and crossed over the lake. Another English mile on land brought us to the banks of the Jeaudisjaure; and from this by a small Elv we arrived at the somewhat larger Calishjaure. Every lake is here taken advantage of to shorten the way, and render it less difficult; and the country is not sparing of them; for the innumerable multitude of small lakes which follow one another in this plain is quite incredible. It seems as if the country were not now intersected by vallies but by tunnels *(flache trichter)*, the bottoms of which are now occupied by the lakes. At the Natgieckjaure we found the hut and the fishing utensils of one of the Quäns: he had been there for several weeks, but he had still hopes of farther success. The fish do not shun the severe climate, or rather they eagerly crowd to it; for from the great quantity annually caught in the lakes, we could hardly expect that in the following summer they should increase to such a degree as always to supply the waste of the former. The principal fishery is always directed towards the sick *(salmo lavaretus)*, which abounds in every lake. The pike is not so common. The former is dried by the Quäns on wooden erections in nearly the same way as the cod and the sey are dried on the sea-shore. When they want to eat it, few preparations are necessary. I have often wondered at the manner in which my three guides produced their dried fish.

fixed them on the points of pieces of dry stick, and surrounded the whole of the fire with these spits. If the fish were roasted so as to appear brown, and approaching to being burnt, they consumed it without any other preparation, and, as it appeared to me, with a singular relish. The potatoe is praised on account of the small preparation which it requires to be eatable. The sick scarcely requires a greater expence of time and labour. It forms the principal diet of the Quäns, and they are seldom able to procure, in addition to it, the Norwegian bread, called *fladbröd*, which is thin as paper, unleavened, and made of the bran of barley or oats, and baked on stones or iron plates.

The fish of the lakes has the reputation of being fatter and more savoury than what is procured in the rivers. The rivers, however, afford an earlier supply in spring than the lakes, which lie higher, and are not visited so early.

The Quäns drew the boat ashore from the Natgieckjaure with great activity, and proceeded with it for a considerable space till we came to a small and rapid stream, the Bojaweckiejock. I perceived from the ease with which it was carried that a particular boat on each lake is not essentially necessary. The nearest way between two lakes lying near to one another is in fact frequently pointed out by round stakes, which lie behind one another at about the distance of the length of a boat. The boat glides lightly by means of these stakes over the inequalities of the surface, and little more is requisite than the strength of a single man to draw it over the gentle hills.

Bojaweckiejock, which Skjöldebrand calls Poiovaivijock *, is one of the principal sources of the Kautokejno-Elv, and consequently of the Altajock, or the great Altens-Elv. It was but a rivulet, however, and in no wise to be compared with the furious Siaberdasjock. We proceeded up numerous turnings and windings through small but close birch and alder bushes. The stream frequently all of a sudden dashes through stones, between which the boat can scarcely find sufficient depth to proceed; but in the serpentine windings the water becomes again still and gentle, like a river in a flat country, in the Ellernbruch. We pushed the boat up for at least seven English miles, till we reached the Aibyjaure, from which it issues. There we already perceived traces of the neighbouring kingdom. Notwithstanding

---

* Voy. au Cap Nord, 135.

we were still on the Norwegian territories, the Swedish Finns had however set their nets in this lake. The access to it is, in fact, easier to them than to the Quäns of Kautokejno; for as we got on land on the opposite bank of the Aiby-jaure, in moonlight, we had scarcely to proceed more than two English miles over hills, which were white with rein-deer moss, when we came to the banks of the Jedeckejaure, whose waters touch the boundaries of the kingdoms. The fishery of this frontier-lake has always, however, been the undisputed property of the Swedish Finns.

We took up our quarters here at the foot of the steep hills which environ the lake. The moon shone bright and full on the unruffled water, and threw a pale light on the points covered with birch bushes, which projected into the lake. The leaves of the bushes above us scarcely seemed to move; the boughs waved softly towards the water, as if they were desirous of being reflected in the moonlight. How easily could I have here dreamed that I was at a lake in Holstein, and forgot that I was on the greatest heighth which separates the waters of Lapland! As the fire began to blaze among the birches, and the Quäns lay around it busied with the roasting of their fish on the points of sticks, the whole appeared such an original, and such a clear, peaceful, and animated moonlight landscape, that no person, however rapidly it might have disappeared from his sight, would ever have forgotten constantly to number it among the most agreeable dreams of life.

The bright sun of the following morning covered the moss for a moment with a light rhime; but before we were in motion it had again disappeared. The whole night through we had no frost. Such a climate in the middle of September, and in such a latitude, banishes every thought of perpetual snow or uninterrupted winter on these plains. I saw a great way all over the country, where there were hills enough, but no mountains; but even on those heights there was no where the smallest trace of spots of snow to be seen. The birches appeared every where as small bushes to the very summit, and their green formed a strong contrast with the rein-deer moss below. And yet we were here at the greatest heighth between the Bothnian Gulph and the Frozen Ocean, and this heighth was in fact by no means inconsiderable. The barometer stood on the banks of the Jedeckejaure at twenty-six Paris inches, 3. 4 lines, and assigned to this frontier-lake a heighth of one thousand three hundred and seventy-eight English feet above the sea. Boiaweckiejock was even two hundred feet above Kautokejno, and one thousand one hundred and fifty-five English feet above the sea. We went through between Jeaurisvara and Salvasvaddo, both on the separation of the waters, the former perhaps five hundred, and the latter scarcely four hundred feet above the plain. These are, however, the greatest elevations of the country, one thousand seven hundred English feet above the sea; and therefore they prove, though merely hills, that the two kingdoms are no longer separated here by a range of mountains, but only by a height, somewhat in the same manner as the Black Sea is separated in Poland from the Baltic.

At the end of the Jedeckejaure the boundary of the two kingdoms runs over the summit of Jeaurisvara, and close past Salvasvaddo. It is very striking; for a broad way cut through the birch bushes allows it to be seen at a great distance; and the *Röse*, or March-stone, appears on Jeaurisvara like a monument. La Tocnaye is in the right, when he asserts that no boundary was perhaps ever drawn with equal accuracy as this, even through the most remote and almost inaccessible wildernesses. For a length of one thousand one hundred and ninety-six English miles it is every where determined, even to an inch. The march-stones rest on firm plates, in which the point of commencement is accurately pointed out; above this the whole has the form of a tower of uncemented stones, three ells in heighth, and nine in circumference. On the top of all there are five particular stones; one in the middle with the King of Denmark's name on the Norwegian, and the King of Sweden's name on the Swedish side. Two lines cut out on the under part of the stone point out the direction in which the next *röses* are to be sought. Two other stones, the *directors*, are situated on the continuation of those lines, some feet from the middle stone, to point out this direction of the course of the boundary line. Two other stones, the *witnesses*, surround the middle stone on other sides, that its place may be still more determinately ascertained, if any accident should remove the stone from its proper place. Between the *röses* the wood is every way cut through for a breadth of sixteen ells, so that the one *röse* may generally be seen from the other. They are seldom more than four English miles from one another. The two governments have besides deposited a very accurate description and plan of all their doings in every parish which the boundary touches, to which there is added a very minute description of the boundary, so far as it is connected with that Prästegieldt. These documents are subscribed by all the Danish and Swedish commissioners who were present at the fixing the boundaries, and they are carefully preserved in the churches. Besides all this, in the boundary treaty of the 2nd of October, 1751, it is stipulated that the limits shall every ten years be investigated, the *röses* kept in order, and the wood again cut ; and notwithstanding it is not mentioned in the treaty that new commissions should be issued from the two kingdoms, this actually took place in 1786 and 1806, and the boundaries were throughout their whole extent from Svinesund at Strömstad to Russian Lapland of new gone over and improved. It is scarcely possible for two neighbouring proprietors, or even two neighbouring householders, to ascertain the limits of their respective properties with such accuracy.

These boundaries were frequently, however, the occasion of differences between the two countries, and the cause of bloody wars. The active Charles the 12th of Sweden, " a great king, a bad citizen, a perfect statesman, harsh, tyrannical and cruel, but who had always the honour and the good of the state in view," * soon after his accession to the government, had irreparably deprived the Norwegian rulers of the old frontiers which formerly stretched to the Bothnian Gulph, and even to Uleoborg, and which, under the kings who governed in common after the Calmar union had been rendered dubious and unfixed; for Charles the 12th first divided the whole of Swedish Lapland into pastorates, sent missionaries and clergymen among the Laplanders, and built churches in the most remote wildernesses. The possession decided the boundaries, and the Norwegians had no settlements on this side of the mountains. In fact these districts might have been considered as lost to Norway after the mouths of the great rivers of Piteo, Luleo, and Torneo, were occupied by Swedes and not Norwegians, in the middle of the 14th century ; for the advantages to be derived from that part of Lapland, above these streams, now fell into Swedish hands, and the produce of the chace was conveyed by the easier course down the rivers, instead of over the mountains to the sea-coast.†   Charles the 12th connected the countries still more intimately with Sweden, when he introduced

---

* The words of Gustavus the 3rd. in his writings, I. Eulogy on *Torstensohn.*

† In the year 1335, Piteo and its environs (dens Oedemark) were confirmed to the prefect (*Drost*) Niels Abiörnsen on account of the labour which he had bestowed in making this country arable, and cultivating it. In the year 1350 the church of Torneo was consecrated by the archbishop of Upsal. *Schöning Gamle Geographie.* 110.

civil authorities there, appointed Härädshofdinger, and sent out Fogde to demand taxes. But his ambition was not satisfied with these internal conquests. He knew the value of the fisheries on the sea-coasts, and to bring these coasts under the Swedish dominion was an idea which he long followed with great perseverance and obstinacy. With this view he procured a cession in a public article in the treaty with the Russians at Teusin, near Narva, on the 18th of May, 1596, of all the countries which lay between East Bothnia and the Varangerfiord; but a secret article extended this cession an immense way farther. The Czar renounced in favour of Sweden all his claims on Finmark, and a great part of Nordland, probably as far as ever the Russians had proceeded down the coast as robbers, or perhaps as far as Charles might determine the bounds of the cession. These Swedish plans would have remained long concealed from Christian the 4th of Denmark, if an accident had not discovered them. For as he commissioned from Antwerp the maps which were then published there by the well-known geographers Ortelius and Hondius, be found to his astonishment on a new map of the northern kingdoms, that Finmark, and a part of Nordland, to the Tysfiord, were separated by a strong boundary line from Norway, and distinguished by the same colour as Sweden. He wrote to Ortelius, and asked him the cause of this singular delineation of the limits. He was answered by the geographer that in this he had followed the directions of a French captain, who, in the last war between Russia and Sweden, had accompanied General *Jacques de la Gardie*, and been employed at the peace, and who, after his return, had published at Rochelle an account of the whole war in French. In this book it was expressly stated as one of the articles of the treaty of peace that the Czar would not prevent the Swedes in any manner from levying contributions on all the Laplanders down to the Malangerfiord*. The King of Sweden did not content himself with empty words; he endeavoured immediately to avail himself of this concession, and if he had found a less active and penetrating opponent in Christian the 4th, he would probably have united the whole of Nordland for ever to Sweden. But Christian, in his journey to Wardoehuus, had made himself acquainted with a part of these regions, and though he

---

* *Slange Christian IV. Historie* 128. *Schlegel Kiöbenhavns Vidensk. Selskab Sckrifter* XII. 59.

had not been able.to drive the Russians from Cola, he succeeded better in preventing the Swedish settlements. His complaints brought about a congress at Flakebeck,in Bleckingen, in 1603, which, however, was productive of nothing; for in 1604 the Laplanders in the interior of the Fiord complained that the Swedish *Fögde* had levied taxes on them the whole way from Waidöe to Malanger: and with the privileges of Gottenburg, which allowed that town a right of fishing in the Varsangerfiord and other harbours of Finmark, Charles ventured at last openly to declare his views on the northern coasts. At nearly the.same time, the 4th of July, 1610, Stellan Mörner was sent as Swedish commissioner with a suite to Tysfiord in the Prastegieldt of Lödinger, with particular instructions not to be satisfied with the half of the contributions which might perhaps be offered him by the Danish functionaries.* Lars Lorsson was to govern as foged all the Laplanders to Finmark, according to the Swedish laws. It is singular enough that in the king's instructions for the foged all the small Fiords and bogs of the sea along the whole of the coast are minutely and accurately described, and more properly named than we find them since in any descriptions of these countries, or in any map or chart, not excepting the latest and best which we owe to the diligence of Pontoppidan. This is a sufficient proof of the zeal with which the king had investigated these coasts, and how much he had at heart to plant the Swedish colours there. The attention of Christian destroyed his extensive plans. The Calmar war was to be decisive of the sovereignty of Nordland ; and it was unfortunate for Sweden. Charles the 9th died, and Gustavus Adolphus, occupied with other plans, was early induced at the peace of Sjörud, in 1613, to renounce all the supposed claims and designs of his father; and since that period the Kiölen mountains have always been considered as the boundaries of Sweden. If the later emigrations from Finland, and especially on the Tana down into Finmark, might be supposed to throw any new uncertainty over the boundaries, or to excite any new claims on the part of the Swedes, fortunately such misunderstandings have been timely prevented by the boundary treaty of 1751. Should this boundary not be

---

* Jonas Werwing Konung Sigismund och Konung Carl. IX. Historier Stockh. 1747. II. 216.

altered by a treaty of peace, or by a cession of the provinces, it will always
with the greatest ease be distinctly found out.*

I had now entered Sweden. Danish Finmark was now changed into
Swedish Lapland, the Finns of the Norwegians were become Laplanders,
and the Quäns were changed into Finns or Finlanders. The name of Finn
becomes from henceforth honourable, and no longer an object of contempt
as on the northern coasts; and the hopes excited by the industry and activity
of the Quäns in Alten were now to appear in full maturity. What does not
hope accomplish! The view from Salvasvaddo down towards Sweden over
morasses covered with dark dwarf birches, and over plains covered with green
birches, and at last Scotch fir, seemed far from unpleasant, and comparable
with many of the views of the plains of Brandenburg, as for instance, with
that of the hills at Mittelwalde and Zossen. The uniformity of the level is
broken by the shining lakes between the thickets and the small mountains
in the distance; and the Palajock, which flows through its whole length,
guides the view through what would otherwise be an unconnected detail of
marshes and trees. In the Palajaure, the source of the Palajock, our attention was turned to the small island of Kintesari with the huts on it. We
saw distinctly the men who inhabit there in summer; and we felt that we
were not altogether in a desert. The Finns come up from Palajoen suu,
and fish for many weeks in the lake, and then in the nearest Norwegian
lakes: the island becomes the central point from which they spread themselves out; and at the end of summer they return back to their dwellings
loaded with the fruits of their toil. We went along the height at about
a distance of two English miles from them. The small hills at a distance look like sand-hills, or as if they were covered with snow: the white
rein-deer moss had supplanted every other vegetable; and now, at the end
of summer, it had grown to a considerable heighth. We fell into it as into
so much wool; and if the numerous foot-marks of the wild rein-deer had
not occasionally furnished us with a firmer footing, and allowed us to proceed at a faster rate, we should probably have been a whole day in reaching

---

* Since the peace of Fredericksham on the 26th of September, 1809, Finmark throughout its
whole extent is no longer bounded by Sweden, but by Russia; but no alteration of the boundaries has hitherto become necessary from this circumstance, or probably will become so in
future.

the banks of the Palajock : the sandy levels of the coast of Pomerania, or the ashes of Vesuvius, are scarcely so fatiguing. How scanty the moss of the mountains of Alten and Kautokejno appear in comparison of this interminable mossy level ! There the moss has to struggle with the effects of the climate ; but here it seems to thrive most vigorously. We were not long in descending the plain till we saw the first parched up firs (*pinus sylvestris*) which, under the guidance of some warm summers, had ventured too high up. We were consequently quite close upon the upper limit of Scotch firs, and we here again witnessed what has been often remarked in the north of Sweden ; namely, that the moss never grows so large and thick, and upon the whole so luxuriantly, as when the further ascent of the Scotch firs is prevented by the cold ; and this vigour and abundance extends to the upper birch region about three hundred feet of perpendicular heighth. Farther up, the moss becomes again thinner and weaker, and in a distant view the naked stones of the hills are more prominent than the white covering of moss. This circumstance has an essential influence on the political condition of the two neighbouring kingdoms. In Sweden, the country rises so gently from the Bothnian Gulph, that we frequently can only discover the ascent from the course of the rivers. The plains therefore are many miles in extent before we reach this upper limit of birches from the upper region of firs, and hence many miles are covered with this abundance of moss. It is not so in Norway. The mountains there ascend so very rapidly that they soon, on very short bases, rise above the vegetation of trees. There is not sufficient room for the extension of the moss in the singularly contracted climate ; and instead of covering whole square leagues as in Sweden, it seldom occupies more than two English miles in Norway, between the firs and birches, and then only in narrow enclosed vallies. Hence the rein-deer Laplanders can never settle in Norway, and Wahlenberg remarks, that before the cession of Kautokejno Norway did not possess a single Lapland congregation, for the Laplanders never imagine themselves at home till in their winter quarters.* They could not, however, prosper in Sweden, if they were prevented from annually roaming over the mountains on the Norwegian sea-coasts. For the plains

* Wahlenberg Kemi Topographie. 24.

Y Y 2

in Sweden are seldom high enough to protect the rein-deer from the heat and the insects of summer. The animals grow weak, decline, and die. Norway on the other hand gives them an opportunity of reaching snow even in the warmest month of the year. Both kingdoms are connected together by the Laplanders. Whoever should attempt to restrain the Laplanders from roaming over the mountains, would not affect the prosperity, but complete the ruin of that people.

The rein-deer moss on the other hand is infinitely more accommodating in its nature than the rein-deer which it nourishes. Although it loses among the Scotch fir woods something of the vigorous growth which it exhibits among the birch bushes of the heights, it still remains, and that for a great way down, the most efficacious and striking covering of the soil. In Westerbothnia or Jämteland we are a great way below the region of rein-deer, but still the naked cliffs in the woods are covered with the rein-deer moss. Rein-deer moss is not a stranger to the rocks, not only at Stockholm, but even as tar down as Schonen. In such a climate a rein-deer could scarcely drag out a year or two. The moss even ventures over the sea, and we find frequent traces of it among the sandy plains of Germany.

After labouring for about nine English miles through the moss and through the small dwarf birches (*Betula nana*) in the morasses, we at last reached the banks of the still slow and deep running Palajock. We found it impossible to wade through it, the more so as we had before been obliged to ford the Salvasjock, a pretty strong stream. This stream flows out of a lake on the frontiers beneath Salvaslopel, and adds not a little to the size of the Palajock. But my nimble Finns were never at a loss. They threw down their burdens and clothes, and run about among the bushes, and in a short time they returned with large dried firs. They bound them together with rods of mountain willow, with which the banks of the river were here covered, and formed a raft of them, and in half an hour after our arrival the raft was in a condition fit to transport the navigators. The cords and rods bound together with which we drew it back, when one of the company was ferried over, measured ten fathoms. This was the breadth of the stream at the place, and its depth in the middle was about eight feet. It had been swelled by the rain; for its breadth in summer is generally not so great. The banks are here pretty high, not rocky, but composed of strata of sand and clay, which are concealed by willows, birches, and rein-deer moss.

The whole country resembles a flat land, and has nothing of the mountainous in it. On the opposite side of the Palajock the firs stood closer together, and formed a wood. We had seen nothing of the kind since our leaving Alten. We soon entered on a trodden path: a hedge next glistened through among the trees which was drawn round the small fields: and we at last saw houses at a distance. Again cultivation and agriculture! This was Lippajärfwi. The sun was still in the heavens, and lightened up beautifully the declivity from the houses towards the little hospitable lake, and the bushy sides of the Lippivara beyond it, the highest mountain of the country, which had been our constant guide from the frontiers along the great plain. Here we again had a prospect which would have been distinguished in other countries as well as Lapland.

The Lippivara is wholly insulated and seems high, because for many miles around there is nothing to equal it in heighth. This mountain is not only covered at bottom with Scotch firs, but almost to the very summit with birches; and it can hardly reach more than six hundred feet above the level. Hence we reckon it about one thousand nine hundred English feet above the sea; for I found the ground of the houses at Lippajärfwi one thousand two hundred and eighty-five English feet in heighth, and the banks of the Palajock might be about one hundred feet deeper. The first Scotch fir about four or five English miles above Lippajärfwi, stood at a heighth of one thousand three hundred and twenty-seven English feet, where consequently it was little higher than Lippajärfwi itself, at which place, however, the firs had assumed the appearance of woods. Still, however, they betrayed evident symptoms of the difficulty they had had in preserving themselves so high and in this climate. I saw nearly the half of all the trees dry and withered, and on the part which still continued to live, all the under branches for more than half way down the tree were completely dead, killed probably by the frosts; the needles were conglutinated together like spiders' webs, and gave an inexpressibly dead and dreary character to their appearance. The snow may have prevented the expiration of these branches.

The possessor of Lippajärfwi derives his principal support from fishing, and from the produce of some cows. The oats sown by him round the gaard are merely a trial; they do not always succeed. An arable field is still a rarity.   As Lippajärfwi is probably the highest fixed residence in these latitudes, so this is also probably the highest spot on which oats have yet been cultivated.

———————

KAENGIS, the 18th of September.   The man who resides at Lippajärfwi and his daughter conducted me on the fourteenth through woods along a great multitude of little lakes for about nine or ten English miles down to Suondaijärfwi, on a height above the lake.   The habitation here had less the appearance of nomade Lapland.   I found here a proper room, very clean, with a chimney, and glass windows, which were unknown in Lippajärfwi. The walls were lined with milk-vessels and butter-casks, destined for sale at Alten.   The tables and benches looked dazzlingly white from scouring, and the floor was boarded.   This was by no means to be accounted a wretched habitation.   There are great numbers of a much worse description in the most cultivated countries.   They shoved me in a boat down the little stream which flows in numerous windings into the Palajock; the banks were alternately lined with hay cocks, birches, mountain willows, and alders. After a distance of about an English mile, the stream entered the Palajock, and the boat now flew more rapidly over a number of little falls or rapids. For the last four or five English miles we went a foot through the bushes to the banks of the Enontekisjock, and reached Palajoensuu in the evening by moonshine.   It is a small village of four or five habitations, situated at the influx of the Palajock into the Enontekisjock.

We had now left the wood, and were within the enclosed grounds; and I rejoiced to find myself again amidst cultivation, and by the side of the great river.   The inhabitants came across the meadows to meet us, with burning torches in their hands, like so many spirits of the night.   They conducted us into a habitation.   I again was assigned a cabinet separated from the head *gaard* with a bed of rein-deer hide stretched above hay, and a woollen bedcover, which was altogether better than many a well-decorated bed in large inns.   The men then threw themselves into their boats with their torches, and remained the whole night through on the river for the purpose of striking salmon.   I looked on them a long time from the banks.   The movement is easy and beautiful, and the figure of the striker is highly picturesque, as he remains immoveable in the forepart of the boat, completely lighted by the fire, with the murderous trident in readiness for the blow, and his looks and attention immoveably fixed on the stream and the approaching salmon. But how beautiful was the salmon striking on Muonioniska! Scarcely had the evening commenced when these large and brilliant fires were everywhere seen floating on the clear surface of the water. They crossed one another in all directions, and nothing was to be seen but the immoveable figure before, with the frightful trident, as if the fires were driven about by some unknown power. Suddenly an electrical spark of life darts, like lightning, through the figure. In a moment the trident is driven with force into the water, and the struck salmon by its windings only fixes the barbs deeper into his head. In Ketkesuvando I saw all the preparations for this singular and beautiful mode of fishing. On the front of the boat there was an iron basket, like a chest, fastened to the end of a long crooked stick, in such a way, that the pieces of burning fir which it contained reached far above the boat. Immediately behind the fire stands the fisher with his trident; behind sits the pilot, who with gentle and imperceptible strokes moves the boat along. The murderous instrument is very long, with five or six strong iron-barbed toes or points. There are other tridents in readiness, of the same form, but larger. The salmon is attracted by the blaze of the fire, and, suspecting no danger, he raises himself slowly to the surface of the water. If he is too large for the first trident, the pilot assists with the others which are in readiness. When we see on an evening the numbers with which the river is so majestically lighted up, we may well believe that this fishery is here very profitable. And yet what a number of contrivances between Enontekis and Torneo to catch every salmon that attempts to ascend the river! The salmon still finds ways of getting up, notwithstanding the numerous stakes which run across the river, and notwithstanding the numerous falls, and a distance of almost three degrees of latitude from the sea.

The evenings were somewhat cold; in the morning there was a slight frost; but the whole day through the weather was excellent, and far beyond my expectation of September in Lapland. We had a bright and clear sun, with a still and calm air, and the thermometer at mid-day up to 50° Fahr. This is certainly a great deal for a place which lies above 68° of latitude, and as high as on the mountains, one thousand and sixty-nine English feet above the sea:

The woods of Scotch fir *(fichten oder kiefernwälden)* were now in the most flourishing state. They had no longer any dried and frozen branches ; and I saw a number of stems which would have furnished excellent wood for building. We had the climate of Alten again. What Palajoensuu loses in heat, from its high situation, it gains from its more southern latitude.

On the fifteenth I proceeded from Palajock, or Palajoensuu (*suu*, the mouth ; *Palaj oensuu* the mouth of the Palajock), in a boat down the great Enontekisjock. We went rapidly over a number of small waterfalls, in which they contrived to guide and govern the boat in such a dexterous manner, that it never once grazed any of the numerous stones which lay in the falls like mounds. These fishers are exceedingly expert at this, and few would be able to imitate them. But without the attention perpetually on the rack in the guidance of the boat, all communication by the river would be quite impossible ; for the dashing waterfalls are innumerable. After a course of upwards of twelve English miles, my *wappus* (or *oppus*) of Lippa-järfwi called out suddenly : *here is the first gran.* He was himself rejoiced at the new sight. It was in fact the first spruce fir (*pinus abies*) on our way, and it had the same dreary and desolate appearance of the first Scotch firs above Lippajärfwi. It was small ; the branches hung down black, dried, and frozen, and the needles were also here drawn together like spider-webs. Several others soon appeared on the banks of the river, but still with similar branches. We landed not far from thence at Songa Muotka, a miserable fishing hut on an island under the foot of the Songavara, the highest hill of the country. I then observed the barometer, and determined the heighth of the place at eight hundred and fifty-one English feet above the sea.

Hence the first spruce firs stood four hundred and seventy-six English feet of perpendicular heighth below the first Scotch firs, and appeared also more than half a degree farther south, and consequently had gained not a little in point of climate. If the climate of Palajoensuu is like that of Alten, in the productions of the earth, the mean temperature here consequently gains $0°.9$ R. or $2°.$ of Fahrenheit, for every additional degree of latitude ; and hence, for thirty-five minutes, the difference of latitude between the appearance of the first spruce and Scotch firs, we must allow $0°.5$ R. or $1°.12$ F., supposing places to be situated at an equal heighth. If the temperature diminishes $2°.25$ F. for every five hundred and eighty-five

English feet of heighth, the proportion for the $1°.12$ F. will be two hundred and ninety-two. The Scotch firs would consequently gain this much ; or if we add this heighth to the above difference between the elevations of four hundred and seventy-six English feet, we should at Songa Muotka, where the first spruce firs make their appearance, have to ascend seven hundred and sixty-eight English feet of perpendicular heighth before the Scotch firs would disappear. This difference of heighth coincides very well with the observations in Norway. Deeper down we dare not, however, calculate on such a rapid increase of mean temperature ; for the cold of winter increases with the distance from the Great Ocean, in almost the same proportion that the heat of summer is increased with the southern latitude. If we wish, therefore, to determine the developement of vegetable life from the temperature, which we are entitled to do, as life is alone produced by a higher temperature, we must only compare the summers together, and not the winters ; or more accurately still, perhaps, the heighth of the limit of snow above different places ; for the heighth of the limit of snow is determined by the heat of summer, and not by the winter's cold.

Between nine and ten miles below Ketkesuvando, we come to Oevre Muonioniska, a large village on the hilly banks of the river, surrounded by corn-fields, consisting of a number of neat and smart-looking houses, which have an agreeable appearance, and the more so when descending from upland deserts. About five miles lower down lay Lower Muonioniska, separated from the upper by a long succession of waterfalls. This place also was large, and had a comfortable look from the number of painted store-houses, and the small enclosures in every direction along the sides of the hills. "The people here are all very rich," said my wappus to me, "and especially at the inn." I could not but give credit to this. They showed me into a separate room, with large panes of glass, and gave me silver spoons with my milk *.

When we consider that little more than half a century has elapsed since these regions were first cultivated ; that formerly they were only occupied by a few families of Laplanders, who wandered about from place to place with their herds ; and that now the industrious Finns are thriving in every direction, we cannot help feeling an internal satisfaction that cultivation and

---

* Both the Muonioniskas belong now to Russia since the peace of Fredericksham.

agriculture are thus spreading over parts which were believed to be necessarily doomed to perpetual sterility, and to be inhabited by wandering nomades. There was no doubt a time when Lapland extended itself almost to the town of Torneo. Its boundaries have been gradually removed more and more northwards. They at last became stationary at Muonioniska, more than one hundred and forty English miles above Torneo. They would, however, have been advanced also beyond this place if the numerous emigrations of the Finns into Norway had not occasioned a momentary cessation in the cultivation of Swedish Lapland. When once the Norwegian Fiords are settled, if the active spirit of the Finns is not oppressed by the unrelenting hand of fate, the habitations will again continue to be extended along all the streams and lakes connected with the Muonio river, and the Laplanders will be confined within narrower and narrower limits, till at last the whole race will be exterminated.

Although the Laplanders and Finns may have the same common origin, they have, however, been long separated, and probably long before they came to inhabit the north. We find it extremely probable, when we compare together the old accounts, the customs, and probabilities, that the Laplanders descended from the White Sea towards Norway and Sweden, and that the Finns, on the other hand, ascended from Estonia, through Finland. The two nations are not only at present *toto cælo* different in their mode of life and cultivation, but they have not even the smallest national physiognomy in common. The Laplanders, as is well known, are in general small; tall men are every where exceedingly rare among them; and such a a person as Niels Sara, at Kautokejno, who measured five feet four inches Paris measure, or five feet eight inches English, may not be again found among many hundreds of them. However, such individuals as the two married women whom M. Grape measured would every where be reckoned dwarfs among them. But the Finns, though they remain for centuries in the same country, do not appear to become in the least smaller in size, either at Kautokejno or Muonioniska, than the Swedes or Norwegians. The cause of this is easily discovered; it lies altogether in the difference of cultivation. The polar tribes are small, like all animals and all organic substances which surround them; because they are almost in like manner fully exposed to the oppression and contracting influence of the severe climate, and have not learned to secure themselves from it. The Finn, on the other hand, procures a tropical heat in his *perte*, and what is contracted by the cold is again extended here and stretched into new activity. The Laplander seldom or never keeps himself, even in his winter gamme, in such a temperature as nature requires for the developement and advancement of the functions of life in the physical man; and though this may not be felt by his nerves, yet it must be felt by his constitution and his conformation. The Finn, on the other hand, finds a compensation for the unheard-of cold in warm baths of an equally unheard-of heat; and the advantages he derives from thence is demonstrated by experience in Lapland.

In the year 1799 there were five thousand one hundred and thirteen Laplanders in the Swedish division of Lapland; if we reckon an additional three thousand Laplanders for Norway, though it scarcely contains so many, and one thousand for the part belonging to Russia, where, upon the whole, but few Laplanders remain, the total strength of this people, however widely spread, will consist at most of ten thousand souls. But in Finland alone the Finns amount to nearly a million.

———————

I had not again seen any gneiss from Kautokejno to the other side of the frontiers; the little stone which is visible in these plains bears the character of granite. No block, no piece displayed a slatey or even a striped texture. Here and there we saw low rocks in the shape of small round wool-sacks, such as very much distinguish granite, but seldom appear in gneiss rocks. I doubt much, from what has hitherto come to our knowledge respecting the northern lands, if this granite will ever be found again in greater extent. Even at the frontiers, at the foot of Salvasvaddo, it is very small granular; the felspar is pale flesh-red, the quartz grey, and the mica black, and the folia close to one another. Here commences a somewhat streaky formation; quartz and felspar are often separated from one another by mica, but the true slatey structure is not easily recognized even in large blocks. It becomes much easier recognized in the blocks of Lippajärfwi. We are gradually again put in mind of gneiss; the felspar becomes frequently dark red, and the black mica much heaped. At Svondarjärfwi fine epidote needles and grains of titanitic iron ore lay on it; and at last, at the influx of the Palajock, the gneiss nature was no longer doubtful, though the gneiss was certainly here very small granular, and the felspar very pale flesh-red and white. But the mica, in very small, nearly scaley folia, was distributed amongst the felspar in rows or beds, and the quartz was abundant and of a grey colour, and conchoidal fracture. The stratification here could not, however, possibly be determined ; for I found no rocks. I found at Pala-joensu a large and remarkable block of grey coarse granular zirconsyenite, with angular cavities, as at Christiania. If it has come thus far, the formation cannot be at a great distance ; perhaps above Enontekis, towards Raunula. It is singular that it should appear so high and so insulated. At Muoni-oniska the gneiss becomes very various in its composition. The mica is collected, group-wise, in single folia, with reddish white felspar in long streaks between, so that the slatey structure is here no longer parallel with itself.   Quartz lies only in small grains on it.   Other strata consist of gneiss, but with little mica; the scales of mica are very small, and lie alongside one another in a continued series, intermixed with very fine granular fel-spar and quartz ; and all the varieties which can occur between the two kinds of gneiss.   Here we again have the kind of gneiss of the shores of the Norwegian islands, only the mica slate no longer appears ; at least in my whole way down the Muonio river nothing similar ever was seen by me.

" Fortunately," said M. Kohlström, the clergyman in Muonioniska, to me, " fortunately *Johann Von Colare* is still here ; for he is the most expe-rienced waterman, he will take you over the waterfall."   I heard the noise of the fall long before we approached it, while the river still glided on smoothly, and surrounded two islands which were then thickly covered with hay-cocks.   Then followed several falls ; they were not high nor long ; but the stream became rough and agitated. Rocks began now to rise along both sides, and points to appear above the surface. The agitated water presses through between the closely approaching rocks. The waves began to rear themselves up, to foam and dash over one another. They drove the boat with incredible rapidity down the abyss ; they dash over in the most wild and alarming commotion ; the sky, rocks, and woods, all disappear, and nothing is seen or heard but the foam and roaring of the water.   The wave dashes the boat with one sweep against the rock ; but the bold pilot guides it with a strong and steady hand, with still greater rapidity than the wave, as if in sport, from one side to the other ; and the next moment it is again floating on the no longer agitated current. A few steps farther on the stream becomes a beautiful lake, and almost motionless.   The first waterman who attempted this alarming fall must undoubtedly have been a man of matchless boldness ; and even yet, after so many and repeatedly successful attempts, this Tartarus passage is never entrusted to any but the most experienced individuals. The two men in the fore-part of the boat have a most frightful appearance. Their fixed looks, their eyes, which seem to start from their sockets, endeavour to read every thought of the pilot, whether they ought to row in the fall more rapidly or more slowly. Their own preservation depends on their correct under-standing of the thoughts of the pilot. Every muscle is stretched in the highest degree, and the arms only are in motion. The boats are as strong as sea boats ; the waves would otherwise destroy them in a moment ; and the huge helm seems made for large ships. These men venture in this way to proceed down with large burdens of butter, teas, fish, and hides, to Torneo ; and they drag the boat along the bank in ascending, when they come to the fall, with incredible labour.

At the foot of the fall, which is celebrated under the name of Eianpaika, lies the solitary and miserable cottage of Muonioalusta. According to the barometer it was one hundred and eighty-four English feet beyond Muonio-niska, and the greatest part of this height, much more than a hundred feet, belongs to the great fall, which in length exceeds an English mile.

It was the last house of this region. For many miles down I could see no dwellings more. A thick forest, without any elevations, runs along both banks without any interruption. But notwithstanding the seeming uni-formity of such a forest, the fancy is however agreeably occupied, one moment with the view of the fresh and lively green of the banks, where birches, willows, alders, and the bird cherry tree *(prunus padus)*, bend softly over the water, with a perpetual diversity and change of form ; and then dark spruce firs rise above the close thickets like so many cypresses. Again, when we cast our eye down the stream, we see, a great way off, projecting capes, and an innumerable diversity of small hills in the interior, and blue mountains wholly in the distance. This is no desert land. The plants and trees thrive in it ; and the black marshes and white rein-deer moss disappear.

The mountains which excite such a lively expectation are only, however, hills when we approach them, of a few hundred feet in heighth. The highest of them is Ollos Tundure, to the east of Muonioniska, about eight hundred and fifty English feet above the level. We did not remain long in Parkajocki, a soli-tary inn on the right bank of the river, nor longer in the also solitary Kihlänge ; and late in the evening we reached Houki, rather more than two English miles above Colare.   The two soldiers who conducted me had brought me that day more than fifty English miles down.   They resided in Muonioniska, but were wholly Swedish in their dress of blue uniform jackets. On the other hand my boatman of Palajocki bore a very striking resemblance in point of dress to the statues of Barbarians at Rome.   The same sort of shoes without separate soles (the *Komager* of the Laplanders), the same hanging breeches, reaching down to the shoes and full of folds, with a robe over all, which is held together by a broad girdle in the middle of the body.   It is said that the Finns are descendants of the Hunns.

Hitherto we had proceeded down a great number of waterfalls; from Kihlänge alone I counted six, none of them it is true to be compared in any manner with Eianpaika, though they required all the dexterity of my soldiers not only to avoid the stones which are visible, but also those which are hid beneath the water, but which, however, may come in contact with the boat. Such a number of falls must have brought us a great way down.   In fact, Houki was only four hundred and fourteen English feet in heighth, and consequently was near three hundred feet below Muonioniska, and this difference, added to the southern situation, produced a very striking change of climate.   Higher up the corn was even this year frozen on the ground, but this was not the case at Colare.   " This is our misfortune almost every year," said M. Kohlström to me, " and then we have no bread, and are obliged to feed on fish and potatoes, which, being in the earth, are not so easily frozen."   The carriage from Torneo up all the different falls is too difficult; but they receive brandy and tobacco in sufficient abundance, and more than they want, from the merchants of Alten.   How much easier it is to live in Colare.   The view in the morning down the river from Houki was extremely gladdening; the banks were lined with a number of neat and elegant houses. The bucket ropes of the draw-wells, which rise high in the clear air, and denote at a distance the number of single cottages, and the elegant storehouses, painted with all manner of colours and figures, proclaim the prosperity of the inhabitants, a degree of prosperity which we should by no means have expected.

We proceeded rapidly downwards with a beautiful and warm sun.   At midday the thermometer stood at nearly 54°. F.   The air was quite still, and there were only a few light white clouds in the heavens.   Between four and five English miles above Kängis a black cloud immediately made its appearance.

It began to hail; the hail-stones fell thick, as large as peas, and shaped like pears, as at Altengaard, and beat alarmingly on the water of the river. This however did not continue a quarter of an hour. About one o'clock, as we came in sight of the saw-mills of Kängis, all was over. The sun again shone clearly forth. The thermometer stood at 50°. F. Such is the local nature of this singular and remarkable phenomenon! A heat of 54°. in this still air, and with this sunshine, was sufficient to produce hail, and such large and singularly shaped hail!

We again proceeded down another large and strong fall of the Muonio river, where the Torneo-Elv comes dashing on the right side through a narrow opening, and tumbles foaming over cliffs and rocks. It was so confined between the rocks, that compared with the Muoniojock it seemed a mere rivulet; but we were scarcely opposite the mouth when it drove us with the rapidity of an arrow over to the other side, and in a moment we reached the bank under the saw-mills of Kängis. The boisterous Torneo-Elv carries the Muonio along with it, throws itself above it, and extinguishes even its very name, notwithstanding in the way downwards it is not Muonio but the Torneo river which changes its direction.

Every thing from the frontiers had been so level, that the situation of the saw-mill of Kängis furnished an entirely new and romantic prospect. The dark woods above, the dashing of the river, the great surface of the slow and proudly descending Muonio, and then the large new building of the saw-mill, with several dwelling houses beside it, formed altogether a surpr sing and a rich and lively landscape. And when I landed and saw all the new undertakings, a canal lined with large granite flags, and a number of blocks lying around for the purpose of carrying the canal still farther, I could hardly be supposed capable of refusing to be convinced that I was no longer in Lapland, and that such undertakings could only belong to the industrious Swedes. I was still farther convinced on walking for about an English mile towards Kängiswerk; the vegetation in the hollows seemed so rich and the trees so beautiful. I soon saw maids and labourers employed in getting in on the height the remainder of the harvest; and in the glen below, immediately above the raging and foaming waterfall, the large and respectable farm-house, the red house of Kängis, with large panes of glass in the windows, shone upon us. The black cottages were smoking on the opposite side of the river, and on this side stood like a mound large frames of at least twenty feet in heighth, closely filled with corn, which here awaits its final ripening, and is laid in a triple row one behind another.   M. Eckström, the *Bruckspatron*, or proprietor of the work, received me with the most cordial hospitality.   The circumstance of finding myself again in the midst of men well-informed of the affairs of the world   and in a house where every thing was not only comfortably but even elegantly fitted up, gave rise to very peculiar feelings ; and to repose in a room which appeared to have been transported from one of the best houses of the capital was enjoyed by me with an indescribable relish after a journey through the wastes of Lapland.

In the waterfall of Muonioniska we clearly perceive gneiss, and the stratification below may also be determined: it stretches N. E. and S. W. (h. 3.) and dips towards the south-east.   The stone is not rich in mica, and the felspar is small granular and white.   Deeper down in the falls, where small rocks also appear, the strata lie altogether horizontally.   At Kihlängi the gneiss contains more mica than before, and is scaley foliated.   This is all which in so great a length of way on the river can be observed of the constitution of the earth.

At Colare I found a block of dark-blueish grey, small granular lime-stone, from a bed which must necessarily be near at hand, existing in the gneiss. This is the more credible as a similar lime-stone bed is wrought at Kalkkepatha on the Tärändo-Elv, not far from the furnace of Torneo-fors, and almost at an equal heighth.   This is not unimportant; for this lime-stone is not a bed on the mica-slate like the lime-stones of Norway, but a bed in the gneiss.

Between four and five miles above Kängis the white gneiss disappeared. The rocks and blocks were wholly brown as if covered over with rust of a granular structure, and affected by the weather.   It was now changed into another kind of stone; red granite, the *Rapa kivi* of the Finlanders, (slate-stone, because the blocks fall to pieces on exposure to the weather, and peal off in scales).   All the rocks at the confluence of the two rivers consisted of this, and the mountains to beyond the work of Kängis.   The granite is coarse granular; the felspar dark *flesh-red*, and in great abundance; mica in single black, but not considerable folia; quartz scarcely at all, or when it is there, of a blueish grey.   On the other hand, black hornblende folia seldom or never fail, and they generally exceed the mica in abundance and magnitude. It was truly an *under-lying* granite. Where it first appears on the Muonio river, we see it distinctly, and still much more beautifully in the large waterfall of Torneo Elv, which immediately above the works of Kängis cuts of all communication by water betweeen the countries above and below.* At the rocks of the fall both stones are very sharply separated from one another, and beyond the church of Pajala, which lies above the fall, nothing but white gneiss appears. The mica is very frequent in it, and is black, and in thick scaley folia one above another. The felspar is very small granular, and yellowish white, and the quartz is frequently small—conchoidal in thin beds; the whole is *dick flasrig*, and very similar to the gneiss of Freyberg. The stratification N. N. W. and S. S. E. (h. 10.) with a dip of fifty degrees towards the south-west.

The works of Kängis are very old, but they have never risen to any considerable heighth. The mines of the works lie between thirty-five and forty English miles higher up the Torneo river, at Junos Suvando, on the boundaries between Wester Bothnia and Lapland, where the furnace is situated in a latitude of sixty-seven and a half: it is the last and most northern furnace in the world. Several miles deeper down, and on the immediate banks of the Torneo river, there is still another furnace at Torneo-fors, which, however, is not always in employment. Both were built in the seventeenth century, shortly after the rich treasures of iron ores in this country were discovered. The iron ores actually form whole mountains which rise above the plain in the neighbourhood of Juckasjerfw, at Svappavara, and a number of other places. What an admirable production! to bring life and cultivation into this desert region. This has not, however, yet fully succeeded, notwithstanding all the endeavours which were made in the middle of the last century to bring the mines of Torneo Lapmark into a flourishing state. The iron-stone is every where very difficultly frisible, and yields an iron which is brittle when cold. I heard even in Alten frequent complaints made of the iron of Kängis that it was good for nothing, and in Torneo also I found it in very low

---

* According to trigonometrical surveys, the perpendicular heighth of the fall from Pajala church to the calm water below the works is seventy-eight Swedish feet (*Hermelin Mineral Historia*, p. 69.) The heighth of Kängis above the sea, according to the barometer, is four hundred and twenty English feet; a number which may be, however, about sixty or seventy feet doubtful, and too high.

▲ ▲ ▲

estimation.   Hence it probably happens that the iron produce of Kängis
has never risen above 5000 cwt. (2000 *Schiffpfund*) annually, and at present
scarcely reaches the half of that quantity.*   The active M. Eckström, who
has acquired great merit in the sinking the mines of Junos Suvando, appeared himself to doubt of the possibility of ever rendering this iron-stone a
marketable article.†   His industry has embraced many other branches, which
if successful, will necessarily injure the production of iron.   The beautiful
saw-mill at the confluence of the two rivers will cut down not a few trees
of the large forests, and the returns of agriculture which M. Eckström strenuously endeavours to bring to perfection, were intended by him to be used
in a distillery of brandy.

The iron-stones in the whole of Lapmark, as almost every where in
Sweden, form thick beds in the gneiss.‡   As they are so fixed, and so indis-
tructible, they remain above the surface like mountains of iron, when the
surrounding gneiss is removed.   When, however, we follow them downwards, their true nature as beds is soon placed beyond all doubt.   These
beds vary very much in thickness; sometimes they are narrow, and immediately again become wider.   The bed of Junos Suvando is in one pit, from
three to four fathoms, and in another from fourteen to fifteen fathoms in
thickness.   It consists almost throughout of magnetic iron-stone, in small
adhering crystals, which causes the whole mass to appear as if composed of
round and nearly small granular concretions.   The magnetic iron-stone of
Dannemora and Utöen is on the other hand altogether fine granular.   The
other fossils which accompany the ores of Junos Suvando are principally

---

* Baron Hermelin Forsök til Mineral Historia öfwer Lappmarcken och Westerbottn, 21.

† Although the iron-stones of Lapland exceed in richness those of Sweden, yet they do not
afford even a tolerable iron, if they are not smelted with some of the better sorts of ore, as that
of Utö or Dannemora.   An accurate chemical investigation of the different ores of iron, particularly those of Scandinavia, is still wanting.   The chemist who shall execute such a work will
render an important service to mineralogy.—*J.*

‡ It was formerly the opinion of mineralogists, that all the iron ore of Sweden was situated
in veins.   Werner, from inspection of specimens, was of opinion that the ores must occur in
beds, and this idea has been amply confirmed by the observations of Von Buch and Hausmann.
These beds of iron-stone are extremely interesting, on account of the great variety of minerals
which they contain : nearly the whole of the remarkable mineral substances, brought of late
years from Scandinavia, are production of iron mines.—*J.*

alc, in green coloured folia, which frequently lie on the iron-stone itself, and not unfrequently a foliated green fossil, which might easily be taken for foliated actynolite ; but we distinctly perceive a double cleavage, and the angle of the cleavage is easily determined to be one hundred and twenty-eight degrees ; an angle which corresponds with that of the tremolite. It melts before the blow pipe to an enamel, and becomes white, like tremolite. Both the planes of the fracture are shining.   The cross fracture on the other hand has nothing of foliated, but is coarse and small splintery.   The fragments are never rhomboidal, although such a form might have been expected from the foliated fracture, but are always splintery or fibrous, even when we apply the file to the mineral.   In this manner, the disposition to the fibrous structure becomes evident, though it was in no wise betrayed by the exterior. The fossil is semi-hard, and emits a few sparks with steel.   It is always leek-green, and to be found in pieces and folia of the size of half a hand.   It is *foliated Tremolite.**

This very remarkable and altogether foreign fossil, to appearance, is also not unfrequent in Gellivara in Luleo Lapmark.   White tremolite in promiscuous radii also frequently accompanies the green masses, and crystals of calcareous spar in druses generally repose above it.   The six-sided prism accuminates with three planes.   Similar dodecahedral crystals of calcareous spar, of the size of a hand, are also to be had at Junos Suvando.   But white fine granular limestone, though it is generally the true companion of the fibrous tremolite, is not to be found in the mines of Junos Suvando.

The furnace of Torneo-fors was generally supplied with iron-stone from Luossovara and Svappavara : both beds of a huge thickness ; that of the former being known to the thickness of thirty-four fathoms, and of the latter to thirty-eight fathoms.   But this is still exceeded by the iron hill of Kiru-navara, about eleven English miles to the west of Jukasjerfivi ; for the breadth of the pure ore has been here seen to the extent of eight hundred

---

* This particular variety of tremolite accompanies the magnetic iron-stone in different parts of Lapland, but it has not been observed in the iron mines of the southern parts of Sweden : there its place is entirely occupied by the white fibrous and radiated tremolite, which is contained in a white primitive lime-stone, situated in the gneiss, which includes the beds of iron-stone.

The green colour, and geognostic situation of this foliated tremolite, incline me rather to consider it as a hornblende or actynolite.—*J.*

Paris feet. All these treasures are however at present turned to no account, and must remain so; for the ore cannot afford such a distant land carriage as forty-six English miles with rein-deer, and in small Laplandish pulkers; and it is impossible to erect the furnaces nearer to the ores, on account of the want of wood. It is reserved for posterity to derive advantage from the immense quantity of iron-stone which nature has deposited in Lapland.

Besides its iron mines, Junos Suvando is also distinguished by a remarkable physical curiosity, which is perhaps the only thing of the kind in Europe. Several miles below the furnace, before we come to Torneo-fors, the Tärando-elv, a considerable stream, issues out of the great river of Torneo, and runs for thirty, or thirty-five English miles, in various windings, through marshy plains, and then throws itself into the Calix-elf, which brings it into a quite different place in the sea from that which nature at first appears to have destined to it in conjunction with the Torneo river. All the land between these rivers, for upwards of ninety English miles downwards, is a real river-island, every where surrounded by water. It is the same phenomenon as that of the issuing of the Cassiquiare from the Orinoco, to enter with the Rio Negro into the Amazons river. This is so singular a circumstance, that it was long doubted, till at length the researches of M. Grape, the clergyman in Neder-Calix, and the excellent maps of Baron Hermelin, have placed it in a full and satisfactory light.

With feelings of gratitude I left the works of Kängis, in the afternoon of the eighteenth of September. We proceeded in a boat down the roaring fall, beneath the works, where the foaming waves again dashed over the boat and the crew. We then proceeded quietly through the wooded straits beneath the saw-mill as far as Kardis, on the now united streams of Muonio and Torneo. The boat was heavily laden with unwrought iron, sent down from the furnace above Kängis to the second furnace at Svanstein. But the river carried us rapidly along, and the same evening, in the dark, we reached the classic ground of Pello, famous from two measurements of a degree there.

I was in the very cottage occupied by the French academicians, and even in the same room in which they remained and carried on their observations. Every building was yet standing, as Outhier has most accurately described them in the account of his journey: the dwelling-house, the stables, the *Staboe* or store-house, the stranger's house, still remained in the same place, and even the small stair before the room of the strangers seemed the same. At Kittisvara, close at hand, a signal post was standing, as if the astronomers had but just quitted this country. This was actually the case. For the Swedish *Savans*, with *Svanberg* at their head, had but finished their laborious operations here within the last two years. They had carried their triangles far above Kittisvara, the fixed point of the French mathematicians, and had even extended them to beyond Kängis, a length which exceeds by many leagues the measurement of the French. They even carried their observations beyond Torneo into the Bothnian Gulf, to the last solitary situated island of Malörn. They were there enabled to find with better instruments than the advancement of that period would allow the Frenchmen, and with greater accuracy of observation, that the degree which cuts the polar circle contains 57,198.7 French toises, whereas Maupertuis and his companions imagined it to contain 57,405 toises. If the labours of Maupertuis first established the idea of the spheroidal form of the earth, the figure of the north, however, was first made known by Svanberg, and through him theory has obtained a new triumph, having ascertained the degree much more accurately by calculation than was done by the earlier attempts of the French by actual measurement.

Kittisvara is the highest, and almost the only mountain of this region. It is but a hill, however, scarcely three hundred feet in heighth, and covered with wood to the very top. Neither does it appear rocky. On the other hand the strata of the rocks appear in the middle of the river, and the water here forms again several foaming though neither very high nor very dangerous falls. It was still *dark-red* coarse granular *Granite* as at Kängis. Can it be uninterruptedly continuous from that place to this ?*

On the morning of the nineteenth I set out by break of day, and proceeded down to Svanstein. The country became now highly striking; the banks on both sides were extremely rich. The rows of houses continue almost without interruption. Inclosed corn-fields lie all around towards the thick wood, and the islands in the river were every where covered with large and rich hay

---

* It probably is, and so says Baron Hermelin expressly, *Mineral Historia*, p. 69. Here and there iron-stone is found interspersed in the granite, or in veins, as at the Rotirovaberg, about seven English miles to the south of Kängis, but the thick beds seem to be in a particular degree peculiar to the white gneiss.

cocks. Here they are no longer under the necessity of recurring in winter to the bark of trees, lees of train oil, sea-weed, fish heads or bones. These cottages would shame those of the best districts of Germany. In a few hours the two high and wooded mountains of Kynsivara and Pullingi raised their heads on the right bank. We proceeded in that direction, landed between them, and saw the buildings of the iron works of Svanstein, in the valley along the rivulet, and up the declivity.

Pullingi was also a station where the degree was measured, and the highest of all; for according to the account of Baron Hermelin, the mountain rises eight hundred and two Paris feet, or eight hundred and fifty-four English feet above the river; and if we allow the river at Svanstein to be one hundred and thirty-seven English feet higher than its mouth, the whole heighth of Pullingi above the sea will amount to nine hundred and ninety-one English feet. Most of the other mountains where the degree was measured scarcely attain the half of this heighth. Kynsivara even, which lies quite near, and opposite to it, is lower, being only five hundred and seven English feet above the river, and six hundred and forty-six English feet above the sea.

I ascended Pullingi through woods of spruce and Scotch fir to the top. There was a signal-post above, and I saw a great way round. It is a singular prospect. We follow the river in its course, both upwards and downwards, for a great way; and it is every where lined with crowded habitations along its banks. Opposite, I had a lively and clear view of the large village of Tuttila. A small way, however, from the river into the country, all ap peared a huge and boundless forest, interrupted by nothing but the empty space, occupied here and there by small lakes, and the small blue mountains on their banks. The tract of the river is alone inhabited and animated; the remainder is dreary and dead. This view considerably lowered my idea of the great cultivation of these regions, and in rather a gloomy state of mind I descended to the works.

Not far from the house I saw the garden of the inspector, and in it, for the first time, peas, which were ripe and pulled. Even in Kängis they would hardly succeed, and in Alten they would not succeed. The yellow turnips were very large, the potatoes already dug up, and horse radishes stood in large bushes around the garden fence. In this climate, therefore, garden stuffs again succeed; and for the comforts of life, they are no longer depen- dent on other places.

On the declivity of Pullingi there was no longer any of the red coarse granular granite, the *Rapa Kivi*. Very well characterised strata of white gneiss abounding in felspar made their appearance, and continued without change almost to the summit, with nearly twenty degrees inclination towards the south-east. On the height the stone became again large granular, the felspar red, the quartz blueish-white, and the mica contracted itself into small folia. It appeared granite again; but not so definite as below at the waterfalls of Pello, and the strata were also few in number. It may be a faint repetition of the granite lying below, as it appears beneath the church of Pajal at Kängis. This appearance of the gneiss on the Pullingi is, however, very remarkable; for it shews that the red granite, the rapakivi, forms every where the fundamental rock here, and consequently is a peculiar formation, and not merely a change of the gneiss, and that this fundamental formation cannot rise beyond a moderate heighth.

The boat had unloaded the unwrought iron in Svanstein, and taken instead of it a heavy lading of wrought iron, which was to be conveyed to Torneo. We rowed slowly down the proud stream. In the evening Matareng appeared in view. The church of Oefwer or Poiki Torneo, here towered above a thick mass of houses; a large village on a peninsula which projects from the mountains a great way into the river. The sound of bells was borne over the water to us. The still air conveyed the sound tremulously to the woods along the banks, from whence it was re-echoed to the water. The sacred solemnity of these sounds, which had not been heard for so many months, produced a strong and powerful impression. We proceeded along, and heard the bells for a long time, till they were at last lost in the distance, and in the hollows of Avasaxa. A number of salmon weirs narrowed the river at this place; and a narrow opening alone remained for the boat, frequently not broader than the boat itself, in which the water pressed through with great force. These weirs, however, do not obstruct the salmon: notwithstanding their number, and the number of salmon caught in them, and notwithstanding what is killed at night by the trident, they still proceed in great bodies as far as Enontekis, and even as far as Raunula on the Norwegian frontiers. And yet those who live high up the Tana Elv in Finmark are in dread of a weir that does not take in more than the half of the river, which they consider as an insurmountable obstacle to the farther progress of the salmon.

It was dark as we reached Niemis, where we missed very much the clean and comfortable inn of Pello.

From Matareng, or from Oefwer Torneo, there is an excellent road down the river side. But my boatmen proceeded with me seven English miles farther to Wetsaniemi, where the passage by water is almost altogether obstructed by new waterfalls, the greatest in the whole passage from Pello to Torneo. A car and a horse, which were bespoke from Pakkila, were there in waiting for me; for from the commencement of the road at Oefwer Torneo there are everywhere post stations, as throughout all Sweden. The roads were admirable, and like those in the south of Sweden. The young Finlander proceeded down the excellent road with furious rapidity. Such regulations of cultivated society, such admirable weather, and such rapid driving, were a pleasure to meet with. This is a rich country. Where is there any thing in Norway, or even in the south of Sweden, equal in point of cultivation? Houses closely following on houses, corn-fields, meadows, and extensive prospects over innumerable villages down the river! I reached Korpikylä at mid-day: the horse and car were quickly changed, and trees, houses, and fields, flew rapidly by. The road was wholly covered with people: they came from the church; youths and maidens hurrying gaily along, and the old people as venerable as Armenian priests, dressed in a long black *talar*, or overall, buttoned from the throat downwards, a sulphur-yellow Swedish scarf round the body, and a small black bonnet on the head. A singular dress for a peasant.* We imagine we see so many ghosts. They are, however, clean-looking, rich, and prosperous people. About two o'clock I passed the beautiful church of Charles Gustavus, surrounded by large farm houses, and at Frankilä, a short distance from it, I again changed horse and car. They were not in readiness, and yet I no where waited a quarter of an hour, notwithstanding I was never fortunate enough to make myself understood to the people, who speak only Finnish, and not a single word of Swedish. In an hour and a half I reached Kockos, and shortly afterwards the large villages of Wojakkala and Kukkola, and about five o'clock in the evening I pro-

---

* Baron Hermelin has given a drawing of it in Retningar til en Bescrifving öfwer Suerige.— Stockholm, 1804.

ceeded round the Peninsula surrounded by the river on which Torneo is situated, and entered the anxiously expected town.

The rapidity of the passage had allowed me very little opportunity of examining the changes of rock on the way; and the advanced season of the year required me to hasten my journey southwards. At Jourengi, where the polar circle runs through the country, the red granite again made its appearance in small rocks, and confined the river. It was more small granular than higher up, and almost stratified like sand-stone.* Gneiss was here out of the question. It was the same at Matareng; the greatest number of blocks, and the largest along the banks, consisted of it, yet white gneiss blocks lay between, probably derived from the heights above Oefwer Torneo : the base was still, however, red granite, with white gneiss immediately above it whenever the banks began somewhat to rise. Hence Avasaxa, which rises opposite Matareng, on the eastern side of the river, to six hundred and twenty-six English feet above the stream, and about seven hundred and twenty-four English feet above the sea, may also be gneiss. Below Matareng the high Luppiovara rises above the road to a heighth of five hundred and thirty-five English feet above the level, or six hundred and thirty-seven English feet above the gulph. It has been described by Baron Hermelin, who has given us several views of its summit (*Mineral Historia öfwer Westerbottn och Lappmarcken, p. 66*). The grey granite (or gneiss) lies on its declivity in completely regular beds floetzwise on one another, and appears like a regular building. The mountain is very steep towards the river and the road, and is only connected on the west side with other heights. In the middle of its height appear several beds with large granular red felspar, almost similar to the summit of Pullingi, which, however, do not belong to the red granite of the base.

Below Korpikylä this red granite was no longer to be seen. It may now be wholly concealed beneath the surface. The large blocks lying around seemed also to be granite with white felspar; but when we examine the stone in small rocks on the low rising grounds over which the road runs, we distinctly recognize the striped and slatey nature of the gneiss.

---

* This example of stratified granite deserves to be particularly noticed, as several naturalists are of opinion that granite is always unstratified.—*J.*

Nivavara, on the other side of the river, one of the stations where the degree was measured, is still more distinctly gneiss. The stone seems slatey quartz, with a very little reddish white felspar, and not unfrequently with red garnets. The strata of the mountain stretch from north-west to south-east, and dip slightly towards the west.—(Hermelin).

I was much astonished, however, when I saw besides at Korpikylä, and much farther down, not unfrequently large blocks of black greywackish looking glimmering *clay-slate*; a principal rock of the transition formation. In the country from Torneo upwards it was not so easily to be looked for. But about four English miles above Torneo, at the waterfall of Jülhä, between Vojakkala and Kukkola, black compact limestone of this formation even appears above the ground *(anstehend)*, and not far off, at Liakala, it is even burnt and made use of. How strange is it to see these rocks so deep in the gulph between primitive rocks which so closely surround them on all sides. Korpikylä is scarcely eighty feet above Torneo, and more than this heighth the transition rocks have not been able to reach. The Torneo river seems also to have interposed a boundary to their extension; for on the west side of the stream I could never find any trace of them in the neighbourhood of the town. But the blackish grey clay-slate appears above the surface eastwards, at the church of Nieder-Torneo, almost at the edge of the river. It shows itself then several times in a farther continuation eastwards, and appears above ground also at Kemi, but is no longer seen at Uleo, on the eastern shore of the Bothnian gulph. It will also hardly appear in a continuous state on the northern shores of the gulph. Is it not quite clear from these particulars that the causes of the origin of the transition formation have operated upwards from the south, and have been but faintly felt in these regions? and that the clay-slate and black limestone are as it were but fragments from the strata which have so often deposited themselves in small receptacles among the gneiss rocks on both sides of the Bothnian gulph; at Abo and Biörneburg; at Oeland and Gothland, and on the extreme islands of Geffle?

How much these regions have changed since we became somewhat better acquainted with them, by the French measurement of a degree! Many villages along the Torneo river, now large and full of animation, were not then in existence, and a map of the river drawn up at that time exhibits wastes, woods, and deserts, where the country is now diversified with carm houses and cultivated fields. Scarcely a country in Europe can be named which has made such rapid and astonishing progress in cultivation and population as this very northern part of Sweden. There is an account of Lapmark by Count Douglas, in 1696, according to which there was then only three families (Nybyggare) settled in all Lapmark. Almost as many Laplanders then lived in the country as at present. In the year 1766, there were found three hundred and thirty fixed habitations in Lapland, containing one thousand six hundred and fifty souls, and nine hundred and ninety-four families of Laplanders, containing four thousand and forty souls (Tuneld). In the year 1799 there were six thousand and forty-nine settlers and five thousand one hundred and thirteen Laplanders. Such was the progress in these waste districts in the course of thirty years. In the last year this spirit of cultivation and the consequent increase of population have gone on at a still greater rate, and we might entertain the hope of soon seeing here the maximum of inhabitants which the polar lands are capable of supporting, if destructive wars had not destroyed the rising shoot of cultivation and advancement throughout the whole nation, and perhaps extirpated it for centuries.

According to the very remarkable tables which Baron Hermelin gave to the world at Stockholm, in 1805, respecting Westerbottnslän, and which were extracted from the government registers, there were living in all Lapmark

	1751.	1772.	1801.
Jordbrukare (cultivators of the ground)- - - -	25,842	30,807	51,997
Bruksfolk (connected with the mines) - - -	180	246	1,108

In the whole province from 1795 to 1800, the new settlers amounted to three thousand five hundred and seventy-nine souls, and the emigrants to four hundred and forty-five.

These are true conquests! The Swedish kingdom has acquired new territories in its interior; for the new settlers were not criminals, exiles, refugees, or wretches weary of life. They were for the most part active and industrious Finlanders from Cajaneburg and farther southwards, who did not roam about the country like the Laplanders, but gained new productions from the soil for themselves and the whole human race. About thirty years ago, when the different parts of Lapland were enumerated, Kusamo Lapmark, between Uleo and the White Sea, always appeared in the list; but for a long time no Laplanders have been known there, and the Finns inhabit in cottages throughout the whole country as high as the uppermost water dividers. A hundred years ago there was not a single Finn in Soldankylä or in Kemi-Lapmark. The Laplanders every where roamed through the woods and waste morasses. In the year 1738, twenty-three Finnish families had settled on the rivers. In 1755, there were already seven hundred and sixty souls in the parish; in the year 1800, the population amounted to one thousand six hundred and seventeen; and in 1802 it amounted to one thousand seven hundred and eighty-six. Nothing but Finns. The Laplanders have now wholly abandoned Sodankylä*. From what parish in Lapland could we not cite similar particulars? The Finns ascend towards Lapmark, because they find their account in the new settlements which they found, and not because they are driven from their paternal hearths by misery and want; for these northern regions teach the singular and remarkable truth, however little it may often be attended to, *that emigration does not depopulate a country.* Cajaneburg and Oesterbottn, from which the inhabitants of Kusamo and Sodankylä, and most of the places above Torneo, have issued, far from declining in population, have increased almost in an equal degree with the newly cultivated districts; for it was not the fathers of families who emigrated, but the sons, who, probably, would not have founded new families in their mother country, and consequently would not have increased the population and general industry. It is on this account that the population increases with such astonishing rapidity in all countries which have been newly settled with any degree of success: there is opportunity and room for new families. But when, by emigration, cottages remain empty, fields remain uncultivated, and the neighbour lays hold of them to increase his possession as at Salzburg, and in 1688, in France, then it may be bitterly felt by the country whose prosperity receives from it such a powerful shock.

In all countries which increase by new cultivation, the number of births to the deaths is large beyond almost all proportion, when compared with other countries, where cultivation and population are at their level. This is not because in the former the life of man is exposed to less danger, because innocence and ignorance of destructive luxuries shorten life, or because vice and enmity are enemies to health; but because every where new families spring up where there were none before, by which the number of births is increased. Life is neither longer in these regions, nor are the families more numerous than in other provinces of the country, except

---

when particular causes concur to increase the mortality. According to the laborious calculations of Henry Nicander* from 1795 to 1801 the proportion of deaths to the births was:

In *Kuopiolän* as 100 : 198.—The new town of Kuopio rapidly increased in population.

In *Uleoborglän* as 100 : 172.—Kusamo and Kemi-Lapmark belong to this.

In *Wasa* as 100 : 166.

In the whole of *Westerbothnia* and Hernösandslän as 100 : 149.

In Upsalalän as 100 : 128.

In *Linkiöping* and in *Oerebrolän* as 100 : 119.—Eastgothland and Nerike.

In *Skaraborg* as 100 : 112.—West Gothland.

In the whole kingdom as 100 : 136.

But in Findland alone 100 : 164.

But who would believe that in Kuopio or in Sodankylä and Kusamo, life is either easier, or the people healthier, more uncorrupted, more virtuous, or the marriages more fruitful than in the fertile districts of Westgothland, the vallies of Eastgothland, or the plains of Upsala. But these provinces had attained their level of population. New families and new houses could not arise, because there was no opportunity for their extension and establishment. On the other hand, how powerfully has Finland increased within the few last years ! Every where towns and villages have sprung up. Kuopio, Heinola, and Uleoborg and Wasa, almost as if wholly new. When the country, however, increases in cultivation till it reaches the bounds which nature has set to it, the proportion of deaths and births will rapidly fall and resemble that of similar districts, or the state of almost the whole of the cultivated surface of the earth.

The whole of Westerbothnia has not less increased in population than Finland ; and what Baron Hermelin relates respecting it is deserving of the highest attention.

In Westerbottnlän, to which Umeo, Piteo, Luleo, and Torneo Lapmark also belonged, there lived in the year 1754 a population of thirty-six thousand eight hundred and sixty-nine. From this year forwards the population has been constantly increasing : at first slowly, till 1771.

---

* Nya Vetensk. Acad. Handlingar 1805. II. Quartal.

In 1775　it was　51,821.
　　1790　- - -　59,777.
　　1795　- - -　67,890.
　　1800　- - -　71,872.

Such facts are not calculated, one would think, to induce a persuasion of an increase of the severity of the climate in which so many believe, and they do not strengthen the opinion that the earth will be depopulated from the poles downwards. And can there be a stronger proof of the good government of the country?

The lists in Hermelin's tables respecting the exports from the towns of Westerbothnia to Stockholm afford us a very interesting, distinct, and clear image of the occupations and the moral condition of the inhabitants of Lapland. For almost all the exported productions have come down from Lapland. We see at one view what these polar regions are capable of furnishing, and what the industry of the inhabitants is employed on.

The numbers in these lists are averages from the custom-house books for the years 1793, 1794, 1795. The exports to other places besides Stockholm are not perhaps of any consequence.

Exports from Westerbothnia to Stockholm.

Victuals and Birds.

		Rix-dol.	Sk.
28,000 Lispund Smör (Butter) at 2 Rix-dol.	-	56,000	0
150　-　-　Ost (Cheese) at 1 Rix-dol.	- -	150	0
970　-　-　Tallow at 2 Rix-dol. 8 Sk.	-	2,101	32
380 Tunnor salt Ox eller Nötkött (Salt Beef) at 6 Rix-dol.		2,280	0
246 Lispund tört Kött (Dried or Hung flesh) at 24 Sk.	-	123	0
1,200　-　-　salt och ferskt Renkött (Salt and Fresh Rein-deer flesh) at 1 Rix-dol.	- -	1,200	0
60 Lispund Rentungar (Rein-deer Tongues) at 3 Rix-dol.		180	0
70 Head of Renar (Rein-deer) at 2 Rix-dol.	-	140	0
80,000 Styck Hierpar (Tetrao Bonasia) at 4 Sk.	-- -	2,500	0
2,800　-　Orrar (Tetrao Tetrix) at 8 Sk.	- -	465	32
4000 Styck Tjudrar (Tetrao Urogallus) at 12 Sk.	-	1000	0
1,400　-　Snöripar (Tetrao Lagopus) at 2 Sk.	-- -	58	16
		66,199	32

Fish.

		Rix-dol.	Sk.
2,800	Tunnor Salt Lax at 10 Rix-dol. - -	28,000	0
	mostly from Torneo-elv.		
1,600	Stycken rökta Laxer (Smoked Salmon) at 30 Sk. -	1,066	32
2,700	Stycken rökta Laxoringar at 12 Sk. - - -	675	0
10,000	- - rökta Nejonögon (Liver Lamprey) - -	50	0
200	Tunnor salt Sik (Salmo Lavaretus) a 6 Rix-dol. -	1,200	0
45	Lispund Sikrom at 1 Rix-dol. - -	46	0
400	Tunnor salt Strömming (without reckoning what is exported to Oesterbottn) at 3 Rix-dol. -	1,200	0
600	Tunnor salta Gädder (Pike) and other salt fish, at 6 Rix-dol.	3600	0
2,800	Lispund torra Gädder (dried Pike) at 24 Sk. -	1,400	0
1,000	Lispund diverse torr Fisk at 16 Sk. - -	333	16
86	Casks of Seal Oil, (Skältran) each cask containing 60 Kanne, at 10 Rix-dol. - - -	860	0
28	Tunnor of Seal blubber of 12 Lispund, at 9 Rix-dol. per Tunnor	252	0
	From the Frozen Ocean, by the Laplanders.		
5,000	Lispund Cod or dried Dorsch at 1 Rix-dol, 16 Sk.	6,666	32
8,000	Lispund dried Grosidor eller Said (Gadus Virens) at 40 Sk.	6,666	32
		52,015	16

Hides and Furs.

		Rix-dol.	Sk.
60	St. Bear skins, at 10 Rix-dol. - -	600	0
3	—— Wolf skins, at 3 Rix-dol. - -	9	0
70	—— Fox skins, (Räfskin) at 2 Rix-dol. - -	140	0
24	—— Järf or Glutton skins, at 3 Rix-dol. - -	72	0
80	—— Martin skins, at 2 Rix-dol - -	160	0
70	—— Schwanenbälge, at 24 Sk. -	35	0
20	—— Castor skins, at 3 Rix-dol. - - -	60	0
6,000	—— Hare skins, at 2 Sk. - -	250	0
1,600	Ton. Growerk or Squirrel skins, at 40 Sk. - -	1,325	0
500	St. Ermine skins, at 2 Sk. - - -	20	20
450	—— Seal skins, at 12 Sk. - - -	112	24
4,000	—— Renoxhudar, (male rein-deer skin) at 36 Sk. -	3,000	0
15,000	—— Renko eller Vajhudar, (female rein-deer skin) at 24 Sk.	7,500	0

		Rix-dol.	St.
1,200 St. Various other rein-deer hides, at 24 Sk.	- -	600	0
100 Decken of young rein-deer skin, at 32 Sk.	-	66	32
9,000 Pair of Lapland gloves and Lapland shoes, at 4 Sk.	-	750	0
600 —— Lapland boots, at 16 Sk.	- - -	200	0
300 Styck Lappmuddar och Muddskin, at 5 Rix-dol.	-	1,500	0

Lapland Furs.

		Rix-dol.	St.
300 Dozen skins of the fore-feet of the rein-deer, of which the Lappmuddar are made, at 6 Sk.	- -	37	24
200 Styck Ox and Cow Hide, at 2 Rix-dol.	- -	400	0
8,000 —— Calf-Skins, at 6 Sk.	- - -	1,000	0
2,500 —— Get (Goats) och Risbit samt Killingskin, at 2 Sk.		104	12
300 —— Sheep-skins, at 3 Sk.	- - -	18	36
32 —— Dog-skins, at 8 Sk.	- - -	366	32
10 —— Cat-skins, at 4 Sk.	- - - -	0	40
Different other Skins	- - -	400	0
		24,870	0

Iron to Stockholm.

		Rix-dol.	St.
1,600 Schiffpfund (of $2\frac{1}{2}$ cwt.) wrought Iron (Stongjarn), at 7 Rix-dol.	- - -	11,200	0
126 Schiffpfund Knippjärn, at 9 Rix-dol.	-	1,134	0
43,400 Dozen of various sorts of Deals, at 1 Rix-dol.	-	43,400	0
2,700 Lasts, at 12 Ton of Tea, at 24 Rix-dol.	-	13,500	0
300 Tunnor Tjäruperna eller Tjäruvatten Theer wasser, at 1 Rix-dol.	- - -	300	0
250 Lispund unwrought Rein-deer Horn, at 8 Sk.	-	41	32
950 Lispund Draglim (Glue) at 2 Rix-dol.	-	1,900	0
300 ———— Hornlim			
35 ———— Fnöske (Spunge and Touch-wood, at 2 Rix-dol.		70	0
480 Pound of made Akerbär (Rubus arcticus) at 24 Sk.		215	0
45 Tunnor Vattenlingon (Tyttlebär-Vaccinium Vitis idaea, Whortleberry) at 1 Rix-dol.	- -	45	0
300 Kannen Hiortron (Mountain-bramble, Rubus Chamae-morus) at 12 Sk.	- - -	72	24

		Rix-dol.	Sk.
500	Head of Fir Stools and several *Mollen*, at 8 Sk. -	83	16
700	Lispund Land and Sea Bird Feathers, not torn, at 3 Rix-dol.	5,100	0
10	Tunnor Rein-deer Hair - -	40	0
	Glass from the Glass Works in Westerbothnia -	600	0
	Amount of exports from Westerbothnia -	221,226	24
	Amount of imports - - -	145,862	16
	Among the imported goods		
	Amount of the value of Grain - -	31,285	0

Under which 2,750 Tunnor Rye

29,80 - - Barley

2,700 Lispund Wheaten Flour

Of other necessary articles there were imported

5000 Tunnor Salt, at 4 cwt. the Tunnor

Tobacco for 24,150 Rix-dol. viz.

18,000 Pound of unground Tobacco

9,000 Pound of Snuff.

43,000 Pound of Sugar.

26,960 Pound of Coffee, and

150 Ankers of Wine.

Torneo does not satisfy the expectation which we are led to form of a town, from which, as a central point, the trade with all the polar regions as far as the Frozen Ocean is carried on ; and it disappoints us the more, as it is so much connected with the rich and cultivated country down the river. There are a number of streets, it is true; they are generally straight, and intersect one another at right angles. A plan of the town would exhibit it as large and regular. But these streets are not paved ; they have quite the look of so many fields or meadows, they are so grown over with grass. We see few traces of footsteps in them, and carriages only go through the lower principal street. The upper streets are actually barred. The inhabitants have the exclusive right of feeding their own cows on them, and perhaps also of making hay ; and hence it not unfrequently happens that when the people of the more southern parts of Sweden wish to divest themselves at the expence of Torneo, they assert that the market-place and the hay made from it belong to the regular revenues of the mayor *(burger meister)*. However, the road from Stockholm to the Bothnian Gulph goes through the market-place, and the streets which run from it towards the river-side. In this quarter, therefore, every thing looks better and more respectable. We see several houses two stories high, painted and ornamented; and at the river-side the house of M. Carlenius, the merchant, has a most distinguished appearance. All the other houses are merely low huts, and they are situated at a distance from one another; for every hut has generally its little garden beside it, containing flowers, garden stuffs and herbs, and beautiful mountain-ash trees. The life and stir is confined to the landing-place on the banks of the river, and it is soon lost in the empty streets. The country round is altogether flat, and almost Dutch. Three windmills with the church cover the almost imperceptible hill before the town; and in the adjoining church-yard wall there is probably the same wooden steeple on which Charles XI. once saw the sun at midnight, and from which Manpertuis and his associates began to calculate their degree. The sea is not visible, and the stir in the harbour is not at present very great; I saw only two or three yachts. But the prospect down the river is rather beautiful; for the church of Nieder Torneo, on an island on the other side of the Elv, has a beautiful appearance. It is a shining stone building in the shape of a cross, and the steeple a high cupola rising from the middle with lanterns on it. There is something unexpected and singularly attractive in this view.

This town, as well as all the towns in Westerbothnia, and a great number in Finland, was founded by Gustavus Adolphus in 1620. There was a church, however, and consequently dwellings around it, from the year 1350; for in this year the church was consecrated by the Archbishop of Upsala*. It was probably about the time of the first arrival of the Swedes so far north-wards. The cultivation of the country seems to have long languished in their hands. The Finns got the start of them, and proceeded with such activity, that at present there is not a single Swedish peasant along the whole course of the Torneo river. All are Finns, and the Swedes are confined to the town. In the town, however, according to Tuneld, there are only six hundred and thirty-two inhabitants. This seems a surprisingly small number, when we consider the extensive influence of Torneo, and even when we wander through the numerous though deserted streets. The people

---

* Schöning gamle Geographie, p. 101.

seem here, however, to enjoy life comfortably in their own way. They frequently meet together —in the morning at the apothecaries in the market-place, at midday in the public-houses, in the afternoon again at the liquor flasks of the apothecary, and in the evening the punch flows in streams in the coffee-house.   There is no want there of frequent sallies of joviality and animation.   They are beings who are very faintly moved by the convulsions of the world.   They penetrate but seldom to them *.   This mode of living has long made Torneo a by-word in neighbouring places.   The four towns of Westerbothnia are thus characterised in a Swedish doggrel rhime : *Umeo* the *fine*; *Piteo* the needle making ; in *Luleo* nothing is done ; but *in Torneo they get drunk (I. Torneo drikkes skaaler)*.   Truly the Torneensians still strongly justify the by-word.

The greatest part of September was now past.   I was looking hourly for the snow, and yet I wished to flee from it before the earth should be permanently covered.   But, like many others, I had rated Torneo's climate too low.   They were still living here in a pleasant autumn, and not in winter. The air was calm, clear, and still.   It freezed in the night, though but little, and the first forenoon rays of the sun soon annihilated the ice.   The sun gave out a gentle heat at midday, and with pleasure and enjoyment I went about the country in this temperature.   The thermometer generally rose to 50° F. and sunk in the afternoon very slowly.   The trees were yet in full glory, and no where had lost any of their leaves.   A firm snow tract was not expected here till the end of October, and it is seldom earlier.   September is in Torneo what October is in the north of Germany ; and the polar region does not vindicate her violated sovereignty till the end of November.

---

* Yet Torneo is now a Russian town !

CHAPTER IX.

JOURNEY FROM TORNEO TO CHRISTIANIA.

Woods on leaving Torneo.—End of the Finn Population.—M. Grape, a Clergyman.—Calix-Elf.—Baron Hermelin's Undertakings in Luleo Lapmark.—Raneo.—Luleo.—Piteo.—Splendid Church of Skelefteo.—Decrease of the Surface of the Sea in the Bothnian Gulph.—Wahlenberg.—Umeo, its Church.—Gneiss in Westerbothnia.—Angermann-land.—Linen.—Skulaberg.—Sundsvall.—Changes of the Gneiss.—Helsingeland.—Gestrikeland.—Gefle.—Course of the Limit of the Scotch Firs over the Surface of Europe.—Dal-Elf.—Upsala.—Observations of the Temperature.—Stockholm.—View of the Town.—Collection of Minerals belonging to the College of Miners.—Departure from Stockholm.—Söder.—Telje Canal.—Orebro.—West Gothland.—Ruins of Udewalla.—Storm at Swinesund.—Frederickshall.—Return to Christiania.

On the 21st of September the car was transported in a boat over the sea bay, which surrounds Torneo almost like an island, and which is entered by a considerable arm of the Torneo river, when it rises very much. The horse was in waiting for me at the other shore; and I now proceeded rapidly through the woods towards Nikkala. The great population all the way along the river now disappeared. Here and there stood a solitary cottage. The country was covered with thick woods of Scotch and spruce firs and birches. Whoever believes, what has more than once been printed, that the cold prevents trees from growing at Torneo, let him come and see these woods. The birches before Sangits were higher, larger, and more beautiful than any I had ever before seen. They have evidently found here the most advantageous climate for their growth. They rise a great way over the tops of the spruce and Scotch firs. It is a magnificent tree. We several times left the forest between Sirvits and Nikkala, to cross two sea-bays which interrupted the road, not in boats, as was done in 1736 by the French mathematicians, for that was now rendered impossible. Large and beautiful bridges were built across the whole of the breadth. The sea-bays have become marshes from the continual decrease of the Bothnian Gulph;

and we may soon expect to see fields and cottages on the surface now occupied by that gulph. The decrease here will admit of no doubt. Such sea-bays are not filled with mud by the small streamlets which here empty themselves; and the open sea furnishes mud in no such abundance. The shores also, and the rocks rise *above* the level of the water; and are not buried in any mud and earth brought down by the streams.

At Landjerf, twenty-three English miles from Torneo, the long Armenian dress, with the yellow Swedish scarf, worn by the peasantry, disappeared. The Finnish nation and the Finnish language were now left behind us, and Swedes everywhere made their appearance. In the evening I reached Neder-Calix, and was enabled to have a few instructive moments of conversation with the agreeable and intelligent clergyman, Erick Grape, to whom we owe the excellent description of the pastorate of Enontekis, in Lapmark, in the Transactions of the Stockholm Academy of Sciences, and who surveyed a great part of Torneo Lapmark, with the most minute detail of rivers, streams, and lakes, and drew up maps of it. They are the ground-work of the maps of this country by Hermelin, and of that which accompanies the Travels of Skjöldebrand; and I also have made use of a drawing, with which, in the kindest manner, I was favoured by M. Grape.

After crossing the great Calix-Elf, one of the most considerable rivers of Lapland, I passed the night in the excellent inn of Grötness, where none of the comforts are wanting which we might expect in the first inn of the country; not even bread, which we so seldom find in country-places in either Norway or Sweden; for the cakes, full of poppy of the Norwegians, the Fladbröd, and the somewhat thicker bread of the Swedes, the Knak-kebröd, both of which are baked for months and years beforehand, can only be considered bread in these countries.

———

In the woods from Torneo to this place there were no hills to be seen, and no rocks appearing above the surface. Small granular granite, with white felspar, not gneiss, made its appearance on the banks of the Calix-Elf. I did not, however, follow it any great way; but nearer to the sea there are several interesting geognostic combinations which are cited by Baron Hermelin. The transition formation again commences on the shore, and forms several hills; but even here it does not go farther up the country than at Torneo. At Storöe, a peninsula below Neder-Calix, blackish grey

limestone forms beds of from three to four fathoms in thickness : not far removed from " syenite, which is composed of hornblende, felspar, and quartz,
and blackish grey clay-slate" (this syenite is here consequently in all probability also a species of transition formation). The strata stretch from
north-west to south-east, and dip towards the north-east.

Not far from thence, towards the sea, on Hastaskäret, appears a black and
dark grey *compact limestone*, calcareous slate, and blackish grey *clay-slate*,
following one another in such thin and straight slates that they might be
conveniently broken into roof-slates, as experience has proved, if the custom
of the country were not averse from the use of roof-slate. On Lutskäret,
and other smaller islands, the compact limestone is burnt for lime.—Hermelin *Mineral Historia*, p. 63.

Up the Calix-Elf we again see villages and cottages in abundance, but
not so many as above Torneo. Here also there is a considerable salmon
fishery. I proceeded up the river to Morsby, and then passed the furnaces
of Töreo. The buildings were new, and partly, not altogether, restored
from the 'effects of a flood, which had completely destroyed the whole
works. These works owe their existence to the indescribable activity of
Baron Hermelin, who has laid out princely sums in this country in improvements. He has settled colonies throughout all Luleo Lapmark, endeavoured
to establish mines and iron-works at Gellivara and Quickjock, and opened
a number of new channels of industry. We are astonished, on looking at
Robsahm's map of Luleo Lapmark, to see the number of new undertakings
which have been commenced within these few years. Baron Hermelin has
gained an entire province to the Swedish kingdom in its interior.

The iron of Gellivara was to be smelted in the furnaces down at the sea ;
and it was to be conveyed by a difficult way, partly down rivers, partly by
land over snow, and finally down the Luleo-Elf to the furnace of Selett,
to Welderskin and Strömsund, beneath Raneo. Töreofors was built by
Baron Hermelin to work the smelted ore of Strömsund ; but since the
inundation effected such devastations in the new works, he sold them to
three individuals who live quite near at hand, and who can therefore have
an immediate eye upon the new buildings.

But, alas ! the iron of Gellivara also cannot be purely melted. It is richer,
it is true, than the ore of the southern districts ; but it is so difficult to

melt that it can hardly be mastered; and after all it yields but brittle iron. To improve it, whole ship-loads of ore are brought from the mines of Utö, at Stockholm, up to Strömsund; but this carriage is very expensive. At Utö, in 1807, the *schiffpfund* (of $2\frac{1}{2}$ cwt.) of ore cost twenty-one skillings on the spot. Including lading and freight, the price of this quantity is increased at Strömsund to fifty-two skillings, or one rix-dollar four skillings. In what a different condition would these regions be, if nature had thought fit, with the internal wealth and great extent of the Lapland iron-ore, to combine also internal excellence.

The country of Töreo was sandy and flat; yet the farm-houses were large, respectable, and well-built; and their horses were not bad. The peasants themselves appeared to have attained a degree of cultivation by no means common. Those who conducted me inquired carefully, and by no means unintelligently, respecting the political condition of the world; and they did not reason amiss on the subject of the carrying off of the Danish fleet by the English, which had then just taken place. I could almost have imagined that the peasants had a voice in the Swedish national assembly.

At midday I reached Raneo, where a sort of town is formed round the church by several hundred barracks, through whose dead and forsaken streets we are obliged to pass. The wooden windows were fast shut; nothing living was any where to be seen. The extensive and widely-spread congregation (frequently fifty-five English miles distant) comes here to church, and in winter, when unoccupied, remains for several days. They are assembled here in the church, if they have no other point of union; and they feel in their barracks that they belong to a community.

The Raneo-Elf, which is also one of the considerable rivers which descend from the interior of Lapland, is at this place five hundred and seventy-seven English feet in breadth; on the farther bank the country is hilly, woody, and commands several views of the lakes and deep sea fiords; as, for example, at Pehrsö, where, in like manner, the sea is visibly drying up. No distinguished object appears till we come to *Gammelstad* (the old town) Luleo, where I remained for the night, seventy-five English miles from Torneo. The roads had hitherto continued to run from east to west, along the northern shore of the Bothnian Gulph. Hence Hvito, between four and five English miles above Raneo, lay somewhat farther north than Raneo;

but there the road suddenly turns, following the direction of the sea-shore due south. The roads are every where most excellent.

The old town of Luleo was, like Torneo, built by Gustavus Adolphus; but, by the desertion of the sea, it soon was transformed, as is said, from a sea-port to an inland town; and, on account of the shipping, it was obliged to be moved five miles nearer to the sea. It is certain that many places are here now dry which were formerly covered with water. Baron Hermelin relates, that, at the clergyman's house, where there are now cultivated fields and meadows, there was formerly no passage but in boats. The new town of Luleo is now wholly removed from the road.

The road still proceeds for more than two English miles down the river, when we cross it at a ferry, where it is at least one thousand and nine hundred Paris feet in breadth. The day was beautiful. I hastened on, and saw, about midday, the plain of Piteo stretched out before me. The lofty church of Oyeby rose high above the wooden houses around. This was only the old town. A wood of Scotch fir leads to the new Piteo, which lies between three and four English miles nearer to the sea, surrounded by water, by hills and rocks; and the appearance of which is here highly delightful, where we so seldom see houses collected together. It is, however, when we enter it, a dreary town. The log huts have not even the usual clothing which such houses in Torneo generally have; the streets are not paved; and there is scarcely a spot we can look upon which does not betray signs of poverty. We proceeded rapidly through, and did not stop. A long bridge leads us to the island of Pitholm, on which the road now runs till near the mouth of the Piteo stream. The river is here narrowly confined, and lightens the passage over, which is in truth not great (six-hundred and forty-three English feet), and not a third part of that over the Luleo-Elf; but to make up for this we were here powerfully assailed by the rushing in of the waters from the Bothnian Gulph, at which the horses were not a little frightened. I remained for the night on Kinbäck, eighteen English miles beyond Piteo.

What variety can there be along the sea-coast of this country? Flat districts and woods, with here and there a pleasant and rapidly disappearing view of the sea; a rushing stream from the Lapland mountains; cottages along the banks, which prevent the salmon from proceeding up with impunity; and then woods upon woods without intermission: such is the road through Westerbothnia all the way to Umeo. We grow accustomed to it, and except, especially from the rapidity with which this country is generally travelled through, that we shall never again see any thing attractive. In my way from Kinbäck on the twenty-eighth I was not particularly struck with any prospect during the whole morning. The road brought us at midday to several small hills, Here the woods opened; we issued out of them, and saw the extensive plain of Skelefteo, and the large stream which winds through it; and the church of Skelefteo rose in the middle of the plain like a temple of Palmyra in the desert. This is the largest and most beautiful building in the north. What a prospect! What an impression here in a latitude of 64°. on the borders of Lapland! A large quadrangle; and on each side eight Doric pillars, which support an attica. In the middle there is a large cupola, upheld by Ionic pillars ; and a clock and lanterns above it. Why? By what means? By what accident came a Grecian temple into this remote region ? I asked the peasants who built it, and when it was built ? And they answered with no small degree of self-complacency: *We* built it, the congregation of *Almuen*. It cost us indeed great trouble and labour, seven long years, and an outlay of large sums of money....But who gave you the plan and the spirit to undertake it ? That I could not learn. What are we to think of a congregation capable of erecting such a building ? I learned for the first time in Stockholm, that the Academy of Architecture in Stockholm draws the plans and sketches of every public building in the whole kingdom, and overlooks their execution on the spot. This is noble and grand. We need only the experience of Skelefteo to be convinced of this.

Hogström, well known from his description of Lapland, about forty years ago was clergyman here. Well acquainted with the climate of Gellivara, where he had formerly been clergyman, it appeared to him that the temperature of Skelefteo was capable of yielding more than had hitherto been expected from it.   He laid out two fruit gardens here in the latitude of Näröe and Helgeland in Norway, and as high as Archangel. According to Tuneld, he raised apple, pear, cherry, and plum trees from the kernels, and brought them to bear. This must have been a very ephemeral phenomenon. It is uniformly denied on the spot, and really with some show of reason. At least, at present, as the clergyman's house has been removed to another place, every trace of these fruit gardens has disappeared. The high and richly leaved aspens on the height on which the church is situated, the beautiful alders *(Alnus incana)* along the river, may well deceive us into a belief that the climate of Skelefteo is capable of producing more than it actually can do.

About four or five English miles farther on, I came to Innerviken, situated in a small sea-bay. A few years ago, this place was past in boats; but it is now so dry, that the road might be carried along it, and the inhabitants, who are daily witnesses of the decrease, imagine they shall live to see the surface of this arm of the sea converted into fields and meadows. There is hardly a small spot here which does not bear testimony to this decrease, and to throw out doubts in opposition to the belief of the inhabitants along the whole of the gulph would only excite their laughter. It is a highly singular, remarkable, and striking phenomenon ! What a number of questions crowd upon us, and what a field for the investigation of the Swedish naturalists ! Is the decrease the same in equal periods of time ? Is it equally great in all places ? or perhaps it is greatest and more rapid in the interior of the Bothnian Gulph. Before Gefle and at Calmar, by the endeavours of Celsius sixty years ago, accurate marks were cut into the sea-shore, to determine the future decrease with greater distinctness. These marks were examined a few years ago by the able inquirers Robsahm and Hällström, who found the new decrease corroborated. Their observations have not, however, been made public, and are in the hands of Baron Hermelin. May they not remain long kept back from naturalists! Linnæus, in his travels in Schonen, relates that he had also made a correct marking, about an English mile from Trälleborg, on a block that would not be carried off; and describes the surrounding objects with the accuracy of a botanist.* Would not the investigation of this place, and what has happened there, be worth a small journey from Lund, or from Copenhagen ? It is certain that the surface of the ocean cannot *sink;* this the equilibrium of the sea will not by any means admit of. But as the phenomenon of the decrease cannot be doubted, there remains, as far as we can at present see, no other solution of the difficulty than the conviction, that *all Sweden is slowly rising in heighth from Fredericshall to Abo, and perhaps to Petersburg.* Something of this decrease has also been felt on the coasts of Norway, at Bergen, in Söndmör, and Nordmör, as I was assured by

---

* Skanska Resa, p. 217.

M. Webe, the chief magistrate *(Amtmann)* in Bergen, to whom we are in-debted for the excellent charts of the western coasts of Norway. Cliffs which were formerly covered with water now make their appearance. But the belief in the decrease of the sea is evidently not so generally diffused or so certain on the Western Ocean as in the Bothnian Gulph. The irregular and high tide in the Western Ocean is also an obstacle to accurate observation. It is possible that Sweden may rise more than Norway, and the northern more than the southern part.*

In rain and darkness I at last reached Daglösten, nearly eighteen English miles to the south of Skelefteo.

The twenty-ninth of September was a fortunate day, notwithstanding the incessant and continued rain. In Grimsmark, between nine and ten English miles from Daglösten, I came up with *Wahlenberg*, whom I had incessantly been following from Luleo, at a distance of fourteen or eighteen English miles. That excellent man, to whom botany and mineralogy have almost an equal claim, was returning from his third journey in Lapland. I missed him before by a few hours, on the coast of the Western Ocean, in Kierringöe at Bodöe. He had there descended to the sea-coast by the same way which Linnæus before took from Quickjock. He had employed the summer, in not only climbing the highest of the Lapland mountains, the Södre Sulitjelma, but in determining its heighth with accuracy, and seeing, investigating, and describing the great glaciers of the country *(Geikna* of the Laplanders). We travelled now together, and the journey was a source of rich instruction to me.

---

* M. Wrede of Berlin, in his " Geognostische Untersuchungen über die Sudbaltischen Länder," conjectures the centre of the globe to have undergone a slow change, being transported along the axis from the *north* to the *south*, and hence that gradual *depression* of the level of the sea in our hemisphere, which he supposes to be accompanied by its gradual *elevation* in the southern hemisphere. It would appear that M. Von Buch had in view this notion of Wrede's, when he considered it as highly probable, that the land in the northern regions was *slowly rising above the surface of the sea.* Mr. De Luc remarks : " If these motions (that is the rising of the land in the northern, and its sinking in the southern hemisphere) were real, all the *new lands* which the sea has visibly added to the coasts in the *northern* hemisphere would have an *inclination* towards it, as having been formed by it at *levels* successively lower ; and, on the contrary, no lands of a similar kind should exist in the *southern* hemisphere, since all such as might begin to be formed there would be soon covered by the sea as it gradually rose. Thus all *new* lands, by their *horizontality* in the former of these hemispheres, and their very *existence* in the latter, completely refute this whole system." De Luc's Travels, vol. 1. p. 106.

We flew through the woods, where the dark leaf of the alders was now pretty frequently mixed with the pale and yellow leaves of the birches, and the black needles of the spruce and Scotch firs, which are every where to be found in the woods here in equal abundance. Djecknaboda was most agreeably situated on a lake surrounded with high and majestic aspens, as large as those which grew at Torneo. We were surprised in the evening at Savär in Bygdio Socken by the large and extensive iron-works of Robersfors, with numerous and partly very neat houses for carrying on the different operations, and a large and beautiful house occupied by the *Bruckspatron*. As we took up our night's quarters in the good inn of Täfla, I had travelled since leaving Daglösten in the morning ten Swedish miles (between seventy and eighty English miles).

We were now not far from Umeo, and we therefore reached the capital of Westerbothnia, notwithstanding the rain, early in the morning of the thirtieth of September. It had a more respectable appearance than Torneo : some of the streets were paved ; many of the houses looked quite neat, and there were five or six three-masters and brigs lying in the river ; a stir which we had not seen in any of the other towns. Tuneld determines, however, the number of the inhabitants in 1769 at seven hundred and twenty-three ; Piteo had then six hundred and twenty-one, and Luleo six hundred and forty-four inhabitants. In Umeo, their numbers must be very considerably increased since that period.

We remained here till the evening, in the company of Doctor Näzēn, a reputable physician, to whom we owe a series of good meteorological observations, which the Stockholm Academy have published in their transactions. They are almost the first observations which give us any information respecting the climate of these regions ; for Hellant's Journals in Torneo have never been made public ; and accounts of severe degrees of cold felt in Torneo and other places are curiosities which will enable us to come to no conclusion respecting the temperature and climate of the country. The following table contains the results of fifteen years observations by Dr. Näzēn, compared with those made by Julin in Uleoberg, after making those alterations and corrections in the observations of Julin which I believed necessary.* They nearly agree with the temperature of Torneo.

---

* The reasons for these corrections are more fully developed in a treatise on the heighth of the limit of everlasting snow in the north.

	Umeo 63° 50ᴵ	Uleo 65°
January	— 9. 12 Reaum	— 10. 83
February	— 7. 42	— 7. 752
March	— 3. 97	— 7. 91
April	-|- 0. 898	— 2. 59
May	— 5. 34	-- 3. 955
June	10. 35	10 304
July	13. 72	13. 14
August	10. 97	10. 966
September	6. 87	6. 44
October	2. 72	2. 992
November	— 3. 34	— 4. 155
December	— 9. 26	— 8. 18
	-|- 0. 62 R.	-|- 0. 53 R.

If the observations in both places are accurate, the difference of temperature is very inconsiderable, notwithstanding that Uleo lies much higher northwards. But the vegetation says nearly the same thing. We can, in descending the Torneo river, name many a tree, many a shrub, and numbers of herbs, which gradually make their appearance as we advance southwards, called forth by the amelioration of the climate. The Scotch firs at Lippajerfwi, the spruce firs at Songa Muotka; the *Salix pentandra* at Colare; the first fruits of the noble *Rubus arcticus* at Oefwer-Torneo. But from the town of Torneo to Umeo we in vain seek for a new tree, or any new plants. We cannot find any. We are therefore inclined actually to believe that the climate in the northern part of the Bothnian Gulph, from the straits of Quarcken, between Umeo and Wasa, up to Torneo, undergoes very little change.

We crossed in the evening of the thirtieth the broad Umeo-Elf (one thousand and eleven English feet) and continued travelling in the flat country to Styksiö, as usual, through woods, till we came to the hospitable Sörmyöle, the last place which we had to pass through in Westerbothnia.

The formations which traverse this road in Westerbothnia are limited to a great variety of gneiss, in the changes of which we can find nothing determinate. In ascending the Calix-Elv, however, at Monsby it lost the appearance of gneiss, and so much so, that for many miles I only thought of granite. It was small granular; white felspar and quartz in equal abundance surrounded by mica folia. " But at Prestholmsby, between four and five English miles from the furnace of Strömsund, and not far from Raneo," says Baron Hermelin, " the lime-stone, when broken, is greyish white small granular, and mixed with schorl." Such lime-stone can only be formed in gneiss.

At Luleo, a number of red felspar crystals appeared in this granite surrounded by mica folia, and frequently small veins of red and very small granular felspar run through the stone. In the neighbourhood of Piteo every thing again looks like gneiss, and on Pitholm it re-appears with great distinctness. Almost all the rocks are divided into stripes and strata, and at Kinbäck the stratification was quite distinct from north-east to south-west, dipping strongly towards the south-east. It was singular that hornblende beds, which appear almost at every step in Norway, should be so seldom here. At Oby there appeared magnificent gneiss with large white felspar crystals, which all lay in one direction, with a great deal of mica, which surrounds the felspar fibrously (flasrig) as at Freyberg. It continues in this way from beyond Froskoge to Sumana in Skelefteo. It is delightful to see these blocks and the large shining felspar on them, lying along the road side. Every thing is so fresh and animated. Notwithstanding this beautiful felspar, lime-stone still appears to lie subordinately as a bed in this gneiss. About an English mile from the church of Skelefteo, and between Kusmarck and Kogeo, it is quarried and burnt, and the lime sent to a great distance round all the way to Easterbothnia.* The lime-stone is generally ten fathoms thick, immediately surrounded by mica-slate (or gneiss), and on the surface is mixed with iron pyrites, quartz, actynolite, and lead glance, or galena. Moreover, the gneiss here appeared to me to be every where purer, more determinate, and to alternate less with beds of mica-slate than in the Norwegian mountains. Hence the lime-stone beds may be in general unfrequent here; for they every where belong more to the mica-slate than the gneiss.

From Daglösten, above Gumbodan to Djecknaboda, gneiss appeared with white felspar crystals, which the scaley mica encloses in concentric lamina, like the coats of an onion. This scaley mica becomes then characteristic for it. The felspar also stretches lengthways; the mica surrounds it undularly, and forms saturn's rings parallel above one another of a foot in magni-

---

* Hermelin, p. 60.

tude.   At Stycksiö, before Umeo, quartz beds appear in it with large black six-sided and nine-sided crystals of schorl.   But at Sörmyöle the gneiss became again pure, and felspar crystals are surrounded with concentrical mica.

Near to Sörmyöle there is a large wooden triumphal arch for King Adolphus Frederick, who once travelled through this place in his way to Westerbothnia.   It is not far from the boundaries of the province; Lefwar lies in the mountainous Angermannland *(Angermanniae Montes)*, as Linnæus says.   In fact Westerbothnia, so far as the road runs through it, cannot be called mountainous.   The roads run over hills, and are every where excellent.   Every quarter of a Swedish mile is marked on a *wooden mile-stone;** it contains the king's name and the name of the chief magistrate of the kingdom *(Landeshöfding)*, Stromberg, and then the number of miles.   What a number of such mile-stones there are on such a painfully long road as that from Torneo to Stockholm!   And yet there is scarcely a single quarter of a mile which wants one.   The hills begin to rise behind Oenska, and the road runs pleasantly along a lake of considerable length.   Our common German marsh alder *(Alnus glutinosa)* gradually began to make its appearance among the bushes, in a sickly and miserable state, and a short way above the surface. The Lapland alders, with the white bark *(Alnus incana)*, yet rose high and proudly above them; still, however, we gladly greeted their appearance; for they betrayed our approach to a better climate.   Both species of alders in the way downwards from this long continued to struggle for ascendancy; but the round leaved was visibly increasing; and long before coming to Gefle, nearly at Hamrong, it entirely banishes at last the cold alder with the pointed indented leaves of an ash grey colour beneath.

We did not reach Bröstad till it was completely dark, about nine o'clock. The place was somewhat more than forty English miles from Sörmyöle, and one of the best inns on the road.

The second of October.   We were introduced into a large hall, where we saw excellent linen spread out on the table, table cloths and napkins, all fine and beautiful to appearance.   This is manufactured in the mountains of

---

* The expression in the original.—*T.*

Angermannland, and Bröstad is one of the most considerable magazines of these industrious workmen. So high in the north, can such industry be produced and attended with success? Tuneld, who is always out of all bounds in his praise, affirms that this linen is equal to the best Holland. This may well be doubted, but what we saw on the tables was certainly very prepossessing.

We began now to ascend higher and higher the small ranges of mountains: somewhat like Kullen in Schonen, or Hallandsos, and the mountains of Laholm. We had frequently to proceed down into steep vallies several hundred feet deep. All the churches we pass are true ornaments of the country; simple stone edifices: the heighth of the walls, the arched windows, and the gentle inclination of the roofs, are all in the most beautiful porportion. They combine simplicity and sublimity. These churches frequently produce a noble effect in the vallies.

Beyond Spiutha, and after passing Nätra, the hills become higher, and are covered with wood. The road winds laboriously through the vallies and beneath the mountains. The picturesque form of the *Skula* mountain now makes its appearance: this is the highest mountain on the road, and a signal for mariners at sea. It rises with a long perpendicular precipice immediately above the road, a smooth wall of rock of more than eight hundred feet in heighth. We ascended the mountain from Dolstad, where it is rendered accessible by bushes and a gentle declivity. I found its heighth at top nine hundred and fifty-two English feet above the sea.* We see a great way out at sea over woody mountains, which lie intersected around like so many islands. Several sea-bays pass through the narrow openings between these snow covered mountains, and almost to within a little of the Skulaberg itself. But the view backwards into the country is very different. There the streams descend from the height in long vallies. There we do not see insulated mountains, for the whole country, for a distance of eight or ten English miles, rises higher and higher to the elevation of the Skulaberg, more than a thousand and fifty English feet above the sea, and on the height it runs on

---

* 2nd. Oct. h. 4. p. m. Skulaberg 27. 3. 1. Therm. 4. 25. R. clear, south wind.
        h. 6 ...... Dolstad.. 28. 2. 2. Therm. 4. 5. clear, faint south wind, 40 feet above the Fiord.

almost like a plain through Jämteland to the foot of the Norwegian mountains. This is very different from the gentle ascent of the country from
Torneo towards Norway.

The formation of the mountain and its internal composition had also
something remarkable. The gneiss, with frequent scaley mica, had already
left us in the neighbourhood of Bröstad, and given place to those large white
parallel lying felspar crystals which appeared so beautiful on the side of Piteo.
At Horness, at Spintha, where the mountains are heaped together, the
texture of the gneiss altogether disappeared. It became wholly small
granular granite with white felspar, and with several large felspar crystals
between. At the foot of the Skulaberg we had again fine gneiss. A short
way, however, up the ascent the *granite* immediately displayed itself, coarse
and small granular, *with dark red felspar*, such as we had never yet seen on
the whole way ; and this above gneiss and not under, as was the case with
the Rapakivi at Kängis, and at Oefwer Torneo. Grey quartz lay sparingly
in it, and a little mica folia collected in groups, and frequently hornblende.
This was the case to the very summit of the mountain; so that it was not an
accidental bed, but a continuous and distinct modification of the principal
formation of the country. But how far does it continue? And what other
mountains consist of it?

The circumstance of Linnæus having nearly lost his life in a cavity of
this very Skula mountain, as is related by him in the preface of the Flora
Lapponica, has preserved the mountain in remembrance among the Swedish
botanists.

We only returned from the mountain at sun-set, but we proceeded several
leagues farther on our way to Aeskia, where we found fresh cause to be highly
contented with the friendly reception of the people and the internal accommo-
dation of the inn. In fact, such comforts are very unexpected so high northwards, and many days journey from great towns. But all the houses of the
peasantry in Norrland, so far as the road runs through the country, namely,
in Angermannland, Medelpad, and Helsingeland, have an appearance of prosperity which prepossesses us very much in their favour. This appearance is
by no means apparent, for the Norrlanders are actually more prosperous and
substantial than the other Swedes, and more laborious and industrious, notwithstanding their soil and the nature of their country are not among the
most grateful in the world.

On the morning of the third we came to the great Angermannself. It was not like a river here, but like a lake. We crossed over in a large flat-bottomed boat, and rowed for more than an English mile before we reached Weda on the opposite bank. Delightful bushey banks, beautiful prospects and distances down the river, with ships in sail! It is a great and majestic stream. In this neighbourhood one of the ornaments of the north again made its appearance for the first time, the *Norway Maple (Acer platanoides)*. The river is the boundary of its growth; it does not cross it with impunity. In Finland Linnæus first saw it between Christina and Biörneburg, about half a degree farther south.

The road constantly follows the windings of the coast, and never goes to any great distance from the sea. This however increases the number of miles greatly. The views of Fiords which penetrate from the sea are very frequent, but we seldom or never have a view from the road of any extent out at sea. On an island beyond the wood, between three and four English miles from the road, lay Hernösand, the capital of Norrland, and the seat of the chief magistrate *(Landeshöfding)*, who is mentioned every quarter of a mile on elegant mile-stones of cast iron. Late in the evening we came to *Fjäll*, the first place of the small province of Medelpad, and completely adapted to excite in us the most favourable prepossession in its favour.

The fourth of October. Near to Fjäll we crossed the Indals-Elf, the outlet of all the waters from Jämteland, on which account it is a considerable stream. We had twice to cross it, for it incloses a small island over which the road runs. The woods become at last here not so frequent and extensive; the churches crowd closer and closer together; the country seems more inhabited, and the views are more rich and refreshing. The bay of Timmero was astonishingly beautiful. The noble and simple church on a hill in the valley was reflected in the clear unruffled stream, and the bushey declivities of the hills were delightful. The people came flocking down by roads and footpaths from the heights, hurrying to the church, as a central point. Immediately the solemn peal of the bell resounded through the valley. The people on the footpaths now quickened their pace. The groups on the roads began to separate, the whole valley was in motion, and the sound was solemnly borne up the mountains. How grand and elevating is nature!

We were at a short distance from Sundsvall. A valley opposite us descended towards the town; the declivities were green and covered with houses, beautiful and animated like *Tannhausen* in Silesia. The town was burnt two years ago, but was now re-edified, and run dazzlingly from the sea-bay up the river. But internal prosperity did not seem to be yet restored. Many of the streets were only laid down, no where paved, and many houses were not completed; and that the town should contain one thousand six hundred inhabitants is hardly credible at first sight. We see, however, from a few respectable houses along the water, and the ships in the harbour, that there is a stir here which is considerably increased by the linen manufacture. Before the town we beheld again the first fruit gardens: apple trees with fruit on them; and they did not seem sickly, or to stand in need of the greatest care to succeed. High willow trees stood every where round the town, *salix fragilis*, and for the first time descending from the north, in a latitude of $62\frac{1}{2}°$. This is the extremity of the successful cultivation of fruit-trees (apples) along the Bothnian Gulph. On the Western Ocean, in Norway, fruit gardens have been seen it is true at Ertsvogöe, near Christiansund, in a latitude of 63°. full of various sorts of cherries, and even wallnuts bearing fruit, which seldom however ripened.* It is deep in the interior of the Fiord, where the warmth of the great ocean may, but where the fogs cannot penetrate.

Before Sundswall there was a country house, perhaps the most northern in Sweden: a small stone palace agreeably situated on a hill, to which we are conducted through maple alleys. Grefwe Frölick dwells there the whole year through, said the *Sjuts Bonde* (the peasant in charge—*Schützbaüer*), to me, who delivers the horse for proceeding onwards at the station. The climate must have its charms when it is chosen to erect country-houses on.

We crossed the Njurunda-Elv over a beautiful bridge, and shortly afterwards we again proceeded between high mountains betwixt Maji and Grytje. The Norbykuylen is a celebrated mountain in the whole country round, and serves as a mark to seamen a great distance out at sea. It was not so rockey as Skulaberg, nor so steep and perpendicular in its ascent, but it was certainly equally high. It was evening when we descended, and we did not reach Bringstadt, the second station in Helsingeland, till late in the night. The mountains had placed limits to the more southern extension of Medelpad.

---

* From *Bescrivelse ave: Oure Prästegieldt.* Norsk Topographisk Journal. XIX. 107.

It is singular to observe the frequent changes of gneiss into granite. For miles we might be induced to believe the granite to be predominant, and the vanquished gneiss to be altogether expelled by it. At Herskog, and above the Amgermanns-elf, the granite was distinct, small granular, with white felspar, and with insulated black mica folia, and no longer bearing any resemblance to gneiss. But a few miles farther on, at Nässland, fine slatey gneiss lies frequently again as a bed in this granite, and this as we proceed becomes still more frequent. At Westad, on the other side of the Indals-elf, both alternate in equal abundance, and equal extent. And on the beautiful banks of the Bay of Timrod every thing like granite is again suppressed, and distinct slatey and somewhat undular gneiss every where prevails. With gneiss so continuous, so amazingly far extended, and with such a great preponderancy, we can scarcely conceive that these granites have a proper independency. They are only *subordinate* to the gneiss. Towards, Maji, Bringstad, and through the mountains of Norbykuylen, no granite ventures to make its appearance again. The mica lies in the gneiss almost always in thick scaley folia ; this it cannot so easily do in granite ; and hence gneiss is always very easily to be distinguished at a distance, even when the stratification is not so very striking. At Bringstad the gneiss is striped, with white and very small granular felspar, and with a number of small blood-red garnets in the felspar.

On the Njurunda-elf several remarkable new trees again appeared to us, the majestic ash *(fraxinus excelsior)* and the hazel ; with them a number of plants, with which the ground of the woods and meadows were now clothed. The hazel is perhaps detained so far south more from the dryness of the air than the absolute cold of the climate ; as it appears to thrive admirably in a moist atmosphere, if we dare trust to the observations on the Norwegian islands along the western ocean ; for hazels appear there every where in such incredible abundance, that Bergen generally exports a number of casks of them to other countries. And this bush goes as high as 65° in Helgeland before it disappears. The Njurunda-elv is only however in 62° latitude.

Above Malstad and Sauna we proceeded quite close by the gate of Huddicksvall. The town appeared considerable, and new, as it shone over against us. Gardens with large maple trees lead to the place, and the thick green of a great number of other trees rose with freshness and animation out of the mass of houses. It is highly gratifying and delightful to see the better climate daily disclosed to us as we advance.

As we proceeded along the garden walls which ascend from the town, and designate its boundaries, we were greatly astonished at the large red garnets in the gneiss. We had, it is true, seen garnets in the gneiss the whole morning, at Malstad and Sauna ; but here garnets seemed to be almost essential to the formation, in round crystals of an inch in size, between the white felspar, and frequently entirely cochineal red. Other crystals are so interwoven with the felspar, that the whole mass of it is thereby coloured red. It is a beautiful stone, which may be sometimes seen in single beds, but seldom continuing for an extent of several leagues.

At the brisk Iggesund, an entire village run from the red roofed dwelling of the proprietor of the works *(Bruckspatron)* up the woody valley. The houses of the workmen run along like a street, and high ashes and alders lined the road. It is one of the largest and most beautiful iron-works of the kingdom, celebrated for a great steam engine *(Walzwerk)* which Rinman constructed here, and which he described circumstantially in his Mining Dictionary. From the circulation and size of the place we may easily perceive the importance of these works.

Here every thing like mountains at last disappear. The road runs incessantly through unvarying and at last fatiguing forests, and not without difficulty we were enabled to reach Kongsgard in Nörala late in the evening, which is between thirty-five and forty English miles distant from Bringstadt.

The sixth of October. Nörala is important in the history of Sweden. Here Gustavus Vasa assembled the peasants of Helsingeland, and stimulated them against Christiern. On account of this event, a society in Gefle, in 1775, ordered a stone to be erected before the inn with a long inscription on it.

This is of advantage to the strangers who frequently travel this road, and it may also perhaps serve to foster in the peasantry a feeling of their own dignity and importance.

Between four and five English miles farther on, not far below Momiösje in Sörala, the wild and impetuous Ljusne-elv came tumbling down before us, out of the woods, over prodigious blocks, and again entered the dark forest. It brings down all the waters of the province of Herjeadalen into the sea, and it is one of the most considerable rivers in Sweden. Yet they have succeeded, and not long ago, in throwing a large and long bridge over the stream, which is by far the most beautiful in the long way between Torneo and Stockholm. But the forest from Skog towards Hamrong is interminable, three everlasting (Swedish) miles, a sandy soil, and only a single inn in all the way. This great forest was another separation of provinces and climates. When we issued out of it, we were in the small and thriving Gestrikeland. Vallies run down from the level, and small rocks along the declivity. A beautiful iron-work lay on one side, and large alleys of maples run from the houses of the workmen to it. Every thing seemed to betray a new and better nature. The cold alder *(alnus incana)* was now no longer to be seen. There are high and beautiful ashes, such as we had not before seen, and even an elm *(Ulmus Campestris)* growing at Hamrong. It is famous, for Linnæus has immortalised it.* In his time, the inhabitants considered it as a bewitched tree, as it never bore any fruit, and was the only one of its kind. In the whole district nothing similar was known by them. Even limes grow here, and thrivingly.

We did not reach Gefle. We were obliged to remain in Trödje, about nine English miles still distant from Gefle.

Instead of the garnets of Huddicksvall, there appeared at Iggesund hornblende in the gneiss, and at the same time a good deal of scaley mica. This certainly gave it a foreign appearance, but it was far from attaining the agreeable, singular, and striking appearance of the garnet gneiss at Huddicksvall. Towards Bro the felspar of a red colour lay lengthwise, like patches (Flammen) on the stone, surrounded by scaley mica; and at *Sorala* it was undularly slatey. We first saw something resembling granite in the neighbourhood of Hamrong mixed with fine slatey gneiss, but not long, for gneiss with white felspar, which is surrounded by mica concentrically, was traced by us to the gates of Gefle.

The seventh of October. The first view of Gefle did not please me. The black fir-wood approaches too near to it. Every thing is too flat, and cannot be overlooked. Yet large buildings scattered over the plain demonstrate to us that we are approaching a considerable town; and this is not

---

* Flor. Lappon. Proleg.

belied by the interior of the place. The council-house is a beautiful and tasteful edifice, and several private houses beside it are not unworthy of its vicinity. A great number of vessels lay down the quay on the small river, which were a proof of the commerce of the town. It contains, according to Tuneld, five thousand five hundred inhabitants, a great number for so northern a town. We remained here only a few hours, pleased with the bustle and the vessels in the river, and proceeded down the alleys of mountain-ash, maples, and ash along its banks, which run from the town for a great way towards the sea.

The garden walls, and even several of the houses, were built of red fine granular sand-stones. This was almost as strange to us as if we had found an orange-tree by the road, for such sand-stones are very unusual in the north. In the middle of the sand-stone kidneys of black and very soft mineral pitch lay in a very singular manner, which were also frequently disseminated in very small grains in the mass. We saw so many houses built with these stones, and of them alone, that we conjectured there was a quarry in the neighbourhood. We went down the whole of the quay, and asked every person we met where the sand-stone was quarried; but no person understood us, nor could give us the smallest information. We could learn nothing. I was informed long afterwards that these stones were not found at Gefle, but were thrown out by the sea at the outermost cliffs before the harbour. The source of our information was very deserving of belief, but the fact by no means so. It was almost midday when we left Gefle. A few hours afterwards we reached the iron-works of Harness, and at the same time an important place for the determination of the climate in Sweden; for we find here the first oak growing wild, which was known to Linnæus as he was travelling towards Lapland. With the appearance of the oak, it seems as if we had left behind us the last influence of the polar climate. They are now enabled to cultivate as much grain in perfection as they require, and the destructive nightly frosts, which frequently destroy in a few minutes the hopes entertained for several months of a rich harvest, are no longer dreaded.

The course of the boundary of oaks over the northern lands is singular enough. It demonstrates pretty distinctly the manner in which the climate deteriorates with the distance from the great ocean. In Norway the oaks grew vigorously and beautiful in the interior of the sea-bays at Christian-

sund, and at Molde almost as high as 63°, and even at Drontheim they are
not altogether extinguished. Harness, on the other hand, does not lie
higher than sixty degrees forty minutes. On the other side of the Bothnian
Gulph oaks scarcely go beyond Abo; in the Finnish bays they only cover
the coasts to Helsingfors, and on the south side they cannot penetrate into
Ingermannia beyond Narva, so that they do not even reach sixty degrees.
Georgi relates * that the Czar Peter the 1st ordered oaks to be planted at
Petersburg. They grew indeed to the thickness of a man, but irregularly,
and with a straggling decayed appearance, quite unsuitable to their age and
their nature. On the road to Moscow they first appear on the banks of the
river Msta (Güldenstädt); and although the country does not rise to any
considerable extent, their boundaries in an eastern direction towards Siberia
are to be found in the neighbourhood of Ossa, between Casan and Cather-
inenburg, in a latitude of not more than $57\frac{1}{2}$°. †

The rocks at Harness were also remarkable. From Gefle hitherto hornblende beds appeared more commonly than usual in the gneiss; and here,
in the neighbourhood of the furnace, there was nothing in every direction but
fine granular hornblende mixed with epidote, quite similar to the Colmünzer
stone of the Fichtelberg, lying in large blocks, of which walls and houses
were built. Whence? Most undoubtedly out of the neighbourhood. But
in what connexion with the gneiss?

The passage over the great Dal-elv detained us a number of hours. The
ice had broken the bridge, which was constructed with great art over the
stream, in the neighbourhood of the great waterfall. We were obliged to be
ferried over. But next morning was the market or fair in Elfcarleby, on the
opposite side. The number of men, horses, and cars waiting at the river-
side was so great, that our turn to get over did not come till a number of
others had preceded us. The place on the other side looked quite like a
town. A number of itinerant merchants had assembled, and we reached
the houses through a long row of booths. The road was blocked up by
carriages, horses, cows, and men running about and making a deafening
noise; and the agitation and noise in the houses were not inferior. The
whole of Dalecarlia comes down to the fair of Elfcarleby, and half the people

---

* Description of Petersburg, 519.   † J. G. Gmelin Sibirische Reise I. 103.

of Gefle come to meet them, with numbers of people from Upsala and Stockholm. It is one of the greatest fairs in Sweden. On the land side there was also a whole town of carriages waiting before the barrier, and urgent for admission, so that it was with difficulty we could squeeze through, and get into the wood. We were constrained by the night to remain in Mehede. We were now in *Upland.*

Between two and three English miles before reaching Yffre, we descended from the hills, and entered on a complete plain. It is level and flat as the north of Germany. There are no longer any forests, and we see corn-fields without interruption. The houses of the peasantry are now covered with straw, on which, though they may be allowed to congratulate themselves, it is certainly every thing but ornamental. But these peasants eat no *barkebröd.* They might in a case of extremity feed on their own roofs.

From Högstad the road runs for between four and five English miles in a straight line to the cathedral church of Upsala. We hastened on to reach this long expected object, and entered with no small degree of sensation, about five o'clock in the evening, a town which Europe for a long series of years has uniformly named with respect.

There are few places, perhaps none, of which the temperature has been so carefully and accurately ascertained as Upsala. Thanks to the excellent mode of observation which has been constantly followed here, which probably originated with Celsius, and which is still adhered to with equal accuracy; for it has never been deemed here a sufficiently correct method of ascertaining the temperature, to heap observations on observations irregularly in the day-time, or to specify great degrees of heat or cold of a few moments continuance, which cannot possibly determine the general climate of a place. The meteorological journals which commence with the observations of Mallet in 1756, and which have been continued without interruption ever since, are preserved in the observatory. In these observations the thermometer was daily observed in its extremes, as has been done for these eighteen years in Geneva, namely, at sun-rise, and the first hours of the afternoon. This was however more difficult in Upsala than in southern countries, where the sun does not, as is the case there, rise at two o'clock in summer, and nine in winter. Even Mallet's observations for a period of ten years are certainly somewhat irregular and hard to decypher. They appear indeed to have been made very little use of, except for those years the results of which Mallet himself gave to the public in the Transactions of the Academy of Stockholm. But the journals of Prosperin are models of accuracy, method, and perspicuity. They begin with 1774, and come down to 1797, the death of Prosperin. They have also served as patterns to his successor Holmquist, from 1798 to 1801, and Schilling, the present observer in the observatory, who still uninterruptedly continues the meteorological journals with equal accuracy and punctuality.

Upsala may therefore be considered as a fixed point in the north, the temperature of which is thoroughly known to us and which we can confidently rely on in our comparisons with the temperature of other places.

While I remained at Upsala, it was my chief occupation to compare all these journals with one another, to draw a mean from them; for I felt the want of a fixed point in the north, the accounts of the temperature of which might be subject to no dispute.

The following are the averages of thirty years, from 1774 to 1803:

January	— 4°.	21	Reaum.	= 22°.	53 Fahr.
February	— 2.	22	....	27.	
March............	— 1.	26	....	29.	17
April.............	-+ 3.	56	....	40.	
May	— 7.	56	....	49.	
June		11.	66	....	58	23
July		13.	69	....	62.	8
August		12.	63	....	60.	41
September.........		9.	07	....	53.	4
October		5.	17	....	43.	63
November		0.	35	....	32.	78
December	— 2.	975	....	25.	307
	4.	42 R.		41.	94 F.

When we compare the same years with the observations of Wargentin, at the observatory of Stockholm, we find that the annual mean of Stockholm is, upon an average, 0°.423 R. or 0°.951 F. above Upsala. This is produced more by the inferior degree of cold in winter than the greater heat of summer; for Stockholm, which is situated on islands between the sea and the Mälar, does not enjoy so many clear days as Upsala ; but then the severity of the colds is on the other hand modified by the fogs of winter.

I entered Stockholm in the dark, on the night of the twenty-fourth of October. The number of carriages and cars on the road had long forewarned me of my approach to the capital ; but I still believed myself in the thick forest, when we were all of a sudden dazzled by the large lamps of the custom-house, and we stepped at once out of the forest into the town.   The straight Drottnings-gata brought us at once below the observatory, from the heights opposite the Mälar ; and the perspective of the burning lamps for an English mile down was extremely magnificent.   And then what a stir on the great Nordermalms Torg ! and what crowds on the noble bridge, and all the way to the inn, in the Myntgata ! The capital of the kingdom could not be mistaken even in the dark.

It is a singular and wonderful town.   What romantic views of islands, waters, rocks, hills, and vallies ! All that we assemble together in our dreams of distant landscapes is here united in the circumference of one town. Whatever there is grand in nature is to be found in the neighbourhood of the finest monuments of art.   It is true, we do not observe here the astonishing magnificence of Naples ; but we have in place of it such an indescribable diversity of prospects and impressions, that for years, perhaps, we may search in vain before we find them again.   How beautiful is the situation of the citadel, from which we overlook almost the whole of the town, as it ascends the declivities from the water ! How charmingly solitary are the rocky banks at Rörstrand and Carlsbergwick, in which elegant country-houses are hid in the clefts ! How magnificent the view from the rocks of Södermalm, of the interior of the town, the ships in the harbour, the islands, the boats, and of the woods and rocks opposite the menagerie ! The streets are laid out with so much art, that at a great distance, and in the most remote parts of the town, we have always the large buildings or churches in view, and are always employed upon distant objects.   Through the long Drottning-gata we see the church of Catharine in the Södermalm ; through the Storgata we see John's church ; and in the Riddare-gata, the church of Adolphus Frederick.   Such variety is not to be found in any other town of Europe.

I saw but very little of the interior; for it is difficult to tear ourselves from such a rich and instructive collection as that of the minerals in the College of Mines, especially as it is thrown open in such a liberal manner, and with such free liberty to make use of it as we experience under the superintendence of the generous Hjélm, and through his goodness.

This collection is in reality an image of the mineralogical constitution of Sweden, in greater perfection than we find in most countries, where geographical collections of the interior have been formed. All the provinces appear here characterized by their natural produce, from the roughest part of Lapland to the southern Schonen; and from each of them we may expect something new and instructive, even though we may have travelled through and examined them. The great utility of these collections is, that we are enabled by them to continue the series of observations already commenced by us. They are the concentrated inscriptions of nature, which are not only legible, but intelligible to him who has previously studied the history of the province in its sources; and which, from being here ranged close together, frequently throw a mutual and brilliant light on each other.

Baron Hermelin's collection is also one of the most important for the geographical knowledge of the mineralogical treasures of Sweden; but the most elegant, and in greatest repute, in Stockholm, is perhaps the collection of minerals of the zealous and intelligent M. G. M. Schwartz, in Rörstrand, the best mineralogist in the kingdom. He is from Helsingfors in Finland, and known from a manual of mineralogy*.

I left Stockholm on the 12th of November, not without great apprehensions of being stopped in my way to Christiania by the falling snow. Even in the first days of November we had been very much impeded by the snow when I visited Utö along with M. Schwartz and M. Gerjer, the son of the celebrated counsellor of mines. It did not, however, lie. But every thing combined to urge us on. The hope of seeing Wermeland and Westgothland, which was afterwards cruelly devastated in the wars that broke out, in a more favourable season of the year, prevented me still more

---

* Hausmann System der unorganisirten Naturkörper, S. 3.

from staying in those remarkable districts. Hence I saw scarcely any thing of them, and derived no other advantage than that of breathing their autumnal air.

In the neighbourhood of the town we descend incessantly over small lakes with rocky islands, which we imagine to form bason-shaped hollows ; but it is still only the Mälar which winds and curves in such a singular manner. Even at Sodertelje it is not far distant. The sea and lakes here approach quite near to one another ; and they are separated by a ridge of land of small heighth. As the Mälar, of late years, had done much damage in Stockholm by its rises, it was resolved to give it vent here, by a canal of communication with the lake. As the work was actually begun, it was widened in the years 1806 and 1807 to such an extent that it was navigable for small vessels ; this was partly done by the labour of the French prisoners. By this means there is opened a new, shorter, and more secure and comfortable way by water to Stockholm, which probably will be very much frequented.

For now, instead of venturing from Landsort, farther into the open and strong sea, the vessels proceed immediately, by Trosa, into the great sea-bay of Himmersjö, and never once come out of the Skiärs, towards the lighthouse of Landsort. The situation of this beautiful and useful work is not only distinctly exhibited in an excellent chart of the Royal *Landmäteri Kontoret* *, but we also see from it, independent of the greater security, how much distance is saved by the vessels which go through the Söder Telge Canal.

I was driven from Lägstakrog, where I remained during the night, by a severe storm and snow from the north-east, rapidly passed Mariefred and Strengnäss, to the good inn of Ekesog. The night came rapidly on : I saw Torshälla and the outlet of the Hjelmar in moonlight, and reached the large and elegant hamlet of Smedby at ten o'clock in the evening. In the midst of this place there was, what is observable at many other inns, a large green grass-plot, surrounded by an elegant trellice ; and a large tree rose out of the

* Karta öfwer Segellederne ifron Landsort til Stockholm soväl genom den vid söder Telge tillämnade Kanal, som förbi Daläro och Waxholm. Författad uti Kongl. Landmäteri Kontoret af C. E. Enagrius och N. Kjerner, 1807.

middle of it, the highest spruce fir of the district, ornamented with gar-
lands to the very top, which points out the inn a great way round the
country.

All is flat at the commencement of the Mälar. In Kongssör there was
a great deal of stir from the shipping down from Arboga. We reach the
canal itself between four and five English miles farther on ; and we are told
by gilt letters over the bridge that it was begun by Charles the Eleventh,
and finished by Charles the Twelfth. From Fellingbro, beyond Arboga, the
country rises ; and the road runs for several leagues through woods and
vallies ; but towards Oerebro the woods begin to open. I had an extensive prospect, by moonlight, over the great Hjelmar, and proceeded again
down the whole of the height, at the foot of which Oerebro, the capital
of Nerike, is situated.

Although in maps we see a mountainous range between Nerike and West
Gothland, they are not mountains like those in other parts of Sweden,
scarcely any thing more than eminences. The woods continue without interruption, and the cottages and flat vallies lie like islands among them. The
greatest height may be at Bodarne, and yet it is not more than a few
hundred feet.

On the morning of the sixteenth I proceeded from Hofwa, down towards
West Gothland. The Wenern appeared like a great sea from the height,
and Mariästad like a sea-port along the shore, with several beautiful houses
and vessels in the harbour. In the twilight I was still enabled to see the
Kinnekulle ascend between Enebarken and Kolang, with a crown which
announces a basaltic mountain ; it is however only a mountain in such a
flat country. Lidkiöping, which I reached in the evening, is again altogether
on the plain.

The whole province of West Gothland appears no longer to be Sweden.
We frequently find villages. Every part is not only more inhabited, but the
houses are closer together ; and what is unheard of in the north, except in
the neighbourhood of Upsala, we see several country churches at once.
At the long Hunneberg, near Wennersberg, I thought I saw birches in the
thickets. Several are to be found at Hollandsö, at Lidkiöping, in a latitude
of $58\frac{1}{4}$. But this is the extreme point which birches can possibly reach,
and even a consequence of the vicinity of the Western Ocean. For in
Smoland the birches seldom or never go beyond 57°, a few miles to the

north of Wexiö; and on the coast of the Baltic Sea they scarcely go beyond Calmar.

I had expected mountains from Wennersborg, towards Uddewalla, and I did not see rocks till I came to the Rockne-elv: gneiss, in which other fine micaceous pieces of gneiss lay enveloped here like a conglomerate. Nearer to Uddewalla a singular *skiärry* country again appears; rocks intersected in such a way as if large floods had penetrated through them. The road runs between such rocks down to the place where Uddewalla formerly stretched out before us. But how much was the prospect here changed! When I passed this place in my way to Christiania, a year and a half before, it was a beautiful and flourishing town; but now it lay an entire ruin. The fire, a few weeks after I left it, had not left a single building, and the inhabitants were still living among the heaps of rubbish: they had not yet constructed a single house. The king wished, not without reason, the new houses to be built of stone; but the beggared inhabitants were destitute of means, and still more so from the Norwegian war, which broke out shortly afterwards.

The road from Quistrum, northwards, on the nineteenth, was of very difficult ascent. It began to snow hard, and the wheels could scarcely move. But, what was singular, the farther northwards we came there was the less snow. At Hedi there was very little, and at Wick none whatever had fallen. We were told by travellers that the snow was deep, and continued lying below Uddewalla, quite in opposition with the situation of the place. The snowy weather came from the east, and proceeded down the Cattegat.

But the ground could not be long free from snow. At Hogdal, in the neighbourhood of the Swinesund, a deep snow and violent storm came again from the east. We ascended and descended the heights towards Swinesund with great difficulty. As it grew dark, the storm became more and more violent, and all passage was impossible. About nine o'clock rain fell: the storm was allayed; the snow softened, and melted in great streams from the mountains. About midnight we were enabled to get over. In the mean time a violent thunder-storm arose: majestically large lightning, from the dark clouds, darted frightfully over the whole hemisphere, and the rolling of he thunder was dreadful. Every thing proceeded from the south; several claps of thunder descended to the ground; the storm raged anew, but from the south; local rain, uproar, and commotion in all nature. What a night!

and what a passage in such a night, till we reach Westgaard! and such a thunder-storm, so late as the twentieth of November!

How necessary a repose of several days seemed in the noble house of Niels Ancker, in Frederickshall! and how grateful the conversation of such distinguished and excellent people!

On the twenty-seventh of November I again reached the house of General Von Wackenitz in Christiania, after an absence of seven months, a house which, from the reception I experienced in it, had been long anxiously looked to by me as a fortunate and wished-for home.

CHAPTER X.

RETURN FROM CHRISTIANIA TO BERLIN.

Drammen.—Holmestrandt.—Remarkable Range of Rocks at Holmestrandt.—Basalt, Porphyry in Sandstone.—They belong to the Transition Formation.—Jarlsberg.— Laurvig.—Beeches and Brambles.—Bridge over Louwen-Elf.—Porsgrund.—Cloister of Giemsiö at Skeen.—Boundaries of the Transition Formation at Skeen.—Beauty of the Zirconsyenite.—Limestone with Petrifactions.—Quartz.—Amygdaloid.— Zirconsyenite on Veedlösekullen.—Porphyry beneath.—Porphyry Veins in the Limestone.—Road from Konsberg to Skeen.—Zirconsyenite at the Skrimsfieldt—At the Luxefieldt.—Beautiful Situation of Skeen.—Departure.—Söndel-elv.—Iron-Works of Näss.—Arendal.—Christiansand.—Sources of Industry.—View of the Town.— Windmills.—Passage to Nye.—Helliesund.—Storm.—Situation of the Island.— Signals.—Sea Crab Fishery at Helliesund and Farsund.—Gun-boats.—Unfortunate Endeavour to reach Jutland.—The Lugger Privateer Virksomhed.—Fresh unsuccessful Attempt.—Kumlefiord.—Danger of Grain Ships.—New Attempt, Passage to Brekkestöe.—Pilots.—New Attempt.—Dreary Prospect of the Coast of Jutland.— Arrival in Lycken.—Difficulty of landing on the Northern Coast of Jutland.— Vendsyssel.—Aalborg.—Randers.—Aaxhuus.—Desert Heath.—Flensburg.—Schleswig.—Kiel.—Berlin.

THE war between Denmark and Sweden broke out in April, 1808. The way by land through Sweden was now altogether closed, and there was no other remaining communication with Copenhagen and Germany but over the insecure sea, in which English frigates and privateers kept the whole of the Norwegian coast in a state of blockade. Yet many a vessel came over under shelter of the night, which encouraged us also to attempt this passage.

I set off from Christiania early on the fourth of October, 1808, in company with Baron Adeler, a royal page and a captain, one of the most polished and amiable officers of the Danish army. We reached the large town of Drammen, which carries on the greatest wood trade in Norway, at midday, and in the afternoon we were in the small town of Holmestrandt.

This road is inexhaustible in the richest and most varied prospects. The great stream of Drammen is without a rival for splendour : it possesses such repose and grandeur in its course out of the majestic valley into the sea-bay. Add to this, the animation produced by so many vessels at anchor, the perpetual motion of boats in every direction, and the two towns of Strömsöe and Bragernäss running along the banks as far as the eye can reach. It is altogether a prospect deserving of being named one of the most remarkable in Europe.

Holmestrandt also has a singularly romantic situation. There is only room for one street between the perpendicular rocks and the sea-shore, and scarcely so much as to afford room for a *place* or square on a projecting sand-stone stratum. Many of the houses are wedged into and attached to the rocks like so many nests; yet all is neat, elegant, and painted, and indicative of the prosperity of this little town.

The remarkable geognostic relations with which the country round Christiania so wonderfully abounds are still the same on this road: such alternations of the most various rocks and phenomena as we should never before have suspected.

The beautiful red (transition) granite of the mountains of Strömsöe continues for about five English miles to near Oestre, a small mountain range which accompanies the Dramsfiord, in its course into the greater Christiania-fiord. Black lime-stone then shews itself on the mountains, when they begin to sink, compact and splintery, such as we are frequently accustomed to see here. It does not however continue long. Needle porphyry lies above it; and deep down at the church of Sande, on the Sandefiord, the sand-stone appears, on which the porphyry rests. This sand-stone continues without interruption along the edge of the sea-bay, and forms the base of all the following rocks; for a high and black range of rocks advances perpendicularly to within a little of the sea-shore, and runs for a distance of miles. The road to Holmestrandt is made in the sea, for there was no room below the rocks.

I have long and accurately investigated these rocks, when I have always asked myself—Am I then in Italy, or in Auvergne? Conducted here through transition mountains in immediate connexion, these mountains, as well as those of the places just mentioned, appear enigmatical and inexplicable

mountains.   The range of rocks of Holmestrandt is porphyry; but this porphyry becomes basalt through all the imperceptible gradations and changes
of formations in which Auvergne is so rich.   At Holmestrandt itself, small
hills of basalt blocks are heaped up; the basalt is very black, somewhat fine,
granular, heavy, and mixed with numerous greenish black shining augites,
and with nothing else, neither with felspar nor calcareous spar.   The augites
are by no means to be mistaken, and in no respect to be taken for hornblende, for their crystallization, with their characteristic oblique accumination,
is every where prominent.   This basalt is not unfrequently vesicular and
porous; and often red and slaggy when it comes in contact with other strata
of porphyry.   It does not form a cap above the porphyry, or above other
formations, like the basaltic greenstone of the Kinnekulle and the Hunneberg, in West Gothland, but it is a continuous bed in the middle of the
range of rocks itself above needle porphyry, and covered again by other porphyry.   This is very distinctly seen on these steep and naked rocks.   When
the basis loses its blackness, becomes reddish brown, and similar to wacke,
the augite crystals are very beautiful, and their lateral planes and angles
sharp and distinct; and there is also a great deal of white calcareous spar at
the same time, partly in small round vesicular balls, and partly as the filling
up of large and oblong vesicles; very often ornamented internally with small
quartz crystals, and when the kidneys are somewhat considerable with
beautiful quartz druses also on the exterior.   Felspar in needles is also to be
found in the reddish brown masses.

But the conglomerate beds below are very singular balls of needle porphyry
of the size of a head in magnitude, and project from the brown wacke mass
like cannon-balls in wells.   Many of them are broken through, and form
then a strong contrast from the variety of their composition with the basis.
Many of the balls are altogether vesicular, and press closely on one another.
It is one of the undermost strata.

At Angerskleif, a very narrow pass more than two English miles before
coming to the town, we see at a single glance the manner in which the
whole porphyry series rests on the sandstone.   The line of separation is to
be seen for a great way.   The sandstone rises somewhat higher from the
sea, and the rocks remove a little from the road.   The strata dip under the
porphyry to the west and north-west.   The sandstone has a clay basis, and
is much mixed with mica.

On the road from Angerskleif to Revo there appears a thick bed above in the rocks, which from its whiteness is strongly contrasted with the lower masses. It is *felspar porphyry*. The base is bright flesh red, with large thick quartz crystals in it. For the first time I here saw quartz crystals as an ingredient of this porphyry, with the exception of those infiltrations in the calcareous spar of the wacke.

This very remarkable range of rocks turns to the southwards from Revo and Sande, and from the sea-bay, and goes inland in the direction of Hoff. However various the relations which it exhibits, the same order upon the whole as at Christiania and Drammen is not to be mistaken. The sandstone is the same of which the shores of the Holsfiord below Krogskoven is formed. It lies on the black limestone; this we should probably also see at the iron-works of Eidsfoss immediately above Hoff, as well as below Kolaas at Bärum; and the porphyry formation above. All the stones here would consequently belong to the transition formation, and the range of Holmestrandt is peculiarly distinguished by beds and not by formations.

There are consequently basalt beds in the transition formation! And filled with augites; and what is still more, with combinations which bear so strong a resemblance to Auvergne, where the porphyry mountains are far removed from the transition formation.

In fact, when we look on the needle felspar it frequently seems as if we had pieces lying before us from the great valley of *Prentigarde* at the *Montdor*. It is singular enough that at Holmestrandt, as well as at Montdor and Clermont, the quantity of felspar declines in proportion to the increasing blackness of the fundamental mass. In the black basalt there is not a trace of felspar any longer to be found.

_____

The county of Jarlsberg, which we flew through, is a hilly country, with flat vallies like Thuringia. When we are quite near to Holmestrandt, however, there are some admirable prospects from the steep rocks over the Fiord, and the islands beyond the flat district of Moss, and the high lands of Hurum.

At Klaveness I saw amygdaloid in the porphyry a red base of wacke with kidneys of white calcareous spar and steatite *(speckstein)*. In the further progress, however, nothing appeared but porphyry without interruption; generally with a red base, and with small felspars in rhombs and not in needles: rhomb porphyry. We see it it is true not often in rocks, but yet we observe

it distinctly at Söllerud in the valley of Undrum, and on the road from Tons-
bergs-Elv to the church of Stökke.   We may consider more than the half of
the country as a large table land of porphyry, elevated about five or six
hundred feet above the sea.

And what is singular! with the boundaries of the county the porphyry
also terminates, and the beautiful zirconsyenite again makes its appearance.
We may therefore call Jarlsberg the porphyry county, and Laurvig the
country of the zirconsyenite ; for both formations are accurately limited by
the political boundaries, as if by authority, and the one does not intrude on
the province of the other.

One of the former possessors of the county which at present belongs to the
king, Count Daneskiold-Laurvig, pointed out their boundaries by a marble
pyramid with an inscription on it in praise of the king.   But what boundaries does not this pyramid designate!   Where Laurvig and Jarlsberg, and
where zirconsyenite and porphyry separate, dew-berries, not only wind round
the marble with ripe berries on them ; but where dew-berries *(rubus cæsius)*
grow, we are again in the climate of *beeches,* and these are the first dew-berries
on the road from Christiania to this place.   Shortly afterwards this road
actually runs through a majestic beech forest almost to the banks of the
Louven-Elv: a phenomenon no where else to be seen in Norway.   In fact
it is very unexpected; for we should not previously have been led to expect
such an increase of the temperature.   It is exactly in the latitude of 59°, and
at Frederickshall, which is opposite, and in the same latitude, as well as at
Bohus Län, beeches are altogether out of the question.   Warm sea breezes,
shelter from the western sea breezes, and the excellence of the soil, may
have all concurred to forward their growth.   This is the more likely, as
neither at Skeen, nor Arendal, nor Christiansand, which are all farther south,
are beech woods to be seen, notwithstanding the degree of temperature
there is not certainly prejudicial to them.

The bridge over the Louven-Elv, about two English miles before coming
to Laurvig, is beautiful, bold, and elegant, and yet constructed with astonishing firmness and security.   Bridges are in general rare in Norway, and
such large and long bridges still more so.   This was erected in 1807, by
the zeal and activity of the general director of the roads, Chamberlain
Peter Ancker.   The stream which descends from the highest Norwegian
mountains through Nummedalen flows through Kongsberg, and here,

when it has nearly reached the end of a course of more than one hundred and forty English miles, it is extraordinarily rapid and broad. But the stone pillars on which the bridge is erected are suitable to this violence. They are not heavy, but almost all the stones are connected together by strong bands into one solid mass, and the whole presents an edge as sharp as a wedge to the floods and the ice. It is a very beautiful work.

The bay of Laurvig, and the neat town under the rocks are also astonishingly beautiful. The works of Fort Fredericksvärn in the distance seem to rise out of the sea, and the small village of Stavärn is situated on the point of the tongue of land. Add to this, an agitation and bustle in the place itself as in a great trading town, to which, however, at this time a strong garrison to repel English attacks may have greatly contributed. At the end of the place there is a considerable iron-work, where at present guns are incessantly cast and bored; and a considerable number of men were constantly running between the bore-houses, the furnace, and the smelting-houses. The iron ore comes from Arendal, and the charcoal from the upper woods of the country. The latter are in such abundance, that since the royal purchase a new furnace has been built several miles higher up the country in Slemdal, which was to be set a-going in the autumn of 1808, and which will not a little contribute to increase the animation of Laurvig.

We were still accompanied by beech woods from Laurvig on towards Porsgrund down to the sea-bay and up again. There is scarcely any thing in other countries to be compared with the views on this coast: so intersected and deeply indented like canals are the sea-bays of Vass Bottn, and the Landgangsfiord. It is altogether singular, and we cannot get accustomed to it.

In the dark we issued out from among these rocks into the large valley of the Skeensfiord, and descended towards the beautiful Porsgrund. The houses are large and excellent, with not merely a prosperous but a wealthy appearance, and run in a row of more than two English miles in length. The place bears a vivid resemblance to the charming Gemarck above Elberfeldt. On the other side of the Fiord, which is here nothing more than a river, there is a similar place, Westre Porsgrund. Both contain the full population of many a considerable town of Norway.

A good road on the height above the beautiful valley brings us to Skeen, another two English miles farther, when it descends steeply to the little place closely confined between the hills and the water. We proceeded on over an immensely long wooden bridge, over dashing waterfalls, and past saw-mills without number. The roads were again narrow till we came to Giemsiö Kloster, the seat of Chamberlain Adeler, to whom almost the whole of this part of the country belongs.

The singular relations of the transition formation, which are not developed in the environs of Christiania, become striking, distinct, and clear, in the mountains of Skeen, just at the very boundary of the whole formation itself; for beyond the Skeensfiord all these extraordinary rocks disappear, which continue uninterruptedly all the way from Christiania. There is no longer any porphyry, zirconsyenite, amygdaloid, or sandstone, on the other side, not even the black limestone, except a few traces of it for once in small hills at the Westre Porsgrund.

Neither do these rocks spread farther west than Skeen. They are bounded by the Hütten-elv at Fossum up to the higher mountain range at Kongsberg.

Hence the newer rocks in Norway are extended over a space which is fully equal to the Christianiafiord; and as the latter forms a long sack in the Cattegat (the Dutch call it the sack of Norway), the transition mountains, with their various members, penetrate like a similar sack between the gneiss mountains: a singular expansion, which has a peculiar appearance on maps, and which promises one day to throw light on the general laws of the extension of formations in the north.

We have but a very faint idea of the magnificence of the zirconsyenite, when we do not see it in its depositaries, and especially in the manner in which it appears between Laurvig and Porsgrund. All the masses and rocks seem as if they had come from another world; they are something to which we have not been accustomed: the splendour and the freshness of the felspar, the large granular planes, the unusual blue colours, and the frequent labrador play of colours on the planes; and in the interior the distinct fresh and shining hornblende crystals, and small brown zircons every where. Every block must be examined and admired: every rock begun, and followed up. We cannot help taking up such stones as we find on the road, and preserving them for the purpose of still farther examining what we are surrounded with here in every direction; and we must again throw them away to make room for new pieces which crowd upon us. All the eminences, all the mountains on the road above Landgangsfiord and Vass Bottn, to the heighth of from at least eight hundred to one thousand feet, and the singularly picturesque rocks above the Fiords, are composed of this noble stone.

On the western shore of the Eidangerfiord, the older rocks make their appearance from below: no longer in similar rocks and mountains, but in hills which are covered with thick woods. The church of Eidanger stands on them. It is brown, fine-splintery hornstone, similar to the flinty slate which we find under the Greffsen at Christiania. White stripes of similar stone frequently run parallel in it like jasper bands. Deeper down, before reaching the marshy valley of Leerkotten, and the water-like plain of Porsgrund, the hornstone is changed into pure white quartz, which conceals itself in the valley.

When we ascend from Skeen, the mountains which separate this valley from Slemdal and Faritzvand, the hills and the strata which confine the town in every direction are the first objects with which we are struck. The streets are paved with the stones, and thus converted into a most interesting cabinet; for the petrifactions of the black limestone shew themselves beautifully and distinctly, and with great variety on the plates:—a number of large madreporites; a great many trochites, and a wild chaos of entrochites scattered together; several patellæ, and now and then an ammonite among them, and very frequently an univalve in the midst of these, which, according to the assertions of Ström and Esmarck, have been hitherto unknown. These hills are about three hundred feet in heighth, and their strata incline towards the north east. They spread out into a plain at top, a broad valley which separates the gneiss mountains above Fossum from the zirconsyenite mountains of Slemdal.

I went to the latter, and reached the Borgensiö, situated at their base, from which the stream of Leerkotten takes its course; that between four and five miles farther down, below Porsgrund, unites with Skeensfiord. All was hitherto limestone, with a few, and but a few, orthoceratites in it. However, as soon as the ground begins to rise on the other side of the small lake, the stone becomes immediately grey coarse splintery quartz, in distinct strata, of two feet in thickness, which dip into the mountain towards the northeast. Very thin slates of black fine mica folia frequently lie on the quartz; greywacke slates; but still however in thin beds. The strata rise several hundred feet upwards. They are covered by a distinct amygdaloid; a blackish greywacke base, small grained uneven, yet fine granular in the sun, with an incredible number of white calcareous spar amygdala and nuts; the most as large as an almond, others an inch in length, and full of white points all the way down. All are not fully filled, but druses and cavities remain in the midst of several. Moreover, a number of black crystals lie in the mass, which have a greater resemblance to augite than hornblende; the foliated fracture is too little seen. This is a rock which seems immediately connected with the basalt formation; and here it rises into high mountains; for it continues through small vallies to the summit of the Vardekullen, which commands the whole valley of Skeen, and an excellent view of the town, the charming Giemsökloster on the other side of the water, and over the animated Fiord down to Porsgrund. Vardekullen however is but nine hundred and thirty-two English feet above the Fiord.

This summit is separated from the higher mountains by a deep valley down towards Faritzvand and Slemdal. The amygdaloid continues down the valley to the bottom. But on the opposite declivity, ascending the Veedlösekullen, the zirconsyenite appears: fine granular below, with distinct hornblende in angular crystals, and with felspar, which is not unfrequently red; above always more and more coarse granular; the grey labradore felspar more and more frequent, and even here and there surrounded by red felspar. This is no insignificant mass. The summit, where there is a stone landmark for vessels at sea, is at least one thousand six hundred and fifty-one English feet in heighth.* We actually see at top an immense way out at sea, notwithstanding it is several leagues distant, over the mouth of the Fiord towards Langesund, and over Friedrichsvärn, to the opposite side into Sweden. We see the whole of the deeply intersected landscape, over which the zirconsyenite is spread, and all the summits which either rise above or fall below Veedlösekullen. We learn here at a glance the heighth to which this singular formation can ascend. However singularly high and steep the rocks appear to be between Laurvig and Porsgrund, there was no mountain there, however, which much exceeded Veedlöse in heighth. A round isolated Kullen between Faritzvand and Louwen-elv northwards, in the neighbourhood of Laurvig,

---

* 24th of August, 1808.

Giemsökloster, h. 7. Bar. 28. 2. 1. 30 P. feet above the Fiord.

Vardekullen, ...h. 9. Bar. 27. 3. 5. Therm. 13. 5 R.

Veedlösekullen, h. 12. Bar. 26. 7. 6. Therm. 15. 6.

Giemsökloster, h. 5. Bar. 28. 1. 8.

was higher; it is seen by vessels far out at sea: they call it Lövesnyta, and avail themselves of it as a distinguishing mark of the coast. It may reach as high as one thousand nine hundred and seventeen English feet. The view from Veedlöse northwards stretches over gently rising mountains to the three caps of the Skrimsfieldt, not far from Sandsvär, in the neighbourhood of Kongsberg, which bound the horizon in that direction. These are also the greatest known heights to which the zirconsyenite has here risen, and they are on that account very remarkable. M. Esmarck gives the highest two thousand five hundred and thirteen Paris feet, or two thousand six hundred and seventy-five English feet, about two hundred feet less than Jonsknuden, above Kongsberg, but almost a thousand feet more than Veedlösekullen.

Ulfskullen, above Fossumverck, the highest gneiss mountain of Skeen, is equally high with our zirconsyenite summit.

Other lower summits connect Veedlöse (the treeless) with the range of Vardekullen. I followed along the ridge of this chain, and soon found myself surrounded by porphyry instead of zirconsyenite: a brown compact base with a number of white small felspar crystals, quite similar to the porphyries of Jarlsberg. It continues in this way to the amygdaloid above Borgensiö.

The stratification of these formations is therefore quite distinct here, and not to be mistaken.

Low down the black *transition limestone* of Skeen above the gneiss: over this quartz, which here supplies the place of the sandstone at Holmestrandt and Krogskoven. It may actually be nothing else but very fine granular quartzy sandstone, in which the grains disappear in the basis or ground. Above this the amygdaloid, of four or five hundred feet in thickness: higher than in any other place of Norway. Now porphyry. Lastly, above all, the beautiful, splendid, purely crystallised zirconsyenite.

But this was also the case at Christiania. Only the amygdaloid here entered into the series of formations. At Holmestrandt it approximates itself to the porphyry; at Christiania it sinks altogether into the porphyry formation.

Here there is great distinctness in the order of these formations. They are true accompaniers of one another, and they only rise to a considerable heighth in those instances where the black lime-stone appears at a low level. We have here a whole family of rocks supported by black lime-stone, with which Norway enriches the list of transition formations.

Greywacke-slate and clay-slate are also not wanting in the country round Skeen. They appear at the Fiord, on the eastern road to Porsgrund, from beneath the lime-stone. The greywacke-slate is fine micaceous thin slatey, and breaks into angular fragments.

And what is very singular, we again find in the lime-stone the same thick veins of porphyry and green-stone which were so striking in the country round Christiania : with altogether the same relations and in the same composition ; and with as much epidote between felspar and hornblende. These veins may be traced a great way at Mäla, many hundred yards through rocks which appear above the surface ; and the Kloster bridge over the dashing stream near to Giemsökloster stood once in one of these veins of porphyry, of which the remains still protrude at both sides of the water-fall. These veins of crystallised stones in the transition lime-stone impress the more strongly on us, lest we should be disposed to forget it, that porphyry and zirconsyenite, notwithstanding the extraordinary nature of the phenomenon, still lie on the lime-stone.

On the twenty-second of August, 1808, I went from Kongsberg towards Skeen over the mountains, a road only practicable to riders or pedestrians ; but it is instructive respecting the extension and boundaries of the transition formation towards these districts, and is therefore deserving of a short mention.

The primitive mountains which surround Kongsberg stretch much less southward than we might well believe. Scarcely two English miles down, beyond the Dal-Elv under the church of Hedinstadt, and before we come to Hellestad, the gneiss disappears under the dark bluish-grey fine granular lime-stone. The mountains rise abruptly ; the strata all dip towards the south-east into the mountains. A number of blocks lay by the road at Hellestad, and several were traversed with small veins, of from two to three inches in thickness, such as are only to be seen in this country. For the side of the vein was formed of coarse granular black calcareous spar (containing carbon), similar to what is known as Madrepore stone. Red felspar and blue quartz in coarse granular mixture lay above it down the whole vein, and above the quartz, in abundance, pieces of black very shining completely conchoidal and extremely easily frangible anthracite ; all this through a lime-stone, in which very frequently orthoceratites of three or four feet in length lay here, and patellœ without number.

The zirconsyenite of the Skrimsfieldt, beneath which the road runs, first appears above the Sätergaarden (Alpine huts or shiells) of Breystol and Grönlie, where the ascent is already about one thousand seven hundred English feet, and consequently equal to the whole heighth of Veedlösekullen. The formation begins here at the same heighth in which it ceases at Veedlöse, which is very remarkable. It does not always lie immediately above black lime-stone, but is frequently separated from it by a layer of snow-white small and fine granular lime-stone, particularly at Grönlie, quite similar to that which is used as marble at Giellebeck :—a new depository of the white granular lime-stone in the transition formation. In the zirconsyenite itself, even up to the summit of the Skrimsfieldt, the felspar is, as every where else, not red in the whole, but grey, frequently in long crystals, which shoot through the hornblende and zircon points. Quartz is but seldom found in it, the mica still seldomer. The formation is therefore even in pieces, not to be mistaken for either gneiss or granite.

Above the Skrimsfieldt towards the west there is a lake, the Ravads-siö, quite solitary, on the eminence above the cottage of Linaas. The road runs along its banks towards Skeen, and these banks limit both the lime-stone and zirconsyenite. From Linaas hitherto I had seen seldom any thing but white fine granular lime-stone, such as that of Giellebeck, very sandy and soft. But from the southern bank of the lake, this stone continues without interruption, above Langerudsdal, towards the marshy levels of Finvol, the greatest height of this road between Kongsberg and Skeen, and not altogether one thousand seven hundred English feet above the sea. From Bedstul, between two and three English miles farther on, the mountains again fall towards Luxefieldsvand, and the streams run down between narrow and steep vallies. Quartz is not always wanting here in the stone; but still it does not seem essential to the mixture. Mica folia appears also here and there, and the hornblende crystals are at least not so usual as formerly. The felspar also appears sometimes flesh-red between the grey, and not unfrequently in such a manner, that the former surrounds the grey in angular shells, which has a very singular appearance. The whole is very like the stone of which the Egeröe is formed, and the neighbourhood of Egersund on the south-western coast of the country. The whole of the long lake of

Luxefieldt, about three English miles and a half in length, so narrowly hemmed in between rocks, that me can scarcely proceed along its banks, still lies on the zirconsyenite ; and not till we reach the declivity towards Moegaard does gneiss with scaley mica advance over it,—and but for a short space. The lime-stone arises immediately below Moegaard, and continues without interruption down the valley to Skeen. It has not risen to an excessive heighth here, however, for Moegaard is scarcely more than six hundred and thirty-nine English feet above the sea. The whole road from Kongsberg to Skeen consequently lies for the most part in the transition formation ; still, however, very near the boundary where towards the west it is compelled to give way to the universally diffused gneiss.

Skeen is a small but thriving place, with a population of one thousand eight hundred and five souls. There are many prosperous families in it, who are distinguished in a particular manner for their sociality. The sources of the prosperity of the town, as well as of Porsgrund, which lies below it, consist in their foreign shipping ; for a number of vessels proceed from hence to England or the Mediterranean, for the purpose of conducting the Spanish and Italian trade. A great deal of additional animation is occasioned by the considerable iron-works of Fossum, Ulefoss, and Bolvig, which are near to this place ; and still more is occasioned by the number of saw-mills in the town itself. The powerful Skeens-elv which issues out of the Nordsiö, with all the waters of Tellemarken, a little above the town, is precipitated over a number of falls towards the Fiord, and the saw-mills run like a street under the falls. The deal and log trade of Skeen is in fact none of the most inconsiderable in Norway. As the inhabitants of Tellemarken for the most part come down to Skeen for the purpose of procuring grain and other necessaries, this also contributes considerably not only to enable the people of the town to make the two ends meet, but to place them in a comfortable situation.

This place may also in point of agreeableness be reckoned one of the most favoured in all Norway, possessing as it does so many advantages of climate, so many highly romantic environs, lakes and rivers, vallies and mountains. Giemsiö Kloster, situated in an island in the river, and surrounded by thundering waterfalls, with the town on the opposite side of the Fiord in view, and the hills, and the vessels below would be accounted a charming and magnificent country-seat in any part of the world. The broad, large, and animated valley down towards Porsgrund, would seem rather in the south of Germany than in the heart of Norway, and yet but a few miles up the river, on the banks of the Nordsiö, we find such scenes of rocks and wilderness, such grandeur and sublimity, as Norway, of all the countries of Europe, alone possesses. The district of Skeen can only not bear to be compared with the Italian landscapes.

On the eleventh of October we set off in the night from Skeen down towards Brevig, where there is a ferry over the small sea-bay. The water was here, as well as farther up the line of separation, between lime-stone and gneiss, between the newer and older formations ; and after setting foot on the opposite banks at Stathelle, we see nothing but what belongs exclusively to the gneiss, in the same manner as at the Bothnian Gulph.

The country is here intersected with an uncommon number of small lakes between the rocks, as is usual on this point of the Norwegian coast. In the afternoon we entered a ferry, over a long deeply sunk but very narrow lake, the Holte Fiord, and two English miles farther on we came to the outlet of this water into the sea-bay of Söndelev. There we saw a most astonishingly romantic prospect of the church of Söndelev, on the hill in the narrow valley, surrounded by high and woody mountains, and with the Fiord and the sea in the distance—such a view, that were we to see a drawing of it, we should never have conceived it to exist under a northern sky. In the evening, at Röe, the calm and still Fiord was every where covered with people, and small boats in the most diversified groups, as if assembled for some festival. It was the mackarel season, and young and old were busied in catching mackarel. This gave a highly animated and agreeable appearance to the landscape. Night overtook us : we no longer saw our road till it was lighted up to us by the furnace of the great iron-works of Näss, where we found a most gracious and hospitable reception in the house of Jacob Aal, the worthy proprietor.

The twelfth of October. M. Aal not only possesses one of the largest and best contrived works of the country, but he unites with this technical knowledge a degree of taste and scientific acquirement very rare in the north. His house is admirably, even luxuriously fitted up. His collection of pictures contains many excellent pieces, and his library, by no means inconsiderable, is rich in physical and literary works.

We unwillingly and with difficulty took our leave of this house, in which a longer residence would have proved so instructive for us, and about mid-day we came down between rocks to the *Bridge-town* of Arendal.

The rocks have not allowed a foundation for this town on the land. The houses are almost all erected on piles; the streets are bridges between the houses; the places or squares are wooden floors above the water. With this singular situation, and crowded as it necessarily is, the place is, however, very animated. The passage is easily made in one night from this place to Fladstrand in Jutland, and ships go and come between the two places almost daily. Only last night, eight small vessels ventured out at the same time, notwithstanding the harbour is sharply watched by three English frigates. The coasts of Denmark and Norway are certainly no where nearer to one another than at Arendal. The inhabited island of Tromsöe in front, and the bay at the town, form an excellent harbour, in which vessels lie with safety close beside the houses, and may proceed out through one of the outlets of the Tromsöesund. Hence the town is frequently visited by vessels desirous of a secure harbour. It contained one thousand six hundred and ninety-eight inhabitants in 1801.

The wind was no longer favourable, and no vessel could venture out. We were obliged to proceed farther south, that we might avail ourselves of the west winds. In the afternoon we proceeded through between the poles and erections at the iron-mines; and then by Oyestad over the Nidelv, which is another of the most considerable streams of Norway. The advantage presented by a narrow opening between the rocks through which the river dashes and foams with great violence, and falls with a thundering noise many feet down, was here laid hold of. A bridge was boldly thrown over the straits, and carriages may proceed over a place which we can hardly look at without giddiness.

In Möcklestue, at the church of Wester Moland, we waited for the appearance of the moon, and proceeded in the night over a plain completely level, through thick woods to Tvede, and then proceeding over mountains and hills, we reached Aabel on the banks of the Topdals-Elv, at break of day. It was no longer a matter of doubt, that from Arendal onwards we were visibly proceeding through a better climate. The oaks became more and more frequent; and the hollows and ditches by the road-side were

covered by several birches and brambles.   On proceeding over the Topdals-
Elv by a ferry at Aabel, and ascending the steep height on the opposite side,
the whole variety of trees and bushes belonging to the beech climate at once
appeared before us on the heights, and furnished us with many an interesting prospect.   But several leagues from Christiansand, the vicinity of
the great Western Ocean seemed hostile to plants.   The gneiss mountains
were now bare, or covered with very small Scotch firs.   We ascended and
descended the rocks with great slowness, and it was midday before we
reached the large and level plain at the church of Oderne in the view
of the ferry which was to convey us to Christiansand, over the great Törrisdals-Elv.

The town begins whenever we set foot on the opposite bank; long and
broad streets laid out with the utmost regularity, and covered in the middle
with deep sand.   The houses are not built close to one another, but separated by large gardens, and the streets on that account are almost interminable, considering the smallness of the place.   But the houses or courts are
every where constructed very neatly, every where pleasant and delightful.
This fourth capital of the country has upon the whole a very peculiar
character, which is very different from that of the three other towns of
Christiania, Drontheim, and Bergen, and which hardly seems to belong to
Norway.   Situated almost on the southernmost points of the kingdom, in
the bottom of a majestic bay, in which whole navies may lie securely at
anchor, Christiansand is the principal asylum of the many thousands of
vessels which go through the Cattegat, or return from the Baltic.   The
current from the Sound beats against the Norwegian coast, winds round the
whole of the southernmost point, and then flows with great rapidity along
the coast northwards up towards Bergen.   All vessels from the Sound endeavour in view of Norway to reach the North Sea: they are sometimes
carried along by the current, and they are sometimes desirous of avoiding
the returns of it in their passage.   Hence whole Baltic fleets are always to
be seen before the rocks of Christiansand.   It too frequently happens, however, that vessels in the storms of the Cattegat lose masts, sails, helms,
planks, ropes, or whole parts of a ship.   Christiania is then their asylum.
Here they find every thing to the most minute articles which are necessary
to their reparation and refitting: good workmen, dispatch in working, and

consequently little delay in their passage from their misfortune, and advances of money for the payment of the repairs are offered, of such infinite importance to many ship-masters.

Hence we naturally ask:—*What does Christiansand live by? By ship-chandlery.* Several private individuals possess whole magazines of whatever is necessary for the repair of a ship; and the wealthy consul and merchant Isaachsen was mentioned to us as the possessor of such a magazine, where every thing is stored up for the reparation of whatever damage can possibly befall a vessel. In this town they are not only accustomed to strangers, but they anxiously wish for them; for without strangers a great source of their prosperity would be dried up. This has been the subject of many a joke against Christiansand by their neighbours, who say that on the arrival of every stranger their only thought is what they may make of him; as if he could only be driven there by some misfortune or other, and they therefore consider him as a legitimate subject of profit.

But this is far from being the only source of the prosperity of Christiansand. No harbour in Norway is so full of animation; no wharfs are in such continual activity. Even at present, when, from English vessels being constantly in sight, every vestige of trade and navigation must disappear; even now we were stunned with the noise of hammers. Here was the skeleton of a cutter or privateer just commencing: there lay another on its sides veiled in large clouds of smoke from the boiling kettles of pitch with which it was to be caulked. Others again were launched and quite ready for sailing. The activity which could no longer find a vent in the construction of merchant ships was now wholly employed in fitting out privateers. No place in Norway can furnish so easily and of such good quality all that a ship stands in need of. In the upland vallies of Nedenäss, Mandal, and Raaboygdelaug, there are an abundance of oak forests, whereas these trees will not spread above Arendal, and are not to be found at Christiania. The country every where rises too suddenly above the climate of oaks. Those other ship-necessaries which are not the produce of the country can be easily procured by Christiansand from the vessels which pass by, and those which enter the place. Hence most of the vessels belonging to merchants of Christiania and Drammen are built here, and logs and deal are also exported to England. The ships of Christiansand have always been in great repute.

The logs and deals exported from this place are not of so great import-ance; but there are several saws however on the Törrisdals-elv, and the country in the near vicinity.

The unfavourable wind would not permit us to leave Christiansand. The wind frequently appeared to us to come directly from the west, and some-times even from the north. For even on the highest rocks round the town it blew incontestibly from these points, and they are the best winds for a passage to Denmark. But then this was not the sea wind. One or two leagues out at sea we should have found a south-west or a south wind, and the passage would have been impossible; so much is the direction of the winds changed by the land. The sea wind is always pretty determinately when it proceeds over Christiansand, a few points more towards the east or the north. Many ships, therefore, when they wish to sail, proceed between nine or ten English miles to the westwards out of the bay towards Hellie-sund, where the true sea winds can be more easily observed. We supposed that the wind must necessarily change in a few days, and remained therefore in the town, which enabled us to become more intimately acquainted with the place.

When Christian IV built Christiansand in the year 1641, on a sandy flat as level as the surface of a lake, he destined this place for the rendezvous of his fleet; and till a very late period several ships of the line generally lay here. The situation of the town was so happily chosen, that it soon began to increase and prosper, and has never fallen from its prosperous state. It contains at present four thousand seven hundred and eighty-seven inhabitants, without including strangers. Every seafaring nation which was concerned in the northern trade sent a consul to this place, and a number of foreign merchants settled in it. The consequence was a more intimate connexion with foreign countries than in any other of the Norwegian towns. Hence we find a number of regulations, a number of trifles in the appearance of the people and their manner of living, which put us sometimes in mind of England, and sometimes of Holland or Germany. It certainly appears sur-prising to us when we proceed through the large market, to see on the roof of a house of two stories a huge Dutch wind-mill, rising like a tower in the air. It is the only one of the kind in Norway. Common wind-mills are even quite unknown farther north, whereas they are not unfrequently to be met with on the rocks of the southern point of Norway. This prodigious wind-mill in the market of Christiansand is not only beneficial for the town but for vessels at sea. It is from its heighth visible at so great a distance that it serves along with the church steeple to point out to vessels entering the bay the situation of the place.

The twenty-second of October. The wind changed suddenly to the north-west. Every thing conspired to drive us off to Helliesund, where our vessel was lying. It was a delightful morning. A royal boat with six men, which rowed with methodical steadiness and solemnity through the water, brought us quickly out of the bay. Then we were enabled distinctly to see how strongly this important harbour is guarded. Immediately before the town there were floating batteries built on rafts; new batteries on Lagmansholm; and two or three others on the rocky Odderöe, which almost touches the harbour; others still to the right and left along the shore. The gun-shots cross one another here in all directions; and an enemy's vessel would not easily be disposed to venture herself amidst the showers of destructive balls to approach the town or the harbour.

The island of Flekkeröe forms with the main-land a sound before the bay, between four and five English miles in length, into which we entered. The water is there as calm as at Christiansand itself, with the most excellent anchorage at a depth of from eight to nine fathoms. Hence the whole sound is con-sidered as an excellent harbour, in which entire fleets have remained for many months at anchor without injury. A small fortification is constructed for the protection of the harbour on an island quite close to the main-land, which is called Fredrichsholm, or more commonly the Fort of Flekkeröe; and a garrison of a few men was actually kept up here till very lately. When the English, however, after carrying off the Danish fleet from Copenhagen, came to Christiansand in quest of the two remaining ships of the line which were accidentally lying there, they took possession of Flekkeröe, and blew up the works. It is at present an abandoned and dreary ruin.

We reached Nye Helliesund early in the afternoon. But how were our hopes cast down! They would not hear at this place of a favourable wind to Denmark. Our brig lay solitary and abandoned in the narrow channel without hands, and the master was deep in the country. They prognosticate the weather; and what they had observed in the air did not seem to them worth the coming down for. They were in the right; for the sky began to overcast in the south. A storm from the south-west was announced to us for that very evening; and it soon made its appearance. The wind raged furiously between the rocks; the rain struck the water like hail, and the islands were covered with an alarming darkness. We remained prisoners on a rock.

The twenty-third of October. The wind grew less violent, but gave us no hopes. We looked round the place. The road is formed of a long crooked channel of little more than the breadth of a river between steep rocky islands. The houses, about twenty in number, lay scattered along both sides, pendent among the rocks, and concealed among the windings of crevices. They are however neatly built, almost all of them painted red, and have a pleasant appearance to the eye. We ourselves were not badly off in our quarters with the *Lootsoldermand Langefeldt*, the first person of the place. It is hardly possible to conceive a narrower space than that round the house. Before the door there is a *Brygge* of two paces in breadth, supported on piles driven into the water; and again at the back door there are ladders and steps up the rocks. Neighbours are quite near, and may be easily seen by one another; but there is no land road, except by crawling over the rocks.

We ascended however by the ladders and stairs to the height of the island. It is hardly separated from the main-land. In the middle there is a small watch-house to observe the motions at sea and the signals on land. It is singular enough that with these signals, which are repeated at every ten or fifteen miles distance, whatever takes place on the coast may be learned in one day from Christiania to Hitteröe beyond Lister, a length of two hundred and thirty English miles. This day, about two hours ago, two English frigates were seen before Oester Riisör, almost one hundred and thirty-eight English miles off. Flag-signals are used, consisting of three—a Danish, a blue, and a white, and two stripes called *ständer* between. This is simple enough; and yet by such means every thing is communicated which can happen along the coast in time of war, from the first appearance of the smallest enemy's vessel to the landing and cutting off all communication of every other description.

The rocks of the island consist of gneiss, the universal formation hitherto. The stripes which run parallel over the surface soon banish every idea of granite, which has been often described as existing on this coast. But red coarse granular felspar frequently runs in veins of an inch in thickness through the gneiss. In the formation itself it is almost always white.

The twenty-fourth of October. A raging storm from the south and south-west. The house rocked the whole night through like a ship. We frequently believed the rocks and the whole of the island to be in motion. "Numbers of men are not in life to-day who were yesterday in perfect health," said the dry and very precise Lootsoldermand to us very coolly. It was unfortunately but too true. The storm had driven into the Tregfiord, between nine and ten English miles off, a Swedish galley from London to Gottenberg, which is now a prize. Three other vessels laden with sugar and hemp were obliged to put into Humbersund, Hemness, and Ripervig, to save themselves, not-withstanding their being confiscated and the men being made prisoners. But who knows how many have been devoured by the waves?

We live on sea fish and crabs; for this coast seems the paradise of crabs. Their numbers here are astonishing; and the fishery is a very considerable source of gain for the inhabitants of Helliesund. Baskets, something like those used for catching eels, are placed in the water; the crab creeps in, but cannot make his way out again. The crabs went to London. They were taken away from time to time, and our Lootsoldermand, who collected and kept an account of them, delivered them out, and divided the payment among the captors.

The crab ships had their inside fitted up like a large pond: a large water-proof wooden vessel filled with sea water. Into this the crabs were put, and they arrived living and fresh in the market in London. The English used to pay on the spot five schillings a-piece, and sold them in their market for five marcks, or one hundred and twenty schillings. Hence it was well worth their while to bring crabs from this place; and that this branch was by no means a mere incidental concern is shown by the accounts of the Lootsoldermand.

In the year 1803, the vessels took,

On their first voyage	-	-	12,423	head.
On their second	-	-	6,435	
			18,858	
In the year 1804, first voyage	-	11,923	
Second	-	-	5,234	
Third	-	-	10,766	
Fourth	-	-	7,156	
			35,079	

In 1805, first voyage	- -	5,242
Second	- - -	14,092
Third	- - -	8,521
		27,855

In 1806, first voyage	- -	14,000
Second	- - -	13,028
Third	- - -	8,641
		35,669

This branch of trade has been annihilated by the war. The crabs now multiply undisturbed on the rocks of Helliesund, and the English gold no longer enters the houses of Helliesund.

The crab fishery was equally considerable on Listerland to the west of Lindesness. There four ships used to come yearly from Holland to Luse-havn at Farsund. The Dutch used to enter into formal contracts with the fishers. All the crabs caught from the eleventh of December to the end of May, between Lister and Lindesness,* a space of eleven and a half English miles, were to be delivered over to them. They were bound to take all that were brought to them. They were to give two schillings a head provided it was eight inches long. If a claw, however, was wanting, they were bound to deliver two, which were only to be reckoned as one. Then the profits of that fishery were estimated at between three and four thousand dollars.†

Countries surrounded by the sea can only exist by trade and exchange; but then war destroys every sort of communication.

The twenty-fifth of October. We proceeded over the channel to the opposite island, which lay quite open and free to the great ocean. We wished to see the fury with which the waves in a storm break along the shore and dash over the rocks. On the island there are two stone towers,

---

* Lindes-ness (Lime-Cape), the southernmost point of Norway, and known to vessels at sea from a large and small fire kept on it by night, is on that account celebrated far and wide. But it is singular enough to see how the different nations have altered it according to the genius of their language. The English call it *the Naze* (the Cape); the Dutch *Ter Neuss;* the French *le Cap Derneus.*

† Etatsraad Holm Bescrivelse over Lister og Mandal. Norsk Topog. Journal, XIII. 38.

which were covered on the side next the sea with boards painted white. They were seen many miles out at sea. On account of the war, and lest they should have been serviceable to the enemy's ships, they were taken down. As we descended, and looked without interruption into the open sea, the waves, contrary to our expectation, were very trifling; they were broken and small, and not like mountains rolling over one another. We mentioned this to the lootsoldermand. He told us that a sudden calm of the sea was a certain sign that the storm would re-commence from the same quarter. It came, accordingly, in the evening, and raged furiously the whole night through. This sea-calm is a singular phenomenon. Does it originate in an opposing current?

The twenty-ninth of October. All our favourable prognostics vanished. The water rose high in the sound: this we imagined to forbode a west or north-west wind. But it came a storm, and the water again fell. The temperature also would not allow us to entertain any expectations of a change in the south winds. Night-frosts are yet out of the question; and even in the early part of the morning the thermometer stands at 50°. F.

The thirtieth. We went to the church of Sogne in the gun-boat Berndt Ancker. It was an exercise of the crew under the command of the loot-soldermand. Necessity procured this noble coast-fleet, as if by magic, in a few weeks to the government of Denmark, after the loss of their navy. At present there is in every sound, and in almost every harbour on the coast, one of these boats, which is like a battery. The crew are soon assembled from the neighbourhood; they are generally exercised in the different manœuvres; and several engagements with English sloops of war have shewn them what gun-boats are capable of under good management. These boats may be compared to large ferry-boats over rivers; the breadth is equal to about half of the length. The crew are disposed around both sides, with strong oars, as in a galley, and before and behind there is a gun. In an engagement the fore and hind parts are opened almost to the surface of the water, and the guns are brought forward. The shot, when it takes place, always hits the great enemy's ship in the most sensible part, that is, in the hull. The boat turns rapidly about. The other gun discharges its shot where it is believed to be most effectual. They then retreat, re-load, and re-commence the attack. The large vessel may fire off whole batteries.

A small boat, which scarcely rises above the water, and which is also in perpetual motion, is very difficult to hit. The balls pass over it. Hence it frequently happens that the coast-fleet returns victorious from the engagement without losing a single man. It is clear, however, that such gun-ferry-boats, rising so little above the water, are altogether incapable of encountering any thing like a rough sea. They can only work when it is calm, and among the cliffs and rocks. They dare only venture out into the open sea in summer, during a calm, and with a smooth and almost unruffled water.

Our gun-boat, Berndt Ancker, with two twelve pounders, required thirty men when making an attack. With the half of this crew we proceeded slowly along for the five English miles to Hylle, between two and three English miles from the church of Sögne, which is pleasantly situated on a large plain by the banks of the singularly winding Sögne-Elv.

The hospitable reception we met with from M. Friedericsen, the clergyman, and the agreeable society of himself and family, contributed not a little to mitigate the unpleasant nature of our residence in the stormy Helliesund.

The thirty-first of October. The wind at noon became W. N. W.: " A noble *kuling!*" said the glad seamen. The packet-boat was to go off in the beginning of the evening, and we were to follow in the brig about eight o'clock. The sea became pretty high; our sickness soon forced us to retreat into the cabin. The galleon Kemnäs, with a cargo for Lycken, in Jutland, was immediately behind us. We went quickly on.

About two o'clock in the night, the north-wind became so violent that the crew were obliged to lower all the sails but the foremast. The sea was in the most violent commotion. They were afraid of being stranded on the naked and harbourless coast of Jutland, where the north winds rage with terrible fury, drive the vessels on the shallows, and forcibly upset them on the sands. To escape is then impossible. We were now more than half way over, and no longer very far off the dangerous coast. They consulted long, and at last veered back for Norway. The sea became less agitated before we had proceeded two leagues back. They made another attempt; and the prow of our ship was directed towards Jutland. But the violent sea soon returned, and after a lapse of three hours they determined on returning to the north.

At break of day the wind was faint, and the sea nearly calm. What a difference from what we had felt in the night! And yet perhaps at that moment the waves off the coast of Jutland might be as furious as during the night. We were in sight of Christiansand. Our vessel brought us but slowly on: an English frigate appeared, gave chace to us, and was very near to us, when we fortunately run in among the cliffs; and about ten o'clock in the morning we landed in Romsvig, midway between Flekkeröe and Helliesund.

The attempt therefore failed. The vessels accompanying us also returned. The galleon lost her bowsprit in the night, and the packet-boat with great difficulty escaped the frigate. It lay again in Helliesund.

We repaired for consolation to the friendly clergyman at Sögne. We reached the place by a very pleasant way, through a narrow valley covered with numerous oaks over a lake surrounded with romantic rocks. The full moon shone majestically in the evening. It was so still and calm that we did not see the smallest rippling on the sea bosom. No ship could have moved.

We returned to our lootsoldermand at Helliesund. The brig did the same. It lay there more securely.

The third of November. The English brig took, before our eyes, a lugger privateer, which had too incautiously ventured out. It was a pretty vessel, and belonged to the consul Moe, in Christiansand. This frightened our crew somewhat: they were afraid to venture out.

The fourth of November. A new lugger privateer entered here, the *Virksomhed*, bound for Jutland. A singular vessel! every thing light about it; three masts, and all three very low and small, that it might not be visible at a distance; and with two six-pounders, one before and another behind, and two *svingbassen* on the sides like pullies. The crew was at least twenty. They advised us to accompany them. It sailed very fast; for in such vessels quick sailing is every thing, and they are not here, as is the case with unarmed ships, exposed to the attacks of small Swedish fishing boats, which are much more dreaded than the English. Like true pirates, they plunder every thing.

The sixth of November. Early in the morning the captain of the lugger wished to go out to sea with the wind at east-south-east; but such a vessel only can venture to run so close to the wind. It became always more and more southerly the farther we proceeded out; and the violent current towards the Norwegian coast drove us westwards in spite of all our endeavours. After proceeding about five or six leagues, with a violent sea, we were obliged to return; but we could not get again into Helliesund: we had gone too far westwards. The pilot brought us to Kumlefiord, and into a very secure harbour about five English miles from Mandal.

The seventh of November. The day before yesterday a vessel was driven in here, the bottom uppermost, and the masts in the water. It was a brig from Jutland, laden with corn and malt. About a fourth part of the cargo was saved, and the grain dried on land. The captain was found in the cabin, and two sailors in the cook-room. They were yesterday buried ashore.

This is the misfortune and the danger to which corn-ships are exposed When the vessel turns with any rapidity, or a violent gust of wind comes from a new quarter, all the grain falls with the vessel to one side. The point of gravity is immediately changed; the vessel cannot longer keep erect, but immediately oversets. On this account it has been frequently recommended to corn-ships to be divided into apartments, that the grain may only fall through small spaces, and consequently the centre of gravity may not suffer any material change. But in the straits to which Norway is at present driven for food, all sorts of vessels are used for the transport of grain, and consequently they cannot be suitably fitted up.

The ninth of November. So many weeks have passed unprofitably, and nothing but forlorn hopes. A boat from Mandal made its way thus far to-day. The sea is high, and yet promises no great change of the wind. Who could have anticipated such a delay? And how much is the most distant land communication to be preferred to this uncertainty by sea!

We are on the main-land, and that is a consolation. We can move about. There are also several additional conveniencies here; for the peasants can cultivate the ground, and maintain a greater number of cattle, whereas, on the outermost islands, such as those of Helliesund, neither hay nor leaves of trees can be procured.

We ascended Eids-Heien, a signal station about eight hundred Paris feet in heighth. In the small vallies through which we passed, grew, not unfrequently, bushes of Ilex aquifolium, or holly *(houx)*. We had never before seen this shrub in Norway, neither does it grow in Sweden. They called it merely *torn-busk* (thorn bush), and were aware that it did not grow a few miles farther north. This is a proof of the goodness of the climate. If it were not for the furious and salt winds, excellent fruit-trees might be cultivated here.

We had on the eminence a very singular view of the double and triple *skiär* or cliffy environs of the coast; so many rocks and islots, among which the raging and foaming sea is gradually calmed, till it becomes, at the land, as still as an inland lake. At our feet, on the green valley below, the church of Hartmarck was pleasantly situated on a lake, and surrounded by cottages. The environs of Mandal were seen in the distance.

The tenth of November. About five o'clock in the morning the lugger at last set sail out of Kumlefiord, with a good north wind, which, in the course of a few leagues at sea, changed into a north-west. The land rapidly withdrew from us. The mountains above Christiansand almost alone re-mained. About noon we had made more than twenty-eight English miles, and had the very best prospect of reaching Jutland before the dusk. A ship of war then appeared in the horison at a distance. It perceived us, and made all possible sail after us. This required no consultation. We were obliged to alter our course and return to Norway. We were followed by the brig till within a league of land ; but with this north wind we could not enter among the cliffs. We were obliged to sail up the coast past Christiansand, Randöe , and the old Helliesund ; and in the evening, by the assistance of the pilots, we run into Brekkestöe, about fourteen English miles to the east of Christiansand, on the northernmost point of the island of Justeroe, and not far from Lillesand.

The pilots are excellent people. In their large and strong boats they venture a great way out to sea; and among the skiärs we everywhere see with joy the blood-red and white striped sail which marks the pilot-boats, and is only carried by them. They shun no danger, and do not wait for the calls of the vessel to guide it among the cliffs. Every rock, every stone on land, and above and below the water, is known to them : the vessel moves under their guidance through these labyrinths, like a substance endowed with volition. But what have they not to perform on a coast frequented by all nations, and on which the vessels so often seek rest and assistance among the rocks to avoid being cast away! This makes the pilots of the south of Norway among the most experienced and able, and perhaps also the boldest in the world. In the autumn of the year 1806 I was at Lungöe, at Oester

Riisöer. A few days before, in a severe storm, a number of vessels more than five English miles out at sea demanded to be piloted in. The pilot sprung to his boat. Every person ran to the shore to see how the affair would end, for the storm was dreadful. The pilot run against the wind, disappeared and again appeared, and on reaching the ship, the boat then upset, and he disappeared for ever. " It could not be otherwise," cried the son, and sprung to his boat : " my father ventured too much ; he cut the wind one point too sharply." He took his course towards the vessel, exposed to the same hazard ; but he avoided the supposed error of his father, reached the ship, and conducted it happily into the port. The father was blotted out from the list of the living. Every pilot lays his account with this ; and his fate overtakes him generally even sooner than he imagines. How many wives of pilots are there on the islands who have been six, and even eight times wedded ; and yet with the probability of its not being for the last time !

The police regulations respecting pilots are altogether admirable. There must be a certain number in every harbour, and every boat is marked with its number. The harbour has its district right and left along the coast, over which the *lootsoldermand* has the care. Above him again there is the *loots-captain* in the town. The whole are under the direction of the *loots-commander* in Copenhagen. Wherever the necessity is urgent, there, in general, the most perfect institutions are easily formed.

The eleventh of November. They have now a great change to make in the masts and sails of the lugger ; they imagine the masts too high for so small a vessel, and that they require to be shortened. But we could not have at any rate ventured out again ; for a large English frigate was lying close to the coast, which took several boats, laden with grain, in sight of the island. To venture at present to sea, would display great want of foresight and rashness, was the opinion of every one. Bräkkestöe is one of the largest harbours of the coast. There are twenty-four houses on both sides of the sound, part of them beautiful.

The fourteenth of November. Our lugger-captain said to us unexpectedly in the afternoon of the twelfth that he would go to sea in a few hours to try his luck once more. The weather was admirable, and the whole day no enemy's ship was in sight. A few miles out at sea, the north wind became stronger, and drove us quickly forward. The night was dark, and nothing was to be dreaded from enemies. A brig rushed noisily past us, and so near to us that the vessels could have spoken. But if they had wished to give chace to us, before they could have tacked, we should have been at too great a distance, and they would never have found us again in the dark. Long before the break of day we were in the neighbourhood of Veksio, the extreme western end of Jutland, so rapidly had we been carried along by the wind. It was in fact a storm. The sea with wind and current was in such a furious commotion as if the whole ocean were boiling.

But this was no weather and no sea to allow us to land on Nordstrandt, or on the northernmost coast of Jutland. The master wished to run along the coast, and pass the point of Skagen, and if possible to reach Fladstrand, where these furious north-west waves and currents do not penetrate. This was another seventy-three English miles, as far as the passage from Norway to where we were.—But there was no alternative. We proceeded on, favoured by the wind and tide, and about midday we saw the point and steeple of Skagen.

In proceeding along the coast we saw in the distance the interminable rows of stranded vessels. Thousands of masts and skeletons of vessels run like an alley or range of palisadoes the whole seventy-three miles of our course along the coast. This is a dismal prospect to the Norwegian seaman. He endeavours to keep at a great distance from a land which is announced in so frightful a manner; for in his own country he is well acquainted with cliffs and rocks, but not with such shallows.

About one o'clock we passed the point of Skagen, and looked on the waters of Denmark. Behind the point lay a large English frigate at anchor, which, on our appearance, immediately prepared to receive us. We were forced to return.

The weather had in the mean time calmed: it became bright and clear, and almost still. We run little more than an English mile along the coast. We distinguished every house, the cattle and people on the shore, carriages passing along, and dogs running. But all our signals to entice a boat out to conduct us so short a way in so calm a sea were made in vain. We called to the villages of Gamle Skagen, which lay so open and distinctly before us, and we called to the desert coast, but all in vain! Our voices were lost. and our signals waved unheeded in the air. It was a severe and impatient moment! So near to the coast, so very near, and yet no means of getting over. And now we had again to expose ourselves to the sea for a whole November night, to be driven back again perhaps to Norway.

The night was extraordinarily beautiful—mild, calm, and clear: a true summer's night. What we gained westwards by the faint north-east wind was immediately lost by the current, and with indescribable pleasure we saw ourselves at break of day almost on the same spot where the evening before we had left the coast. The east wind now began to rise somewhat, and with it our hopes became high. We sailed for Lycken in the middle of the bay. They named already every place on the land, the villages, the windmills and churches. The sea was now slightly agitated, and the coast no longer to be dreaded. We were both of us landed in the small jolly-boat of the ship; they bore us rapidly over the waves, and returned again to the ship; and we trod upon land; the land of Jutland! the main-land, and connected with our own country. We run, we flew along the shore. We were restored from slavery to freedom! No separating sea, no waves, no wind, longer opposed obstacles to us. Every step called out Jutland, Denmark, and freedom! We reached Lycken quite beside ourselves.—Every thing around us was a new world; every thing different, singular, and gladdening. And the houses collected in villages; and straw roofs on the houses—straw roofs! what extravagance in Norway! And the innumerable multitude of churches in the horizon in every direction!

How they looked upon us in Lycken! They had seen the jolly-boat land from the ship. The guards had assembled. They imagined us prisoners belonging to an enemy's ship, which were exposed, and wished to fire at the jolly-boat. But the lugger fortunately soon run also under the battery, and anchored in the roads near to the land.

The difficulty of approaching this dangerous coast is almost incredible. Half vessels, skeletons, hinderparts, and masts, are every where sticking above the water, quite near to Lycken—melancholy memorials of the fate of so many stranded vessels! There is no where the least vestige of a harbour, and no where the smallest shelter. The ships are exposed to the most raging waves and most furious storms from the north-west, and cannot lie a day or even half a night securely at anchor. When a vessel cannot escape the north-west storm, it stretches every sail, gives itself up to the wind and the mountain-waves, and runs with indescribable rapidity on shore. Hundreds of men assemble along the shore. They lay hold of the ropes thrown out, and retain the vessel when the waves return back to the sea. The ship then falls on her side, but lies secure on the shore. We saw several brigs and a number of yachts at Lycken, lying turned up in this manner on the sand. This process is not the most favourable for vessels, but it saves them however.

But when the numerous Cattegat vessels in storms and dark nights miss the narrow entrance of Skagen, or still believe themselves in the northern ocean, when they have not past Jutland, the vessel strands on the Jutland banks, which run in a triple row along the coast. It strikes on the land, sinks deeper and deeper, till the whole of the inside is filled with sand. In calm weather all that is of any use is taken out of the vessel; but the skeleton remains for many years, a frightful warning to posterity. But when a vessel is cast ashore on Norway, it is immediately dashed to pieces on the cliffs, and in a few moments the least trace of it is destroyed and annihilated. Those who follow have no memorial of the misfortunes of their predecessors.

The English war had converted all the places on this dangerous coast into depôts of grain for Norway. In this way it has bereaved many men of life, and numbers of property and prosperity.

We proceeded rapidly from Lycken sideways along the plain to the two high and open-lying church steeples of Hiörring. We see here every where remains of villages, but as if destroyed. The farm-houses are torn from them and scattered over the possessions. This alteration is new here;—but so much land newly acquired for agriculture, the industry displayed in trenching and ploughing, the superior appearance of the farm-houses, from their being longer and more comfortable, are altogether a sufficient proof how fast *Vend Syssel* is improving this province. This is the account of well-informed persons.

The nineteenth of November. We remained but a few hours in the very small and almost open place of Hiörring, and proceeded on in the dark. In the morning we were before Aalborg. The Limfiord, where so many brigs and galleons were proceeding up in full sail, was beautiful. The north-west storms had brought them with great risk, but fortunately in the night, from Norway through the English frigates. The town lying opposite along the shore of the Fiord was very animated. It is built something in the Dutch style, and need not be ashamed at being the capital of Jutland.

But our feet had here no rest. We came in the night to Hobroe, and at break of day we descended from the eminence into the neat town of Randers. The view of Aarhuus, close on the sea in a bay, with an extensive prospect of the sea, was still more beautiful. We proceeded quickly on, and arrived about midnight in Horsens. We had again a prospect of the sea on the height between the town. Several majestic beech woods on this road are quite delightful. How willingly would we have remained in the vallies along the beautifully environed Meyle! much more so than in Snoghöy at the Belt, where the passage over to Fühnen is made. For the Belt is here like a river, and Middelfahrt, opposite, like a miserable inland town.

In the night we passed through Colding and Haderslev, and in the morning, at break of day, we heard German spoken in the streets in Apenrade.

Through dreary and desert heaths we proceeded to Flensburg. Here I often thought, in Lapland it certainly does not look worse than this, and in many places a great deal better. Midway on the highest eminence, with an interminable plain all around, the only inn is situated, which has a large lamp on the roof. It is placed there every night, like a light at sea, that travellers in the dark may find their way through the waste to this light.

But the stir and agitation in Flensburg forms a powerful contrast: it is a pearl in the Danish crown. There is every where cleanliness and elegance in the ornamenting of the houses and shops. It seemed almost as if I were again among human beings. Springs of water in the public places; a superfluity of provisions in the markets; and hundreds of carriages thronging the streets. We proceeded through them with difficulty; but then again we entered on such another heath as before, with an inn scattered only here and there on it.

Schleswig was distinguished from its length and the endless rows of lamps in the dark; and the numerous palaces and Gottorp in the middle on an island in the sea, appeared like spirits of the night. In the morning about seven o'clock the *Cyclus* was completed, and I found myself again within the well-known walls of Kiel.

Here also an embargo was laid on every vessel to convey corn to straitened Norway. It was pleasing to see how much the welfare of the land we had left was at heart here on the borders. They have very much deserved it in Norway.

On the twenty-second of November I returned to Hamburg; at last, on the evening of the twenty-seventh, the well-known gates of Berlin opened to receive me.

APPENDIX.

POPULATION OF NORWAY

On the 1st of February, 1801, according to the public Enumeration.

Extracted from the Collegial Tidende, 1802 and 1803.

The whole of Norway, 910,074.

BISHOPRICK OF AGGERSHUUS.		
Towns.		
Christiania	-	9005
Friedrichshald	-	3842
Friedrichstadt	-	1837
Moss	-	1438
Bragernäss	-	2859
Strömsöe	-	2549
Tönsberg	-	1543
Holmestrandt	-	863
Scheen	-	1805
		26,995
Kongsberg	-	8000

Province of Aggershuus.

Edsvold	-	4026
Näss	-	4794
Ulensager	-	4079
Hursdalen	-	2098
Gierdrum	-	1333
Nannestadt	-	2745
Eneback	-	2611
Sörum	-	2056
Skydsmoe	-	2323
Uskoug	-	2560
Nettedal	-	1458

Fedt	-	2360
Höland	-	3534
Agger	-	6900
Asker	-	4606
Opsloe	-	694
Krogstadt	-	2033
Aas	-	3673
Vestbye	-	2273
Näsodden	-	763
		56,919

Province of Hedemarken.

Stange	-	3935
Vang	-	4955
Ringsager	-	6860
Näss	-	2741
Rommedal	-	2767
Leuthen	-	2503
Elverum	-	3232
Aamodt	-	2729
Reendalen	-	1685
Tönset	-	3021
Tolgen	-	2017
Quikne	-	1053
Tryssild	-	1597
Hoff	-	6009
Grue	-	4706

Vinger	-	- 6149
Oudalen	-	- 5164
		61,223

Province of Christian.

Land	- -	- 5119
Slidre	-	- 3107
Ourdahl	-	6169
Vang	- -	- 2063
Jevnager	- -	- 2921
Gran	-	- 5416
Gusdal	- -	- 3740
Faaberg	-	- 3645
Oejer	-	- 2578
Ringebo	- -	- 3007
Froen	-	- 4780
Lessöe	-	- 4085
Lomb	-	3406
Vaage	- -	- 3410
Vardal	- -	- 1894
Viri	-	- 2619
Toten	- -	7832
		66,281

Province of Buskerud.

Rolloug	- -	- 4045
Flesberg	- -	- 2694
Sandsvär	- -	- 3853
Eger	- -	- 6713
Sigdal	- -	- 4457
Nordrehong	-	- 6360
Aal	- -	- 4086
Näss	- -	- 5499
Hoel	- -	- 2499
Hurum	- -	1260
Rögen	- -	- 1356
Modum	-	- 4504
Lier	- -	- 3970
In Bragernäss	-	- 140
		51,436

Province of Smolehnene.

Trygstad	- -	- 2864
Glemming	- -	- 925
Thunöe	-	- 3445
Skieberg	-	- 2716
Askim	- -	- 1416
Edsberg	- -	- 3645
Aremank	-	- 2627
Vaaler	- -	- 1771
Rödeness	-	- 1584
Ousöe	-	- 1848
Skibtved	-	- 1703
Haabel	-	- 1724
Spydberg	- -	- 1942
Rakkestad	-	- 4026
Berg	- -	- 1768
Mosseland	-	1118
Ryge	- -	- 1376
Raade	- -	- 1651
Bonge and Thorness	-	1349
Hvalöer	- -	- 859
Tole	- -	- 1806
		42,145

Province of Bradsberg.

Gierpen	- -	- 4229
Eidanger	- -	- 1946
Brevig	- -	- 944
Sannikedal	-	- 2375
Bamble	-	- 2595
Drangedal	-	- 1874
Mikesdal	-	- 1971
Sende	-	- 2665
Boe	- -	- 2999
Holden	-	- 2195
Porsgrund	-	- 1883
Solum	- -	- 2723
		28,399

Laurvig.

Laurvig	- -	- 1897
Hedrum	- -	- 2536

Langestrandt	- -	494
Sandefiord	- -	373
Tiörnig	- -	1427
Brunlaugness	- -	2332
Stavärn	- -	470
		11,692

Jarlsberg.

Borre	- -	1812
Vaale	- -	1682
Hoff	- -	1187
Laurdal	- -	1851
Sem	- -	2633
Sehonge	- -	1433
Anneboe	- -	1694
Nötteröe	- -	3245
Storke	- -	3675
Strömmen	- -	1068
Ramness	- -	2067
Bothne	- -	1488
Lande	- -	1794
In Drammen	- -	124
		25,813

The whole of the Bishoprick of Aggershuus	-	370,903

BISHOPRICK OF DRONTHEIM.
Towns.

Drontheim	- -	8840
Christiansund	-	1642
Molde	- -	803
		11,285

Province of Drontheim.

Aafiord	- -	1488
Biornöer	- -	1585
Hitteren	- -	3685
Statsboygden	-	3178
Hevne	- -	2233
Näröe	- -	1888

Oereland	- -	3142
Fosness	- -	2572
Kolvereid	- -	2184
Overhalden	-	3396
Meldal	- -	4261
Opdal	- -	2772
Holtaalen	-	2191
Röraas	- -	3085
Stören	- -	4519
Kläbo	- -	984
Bynos	- -	3243
Strinden	-	3408
Oerkedal	- -	4825
Melhuus	-	3899
Ytteröen	- -	2047
Inderöen	-	2861
Beitstadt	-	2696
Lervig	- -	1568
Frosten	-	2395
Stördalen	- -	6966
Skogn	- -	3537
Värdalen	-	3890
Sparboe	- -	2735
Sneeasen	-	1879
Sälbo	- -	3184
Stod	- -	2177
		94,419

Province of Romdal.

Heröe	- -	2271
Vandelv	- -	1409
Volden	- -	2936
Jiörringfiord	-	1252
Strandt	- -	1671
Nordalen	- -	2452
Oerskoug	- -	2170
Borgen	- -	4864
Ulfsteen	- -	1709
Haram	- -	1540
Oure	- -	2233
Surendal	-	3655
Stangvig	- -	2766
Edöen	- -	1918
Quärness	- -	3428

Tingvold - - -	3489
Sunddal - - -	2197
Grib - - -	167
Aggeröe - -	2734
Boe - - -	933
Nässet - -	1871
Bolsöe and Kleve -	1694
Grytten - -	2429
Vedöe - -	2784
	54,572

Province of Nordland.
Helgeland.

Rodöe - -	3436
Alstahoug -	4993
Näsne - -	2453
Brönöe -	4386
Vegöen -	949
Vefsen - -	3526
Homnes and Moe -	4561

Vesteralen and Lofodden.

Vaage - -	1530
Hassel - -	2574
Buxness - -	1225
Dierberg -	823
Andenes -	201
Oexness - -	1037
Boe - -	851
Borge - -	1162
Flagstad -	905
Veröen and Röst -	344

Salten.

Bodöe - -	2136
Skerstad - -	2367
Gilleskaal -	2274
Folden - -	2195
Saltdalen -	962
Stegen - -	1675
Hammaröen -	1413
Lödingen - -	2257
Ostoen - -	1815
	52,170

Province of Finmark.
Senjen.

Trondeness -	1289
Astafiord -	1526
Sand - -	653
Torsköen -	320
Berg and Medifiord -	366
Quäfiord -	1197
Tranöe - -	1560
Lenvig - -	1549

Tromsöe.

Tromsöe -	3024
Carlsöe -	1880
Lyngen -	1728
Skiervöe -	1975

West Finmark.

Alten - -	1973
Loppen - -	623
Hammerfest -	922
Maasöe -	454
Kistrandt -	764
Kautokejno -	666

East Finmark.

Höllefiord -	999
Vadsöe -	1141
Vardöe -	160
	26,769

In the whole of the former
Bishoprick of Drontheim 239,215

BISHOPRICK OF BERGEN.

Bergen -	18,080
Vossevang -	4032
Hammer -	2778
Houg - -	3199
Manger -	3506
Lindoos -	3775
Evindvig -	4063
Fanöe - -	2117

Sund	- - -	3184
Findoos	- -	2939
Fielberg	- -	2243
Quindherred	-	2603
Skonevig	- -	1784
Tysnes	- -	2400
Stöisen	- -	2577
Ous	- -	3562
Ethne	- -	1405
Strandebarm	-	2101
Kinservig	-	3402
Graven	- -	2566
Vigöer	- -	1761
Asköenland	-	1253
Aarstadt	- -	291

63,745

Province of Söndre Bergenhuus.

Davig	- - -	1985
Sellöe	- -	2651
Glöppen	- -	3351
Indvig	- -	3899
Eid	- -	2521
Kind	- -	3001
Fölster	- -	1811
Förde	- -	3755
Yttre Holmedal	-	2282
Askvold	-	2571
Indre Holmedal	-	2181
Sognedal	- -	2739
Leirdal	-	2885
Leganger	- -	3230
Lyster	- -	2606
Vüg	- -	3257
Urland	- -	1988
Hafsloe	- -	2107
Justedal	- -	444

49,256

Province of Nordre Bergenhuus.

In the whole Bishoprick of
Bergen - - 103,001

BISHOPRICK OF CHRISTIANSAND.

Christiansand	-	4787
Arendal	-	1698
Oester Riisöer	-	1294
Stavanger	-	2466

10,245

Province of Nedenäs.

Hvidesöe	- -	3280
Sillejord	- -	2411
Hjerdal	-	2812
Moeland	- -	1034
Moe	- -	1146
Laurdal	- -	1117
Tind	- -	2423
Vinie	- -	1736
Holl	- -	4595
Oeyestad	- -	6785
Hommedal	- -	2195
Gjerrestad	- -	2422
Wester Moeland	-	2053
Oester Moeland	-	3636
Sondelöv	-	1653
Evje	- -	2553
Vygland	-	1219
Omblie	- -	1851
Valle	- -	1508
Birkeness	-	1106
Aaserald	-	1016

32,612

Province of Mandal.

Sögne	- -	1764
Mandal	- -	3160
Holme	-	2704
Oddernes	-	3063
Tved	- -	876
Undal	- -	4037
Vandsöe	- -	4947
Lyngdal	- -	3366
Bielland	- -	1955
Oevre Quinisdal	-	2253

Nedre Quinisdal	-	2008	Stavanger	- - -	12 0
Gyland	-	1703	Findöe	- - -	116 9
Flekkefiord and Ness	-	1686	Rennesöe	- -	1174
Kitteröe	- - -	1158	Strandt	- - -	2437
			Närstrandt	-	1427
		34,660	Kjelmeland	- -	2485
			Suledal	- -	2483

Province of Stavanger.

			Vigedal	- -	1385
Helleland	- -	2920	Jelsöe	- -	1908
Egersund	- -	2205	Skiold	- - -	2435
Soggendal	- -	1991	Torvestad	- -	1498
Lunde	- - -	946	Skudesnäs	- - -	2810
Bakke and Tonstad	-	908	Augvaldness	- - -	2010
Hoyland	- - -	1469			
Haaland	- - -	1204			40,162
Närum Boe	- -	1707			
Klep	- - -	977	The whole of the Bishoprick		
Lye	- - -	1807	of Christiansand	-	162,044

Some of the numbers in this List are evidently erroneous, but the Translator possesses ne means of rectifying them.

Directions to the Binder.

Author's Route to the North Cape...................To face the Title.

Route from Alten through Lapland and Sweden........ Page 311

THE END.

B. Clarke, Printer, Well-Street, London.

Printed in the United States
By Bookmasters